Public Sector Entrepreneurship and the Integration of Innovative Business Models

Mateusz Lewandowski
Jagiellonian University in Kraków, Poland

Barbara Kożuch
Jagiellonian University in Kraków, Poland

A volume in the Advances in Public Policy and Administration (APPA) Book Series

IGI GLOBAL
DISSEMINATOR OF KNOWLEDGE
www.igi-global.com

Published in the United States of America by
IGI Global
Business Science Reference (an imprint of IGI Global)
701 E. Chocolate Avenue
Hershey PA, USA 17033
Tel: 717-533-8845
Fax: 717-533-8661
E-mail: cust@igi-global.com
Web site: http://www.igi-global.com

Library of Congress Cataloging-in-Publication Data

Names: Lewandowski, Mateusz, 1982- editor. | Kozuch, Barbara, editor.
Title: Public sector entrepreneurship and the integration of innovative business models / Mateusz Lewandowski and Barbara Kozuch, editors.
Description: Hershey : Business Science Reference, [2017]
Identifiers: LCCN 2016055337| ISBN 9781522522157 (hardcover) | ISBN 9781522522164 (ebook)
Subjects: LCSH: Public-private sector cooperation. | Entrepreneurship. | Public administration. | Technological innovations--Government policy.
Classification: LCC HD3871 .P838 2017 | DDC 352.2/66--dc23 LC record available at https://lccn.loc.gov/2016055337

This book is published in the IGI Global book series Advances in Public Policy and Administration (APPA) (ISSN: Pending; eISSN: Pending)

British Cataloguing in Publication Data
A Cataloguing in Publication record for this book is available from the British Library.

All work contributed to this book is new, previously-unpublished material. The views expressed in this book are those of the authors, but not necessarily of the publisher.

For electronic access to this publication, please contact: eresources@igi-global.com.

Advances in Public Policy and Administration (APPA) Book Series

ISSN:Pending
EISSN:Pending

Mission

Proper management of the public sphere is necessary in order to maintain order in modern society. Research developments in the field of public policy and administration can assist in uncovering the latest tools, practices, and methodologies for governing societies around the world.

The **Advances in Public Policy and Administration (APPA) Book Series** aims to publish scholarly publications focused on topics pertaining to the governance of the public domain. APPA's focus on timely topics relating to government, public funding, politics, public safety, policy, and law enforcement is particularly relevant to academicians, government officials, and upper-level students seeking the most up-to-date research in their field.

Coverage

- Government
- Law enforcement
- Political Economy
- Politics
- Public Administration
- Public Funding
- Public Policy
- Resource Allocation
- Urban Planning

IGI Global is currently accepting manuscripts for publication within this series. To submit a proposal for a volume in this series, please contact our Acquisition Editors at Acquisitions@igi-global.com or visit: http://www.igi-global.com/publish/.

The Advances in Public Policy and Administration (APPA) Book Series (ISSN Pending) is published by IGI Global, 701 E. Chocolate Avenue, Hershey, PA 17033-1240, USA, www.igi-global.com. This series is composed of titles available for purchase individually; each title is edited to be contextually exclusive from any other title within the series. For pricing and ordering information please visit http://www.igi-global.com/book-series/advances-public-policy-administration/97862. Postmaster: Send all address changes to above address.

Titles in this Series

For a list of additional titles in this series, please visit: www.igi-global.com

Handbook of Research on Citizen Engagement and Public Participation in the Era of New Media
Marco Adria (University of Alberta, Canada) and Yuping Mao (California State University Long Beach, USA)
Information Science Reference • copyright 2017 • 503pp • H/C (ISBN: 9781522510819) • US $265.00 (our price)

Global Perspectives on Development Administration and Cultural Change
Gbenga Emmanuel Afolayan (Murdoch University, Australia) and Akeem Ayofe Akinwale (University of Lagos, Nigeria)
Information Science Reference • copyright 2017 • 277pp • H/C (ISBN: 9781522506294) • US $170.00 (our price)

Handbook of Research on Managerial Solutions in Non-Profit Organizations
Vojko Potocan (University of Maribor, Slovenia) Mustafa C. Üngan (Sakarya University, Turkey) and Zlatko Nedelko (University of Maribor, Slovenia)
Information Science Reference • copyright 2017 • 609pp • H/C (ISBN: 9781522507314) • US $300.00 (our price)

Achieving Open Justice through Citizen Participation and Transparency
Carlos E. Jiménez-Gómez (Department of Justice of the Autonomous Government of Catalonia, Spain) and Mila Gascó-Hernández (Institute of Public Governance and Management, ESADE Business and Law School, Spain)
Information Science Reference • copyright 2017 • 296pp • H/C (ISBN: 9781522507178) • US $190.00 (our price)

Emerging Strategies in Defense Acquisitions and Military Procurement
Kevin Burgess (Cranfield University, UK) and Peter Antill (Cranfield University, UK)
Information Science Reference • copyright 2017 • 351pp • H/C (ISBN: 9781522505990) • US $205.00 (our price)

Comparative Studies and Regionally-Focused Cases Examining Local Governments
Ugur Sadioglu (Hacettepe University, Turkey) and Kadir Dede (Hacettepe University, Turkey)
Information Science Reference • copyright 2016 • 502pp • H/C (ISBN: 9781522503200) • US $205.00 (our price)

Theoretical Foundations and Discussions on the Reformation Process in Local Governments
Ugur Sadioglu (Hacettepe University, Turkey) and Kadir Dede (Hacettepe University, Turkey)
Information Science Reference • copyright 2016 • 555pp • H/C (ISBN: 9781522503170) • US $195.00 (our price)

Handbook of Research on Public Finance in Europe and the MENA Region
M. Mustafa Erdoğdu (Marmara University, Turkey) and Bryan Christiansen (PryMarke, LLC, USA)
Business Science Reference • copyright 2016 • 618pp • H/C (ISBN: 9781522500537) • US $300.00 (our price)

List of Reviewers

Albena Antonova, *Sofia University, Bulgaria*
Agata Austen, *University of Economics in Katowice, Poland*
Sabina De Rosis, *Sant'Anna School of Advanced Studies, Italy*
Branko Dimeski, *St. Kliment Ohridski University, Macedonia*
Peter J. Friedrich, *University of Tartu, Estonia*
Adam Jabłoński, *University of Dąbrowa Górnicza, Poland*
Marek Jabłoński, *University of Dąbrowa Górnicza, Poland*
Jaana Leinonen, *University of Lapland, Finland*
Regina Lenart-Gansiniec, *Jagiellonian University in Kraków, Poland*
Slawomir Magala, *Erasmus University Rotterdam, The Netherlands*
Aron Perenyi, *Swinburne University of Technology, Australia*
Konrad Raczkowski, *University of Social Sciences, Poland*
Jarmila Sebestova, *Silesian University in Opava, Czech Republic*
Robert Seliga, *University of Social Sciences, Poland*
Marzena Starnawska, *Gdańsk University of Technology, Poland*
Beata Staszewska, *Kozminski University, Poland*
Łukasz Sułkowski, *Jagiellonian University in Kraków, Poland*
Valentina Tarkovska, *Dublin Institute of Technology, Ireland*
Martyna Wronka-Pośpiech, *University of Economics in Katowice, Poland*
Michał Zawadzki, *Jagiellonian University in Kraków, Poland*

Table of Contents

Detailed Table of Contents

Improvement of public services and raising the citizens' quality of life is one of the biggest concerns of public sector. Changing the way public organizations operate could support such an improvement. In this regard, business model framework is an emerging field of research that could shed some new light. Yet, in case of public organizations, business model innovation remains heavily underexplored. The purpose of this chapter is threefold, to indicate and discuss the opportunities of adapting the business model innovation theory to public organizations, to conceptualize its theoretical framework, and to explore the role of public service design in the process of innovating public sector business models.

Municipal cooperation is important for transformation countries, like Russia, which have to develop legal, institutional and political environments for public service activities. The authors recommend FOCJs as an instrument for coordinated municipal public service provision. To determine the suitability of FOCJs the analyst has to investigate the relation between FOCJ theory, their financing possibilities, the fiscal effects, and the legal forms in which FOCJs can operate in Russia. The authors define several forms of FOCJs and sort out appropriate public enterprises of private and public law for Russian FOCJ. To analyse the establishment, the operations, and the competition between FOCJs the authors present three types of models. One relates to the establishment of an FOCJ, the second concentrates on financing service activities, and the third model deals with competition among FOCJs and demonstrates the effects of different ways of finance. The article concludes with elaborating recommendations for financing FOCJ under conditions in Russia.

The authors discussed the dynamic evolution of the public management manifesting innovative and collaborative methods of social service delivery transcending the public authorities' individual boundaries. The chapter purposely presented the inter-organizational cooperation and public social partnership as an effective business model for public/social value creation unleashing the potential of public organizations and empowering social actors in democratic and collaborative policy/decision making. The chapter identified two public social partnerships in Poland and Spain. The research was conducted in the regime of constructivism. The researchers posed a question as to why the research was being conducted and then adapted the research strategy to it (a case study). In order to analyse obtained data, the so-called thick description was used revealing the contextual determinants that influence the architecture of the researched models.

Although research on social entrepreneurship has been expanding in recent years, the literature regarding business models in the context of social enterprises is still limited. This chapter explores the issues of business models in both profit and nonprofit contexts and discusses how value, being both the foundation and the overall outcome of the business model employed, is created. It begins with the background on the characteristics of social enterprises. This is followed by the conceptualization of business models and their components designed to account for their unique. Then a case to illustrate it is presented. The objective of the chapter is twofold. First, it is to analyse how nonprofit business models are different from its for profit counterparts (at the conceptual level) and to introduce relevant analytical framework. Secondly, the chapter aims at providing a case based on the particular model of a social enterprise in Poland (social co-operative) in order to investigate whatever the business models for profit firms could be applied in the social enterprises.

The aim of this chapter is to summarize the theoretical knowledge from the field of social entrepreneurship and the creation of social innovation and highlight the impact EU funds have on the development of social innovation in selected regions of the Czech Republic. The authors assumed that there could be a positive link between the amount of financial support and the number of created social innovations within the chosen EU programmes. Classification of created social innovation according to type, creator, priority axis in relation to beneficiary etc. came under other objectives. Social innovations are created as a positive externality from other social projects. Finally, recommendation for sustainable support evaluation is provided.

This chapter demonstrates that the local public business model is a phenomenon connected with the multiple value creation system. The primary objective of this study is to gain a more comprehensive understanding of how local public companies may build their business models and simultaneously create new values for customers, stakeholders, and society. The study examines the links between the business model, multiple value creation, business ecosystems with regards to public enterprise. The present study analyses the business model of the local public enterprise in Poland. The findings describe the multiple value creation system from different perspectives and allow for delineating the public business model framework in the tourism industry. Profound understanding of values perception from many perspectives may bring the local public enterprise business model as complex phenomenon closer to recognition.

Preface

Entrepreneurship and business models concerning the private sector have been well recognized in the literature. However, not only the for-profit companies are entrepreneurial and use business models. In fact, all organizations are, and have business models because they create, deliver, and capture value. Hence, governments, authorities and all other public sector organizations pursuing the ideas how to improve public services and raise the citizens quality of life, preferably in a more efficient way, need to become more entrepreneurial. Even though, the major challenges of public sector have been approached in many ways, for instance through new public management reforms, there is an increasing need to adopt "an attitude of experimentation and entrepreneurship" by governments as one of four principles of the public sector innovation movement (European Commission, 2013a, p. 6). Generally speaking, for the sake of all citizens public sector needs innovation (Hollanders et al., 2013; Hughes, Moore, & Kataria, 2011), and innovation is at the heart of public entrepreneurship (Morris & Jones 1999; Windrum, 2008). But public entrepreneurship, even though under scrutiny since the 1960-ies (Ostrom, 1965) and issues covered (Mack, Green, & Vedlitz, 2008), still remains unsatisfactorily explored (Kearny et al., 2007).

This edited book aims to increase the understanding of public entrepreneurship, hence the perspective of business model theory has been employed. Such a knowledge addresses an existing research gap, and has significant practical implications for public managers and policy makers. This book has been addressed to scholars, students and practitioners from various backgrounds and settings, all around the global world, who are interested in public entrepreneurship, public sector business models, and public sector innovation, despite their level of expertise. Because of such a wide potential audience we first outline what is public entrepreneurship and how this book contributes to its understanding. Then, we present the structure of the book. Finally, on the basis of all the chapters and inspired by the usefulness of executive summaries, we draw a framework of public sector business model, which we believe sheds some new light on public entrepreneurship.

PUBLIC ENTREPRENEURSHIP

Despite still unsatisfactory research on public sector entrepreneurship, (Kearney, Hisrich, & Roche, 2007; Morris & Jones, 1999), there are several themes already covered by researchers, like: definitions (Boyett, 1996), types (Morris & Jones, 1999; Oborn, Barrett, & Exworthy, 2011; Zerbinati & Souitaris, 2005), personal and organizational characteristics and barriers (Boyett, 1996; Mack et al., 2008; Morris & Jones, 1999), drivers and facilitators (Luke, Verreynne, & Kearins, 2010; Meynhardt & Diefenbach, 2012), links to other forms of entrepreneurship (Hjorth, 2013), measuring issues (Friedrich & Ukrainski, 2012; Główd, 2015), and finally frameworks of entrepreneurship and differences between public and private entrepreneurship (Kearney et al., 2007; Morris & Jones, 1999; Yeazdanshenas, 2014). There are also particular areas of public entrepreneurship, like academic entrepreneurship (Jones-Evans, 1998; Lewandowski, 2013) for example, but paradoxically, general public entrepreneurship seems to be less scrutinized.

There is a variety of definitions of public entrepreneurship (e.g. Ramamurti, 1986; Drucker, 1985; Morris & Jones, 1999; Boyett, 1996). However, the definition given by Kearney et al. (2007), although narrowed to the context of the research, emphasizes the most important component: "Public sector entrepreneurship (…) is defined as an individual or group of individuals, who undertakes desired activity to initiate change within the organization, adapt, innovate and facilitate risk. Personal goals and objectives are less important than the generation of a good result for the state enterprise/civil service" (p. 279). Seeking and using an opportunity related to arbitrage is the core of entrepreneurial activity, which in the public sector is related to a different hierarchy of values, in contrast to private companies (Kearney et al., 2007). This is one of several features which distinguishes public entrepreneurship. The others pertain mainly to objectives, decision making, authority, motivation, funding, independence, time orientation, skills (Kearney et al., 2007; Morris & Jones, 1999). Moreover, some types of public entrepreneurship and its links to other forms of entrepreneurship also constitute its specificity. Apart from applicability to individual and organizational level (Morris & Jones, 1999), there is also political entrepreneurship (Oborn et al., 2011), next to corporate, public-social, and community entrepreneurs (Zerbinati & Souitaris, 2005). Some other classification distinguishes: professional politicians, spin-off creators, business entrepreneurs in politics, career-driven public officers, and politically ambitious public officers (Zerbinati & Souitaris, 2005). By some, public entrepreneurship is claimed to grasp the notion of entrepreneurship desiring social change, hence social entrepreneurship (Hjorth, 2013).

There are at least few conceptual frameworks of public entrepreneurship. For example, Morris & Jones (1999) noticed the dynamic character of public entrepreneurship in a university, and identified several steps of entrepreneurial process. Others, in turn, focused on a more static approach. Yeazdanshenas (2014) identified the relationships between different internal and external factors with organizational entrepreneurship in the public sector. Meynhardt & Diefenbach (2012) showed that the principles of New Public Management and Public Value Management, next to some factors known from business literature, contribute to entrepreneurial orientation on a department level in public organization. Luke's et al. (2010) study advocates to include among drives and facilitators of public entrepreneurship such factors as: operational excellence, cost efficiency, culture, branding and transfer and application of knowledge. Kearney, Hisrich, & Roche (2007) conceptualized more comprehensive model of corporate entrepreneurship within the public sector, and highlighted performance as a key effect of entrepreneurial behavior. Pursuing better performance is in line with suggestion that arbitrage and exploiting opportunities are the fundamental criteria of entrepreneurial activity, relevant for the public sector (Drucker, 1985; Kearney et al., 2007). Although Kearney et al. (2007) used narrow understanding of performance, currently a well-developed framework of public sector performance is known (Dooren, Bouckaert, & Halligan, 2015). What is missing, though, is better understanding of the processes behind this performance, in particular from the perspective of a business model (Osborne, Radnor, Vidal, & Kinder, 2014).

This book builds on the framework of public entrepreneurship (Kearney, Hisrich, & Roche 2007) as a trigger for better public sector performance. In particular, it explores innovation, the core component of public entrepreneurship (next to risk taking and proactivity), from a new angle of business model theory

(Zott, Amit, & Massa, 2011). Hence, public sector business models and their innovation are considered here as a vital area of public entrepreneurship and public management theory. Three questions delineate focus of this book:

1. What are currently the main challenges of public entrepreneurship?
2. How those challenges are being addressed in practice and what do we learn from this?
3. How public entrepreneurship may be improved through new teaching methods?

DECONSTRUCTION OF PUBLIC ENTREPRENEURSHIP: STRUCTURE OF THE BOOK

The Big Picture: Conceptualization of the Main Challenges and Theoretical Components

We begin with a polemical and conceptual chapter addressing the major challenge of public sector entrepreneurship. It is encapsulated in the title of the Chapter 1, *Between Davos and Porto Alegre. Democratic Entrepreneurship as Crowdsourcing for Ideas*. Yes, a vision of democratic entrepreneurship is our main concern, as Hjorth (2013) puts it public entrepreneurship "calls on the citizen, and therefore also maintains that radical democratic force that addresses everyone" (p. 48). The reasons for concern increased significantly recently, because currently democracy, governance and citizens participation, in many cases, serves rather as an arrangement of the stage for political theatre, than is considered as a real operating model. And it's a worldwide phenomenon, manifested in many ways and on various dimensions, like, for example, the campaign and results of president elections in the USA, or in European Union the problems of disunity like Brexit and Hungarian walls against refugees on the national border with Serbia, or constitutional tribunal issues in Poland, or militarization and power manifestation and NGO restrictions in Putin's Russia, or corruption in India. Just to mention a few topics preferred by media. Although there are no simple, only-truth solutions and answers to the problems expressed above, public entrepreneurship remains an important phenomenon to explore from the scholar point of view. The first chapter provides a big picture of this phenomenon, and, rather than giving the answers, provokes to ask further questions. Despite the variety of possible questions, two theoretical ones come to the forefront. The first one comprises the possibilities to use business model theory to better understand the reasons and processes of public management. This issue is addressed in Chapter 2, *Adopting the Concept of Business Models in Public Management*. It first explains why we need business model theory in public management and why it is so complex. Then this complexity is described in details as a balancing act between three notions of value, from the perspective of stakeholders, citizens, and sustainability. In the final step the chapter outlines an extensive framework of public sector business model, consisting 36 components. In turn, the second theoretical question addresses the problem how public sector business models may innovate. Bridging innovation theory and business model theory on the ground of public management the Chapter 3, *Public Organizations and Business Model Innovation: The Role of Public Service Design*, provides some theoretical foundations and a framework for future analysis in this regard. Some examples of public service design, already known from the literature, support the conclusion that service design, based on various forms of cooperation between different entities, may be applied to innovate public sector business models.

Insights from the Field: Cooperation, Social Innovation, Multiple Value, Marketing

Next chapters provide some detailed analysis of various forms of cooperation, innovation, public value and public marketing. The examples from Russia, Poland, Spain, and Czech Republic are helpful in getting some insights into practical operation of public sector business models. First, in Chapter 4, *Options for Applying Functional Overlapping Competing Jurisdictions (FOCJs) for Municipal Cooperation in Russia*, the Functional Overlapping Competing Jurisdictions, as a form of municipal cooperation, are analyzed for the level of their applicability in Russia. This analysis helps to comprehend the specificity of some components of public sector business model of FOCJ, in particular how value in terms of finance is captured, what are the adoption factors pertaining to legislation, and what is the role of partnership. Chapter 5, *The Business Model of a Public Social Partnership: Contextual Determinants*, scrutinizes the business model of a public-social partnership. The evidence from Poland and Spain support the conclusion that Business Model Canvas is applicable in this context. However, it also reveals that actual results of public-social partnership are most frequently determined by interpersonal relationships between individuals jointly performing a task for a local community. Hence, individual level of public entrepreneurship is crucial to achieve better performance in the public sector through this kind of cooperation. In turn, Chapter 6, *Applying Business Solutions to Social Problems: Social Co-Operative and its Business Model – Evidence from Poland*, investigates how the business models of for-profit enterprises could be applied to social enterprises, and to a social cooperative in particular. It proposes some modifications to the original Business Model Canvas, pertaining to social mission of such an organization. In more general terms, this study emphasizes the role of various forms of cooperation between public sector and social initiatives. This voice strengthens the previous ones in advocating for partnership-centered business models in the public sector. Chapter 7, *Support of Social Innovations – Case of the Czech Republic*, outlines how public sector business model may be embedded in a wider policy business model. In the Czech Republic many social innovation projects were supported from European Social Fund (ESF) by co-funding national, regional and local projects that improve the levels of employment, the quality of jobs, and the inclusiveness of the labor market in the Member States and their regions. Hence, this study shows that the process of creating, delivering and capturing value, which is the basic definition of a business model, encompasses also multilevel political dimension. Value proposition, as a key building block of a business model, is further explained in the Chapter 8, *Local Public Enterprise Business Model as Multiple Value Creation System*. This study reveals that the public enterprise business model involves many financial and non-financial performance categories which determine the creation of value. Moreover, consistency of created values is difficult to accomplish, mainly due to the plurality of values expected by various interest groups, such as customers, managers, employees, suppliers and other stakeholders. Multiple value proposition and related diversity of stakeholders require adequate approach to customer relations and delivery channels, which are important building blocks of a business model according to Osterwalder and Pigneur (2010). This issue is undertaken in Chapter 9, *Public Sector Marketing in Poland*. In particular, a mix of traditional marketing, relational marketing and social marketing, is necessary to balance multiple-value expectations and maintain good interpersonal and institutional relations.

Teaching Issues

Finally, Chapter 10, *The Productive Leadership Game – From Theory to Game-Based Learning,* shows how new digital leadership simulation game can foster the leadership competencies and entrepreneurial mind-set to improve team based organizational productivity in the public sector. This chapter not only presents a simulation-based innovative teaching method. In a broader sense, it provokes fundamental questions about public management curricula. What is the position of public entrepreneurship in the first place? How leadership training could underpin entrepreneurial activity of students and practitioners? How to renew curricula to effectively enhance democratic entrepreneurship?

CONCLUSION: DRAWING A BUSINESS MODEL APPROACH TO PUBLIC ENTREPRENEURSHIP

Public sector innovation is at the heart of public entrepreneurship (Morris and Jones 1999; Koch 2008), and literature describes many types of such innovations (e.g. Armstrong & Ford, 2001; de Lancer Julnes, 2016; Koch & Hauknes, 2005; Lewandowski, 2015; Sørensen & Torfing, 2012). This book, through the effort of all the authors, contributes to better understanding of another type of innovation, pertaining to the public sector business model. Hence, the role of public entrepreneur is to innovate the business model of public organization.

Some major conclusions from the chapters of this book help to draw a proposition of such a business model. Firstly, there is no one best framework of Public Sector Business Model, and many types are successfully applied including Business Model Canvas and its modifications, an extensive 36-component framework, or a 4-component business model. Thus, having such a variety and complexity, there is a need for a new, simple framework which would depict its main sense at glance. Moreover, frameworks applied in the chapters are static, and business model is in fact a dynamic process (Abdelkafi & Täuscher, 2016; Jabłoński, 2015). Hence, such a trait should be incorporated too. Secondly, all the chapters more or less explicitly advocate that various forms of cooperation are a central point of Public Sector Business Model. It supports the argument that co-production is central to the realization of effective public services (Osborne et al., 2014). Such co-production should be understood in a broad sense of co-creation, co-delivery and co-capturing of value. Thirdly, behind creating, delivering and capturing value, there is a process of exchanging values (see also Alford, 2002), and in general every business model is a model of exchange. Such a model has many dimensions, pertaining to how the exchange is organized, what is the subject of exchange, and who is involved in the process. And there is no single answer to each of those questions. In fact, the chapters of this book reveal the variety of possible approaches and points of view, supporting some of the arguments already recognized (Alford, 1993). Stemming from those considerations is the fact, that public sector business model is very complex, and there is no one customer value proposition. Rather, PSBM is a multiple-value creation system, with various stakeholders interests to be secured and balanced. Some are the key ones, like citizens as customers and tax payers, broader community with its social issues and non-governmental organizations, enterprises contributing to economic growth, politicians as decision makers holding a legitimacy to represent citizens but pursuing also personal interests, like political career, natural environment which must be protected. Such a variety of stakeholders, and the necessity to balance often conflicting interests, requires implementation of the

Figure 1. Proposed framework

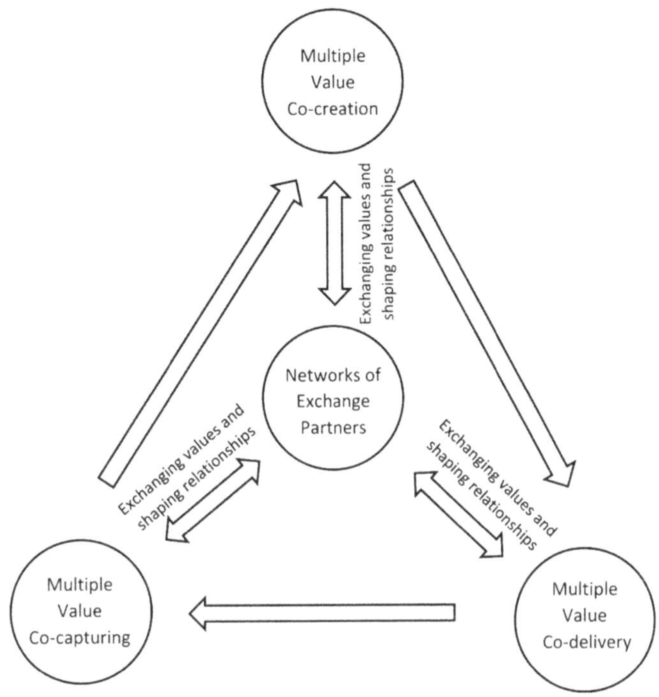

relationship marketing and social marketing methods. Moreover, institutional relations in this multiple-value creation system are often very vital on the interpersonal level.

Taking those conclusions into account, and building on the most general framework of a business model depicting "the rationale of how an organization creates, delivers, and captures value (economic, social, or other) in relationship with a network of exchange partners" (Massa & Tucci, 2013), we propose a simplified and dynamic framework of Public Sector Business Model. Following the above definition it contains such components as: networks of exchange partners as a central one, and three others reflecting the process of co-creation, co-delivery and co-capturing of multiple value. On every stage some partners from the network are involved in value exchange process and the relationships are being shaped, either intentionally or not. The figure below depicts proposed framework.

Proposed framework of Public Sector Business Model and all chapters contributing to this book, could also support teaching process. However, the first and the last chapter argue more explicitly that there is a need for a debate on the curricula renewal, which would enhance making a step forward in teaching public entrepreneurship and public management, a step towards democratic entrepreneurship. So, is crowdsourcing an option?

Mateusz Lewandowski
Jagiellonian University in Kraków, Poland

Barbara Kożuch
Jagiellonian University in Kraków, Poland

REFERENCES

Abdelkafi, N., & Täuscher, K. (2016). Business Models for Sustainability From a System Dynamics Perspective. *Organization & Environment*, *29*(1), 74–96. doi:10.1177/1086026615592930

Alford, J. (1993). Towards a New Public Management Model: Beyond Managerialism and Its Critics. *Australian Journal of Public Administration*, *52*(2), 135–148. doi:10.1111/j.1467-8500.1993.tb00263.x

Alford, J. (2002). Defining the Client in the Public Sector: A Social-Exchange Perspective. *Public Administration Review*, *62*(3), 337–346. doi:10.1111/1540-6210.00183

Armstrong, J., & Ford, R. (2001). Public Sector Innovations and Public Interest Issues. *The Innovation Journal: The Public Sector Innovation Journal*, *6*(1), 1–21.

Boyett, I. (1996). The public sector entrepreneur - a definition. *International Journal of Public Sector Management*, *9*(2), 36–51. doi:10.1108/09513559610119546

de Lancer Julnes, P. (2016). The Study of Innovation. State of Art and Framework for Analysis. In Innovation in the Public and Nonprofit Sector (pp. 12–32). New York: Routledge.

Drucker, P. (1985). *Innovation and Entrepreneurship: Practice and Principles*. London: Butterworth-Heinemann.

Friedrich, P., & Ukrainski, K. (2012). *Measuring Public Entrepreneurship of Quangos*. The University of Tartu Faculty of Economics and Business Administration Working Paper No. 89-2012. Retrieved from https://ssrn.com/abstract=2172801

Głód, G. (2015). Measurement of Public Entrepreneurship in the Polish Health Sector. In *Proceedings of the Management, Knowledge and Learning International Conference, MakeLearn* (pp. 665–673). Academic Press.

Hjorth, D. (2013). Public entrepreneurship: desiring social change, creating sociality. *Entrepreneurship & Regional Development, 25*, 34–51.

Hollanders, H., Arundel, A., Buligescu, B., Peter, V., Roman, L., Simmonds, P., & Es-Sadki, N. (2013). European Public Sector Innovation Scoreboard 2013. A pilot exercise. Belgium: Academic Press.

Hughes, A., Moore, K., & Kataria, N. (2011). *Innovation in Public Sector Organisations: A pilot survey for measuring innovation across the public sector*. London: NESTA. Retrieved from https://www.nesta.org.uk/sites/default/files/innovation_in_public_sector_orgs.pdf

Jabłoński, A. (2015). Design and Operationalization of Technological Business Models. *Acta Universitatis Agriculturae et Silviculturae Mendelianae Brunensis*, *63*(3), 927–935. doi:10.11118/actaun201563030927

Jones-Evans, D. (1998). *Universities, technology transfer and spin-off activities – academic entrepreneurship in different European regions*. Targeted Socio-Economic Research Project No 1042. Retrieved from http://cordis.europa.eu/docs/publications/7864/78645511-6_en.pdf

Kearney, C., Hisrich, R., & Roche, F. (2007). Facilitating Public Sector Corporate Entrepreneurship Process: A Conceptual Model. *Journal of Enterprising Culture*, *15*(3), 275–299. doi:10.1142/S0218495807000150

Koch, P. M., & Hauknes, J. (2005). *On innovation in the public sector – today and beyond*. Publin Report No. D20. Retrieved from http://citeseerx.ist.psu.edu/viewdoc/download?doi=10.1.1.104.3988&rep=rep1&type=pdf

Lewandowski, M. (2013). Introduction to Academic Entrepreneurship. In A. Szopa, W. Karwowski, & P. Ordóñez de Pablos (Eds.), *Academic Entrepreneurship and Technological Innovation* (pp. 1–28). Hershey, PA: IGI Global. doi:10.4018/978-1-4666-2116-9.ch001

Lewandowski, M. (2015). Types of Innovations in Cultural Organizations. *International Journal of Contemporary Management*, *14*(1), 67–78.

Luke, B., Verreynne, M. L., & Kearins, K. (2010). Innovative and entrepreneurial activity in the public sector: The changing face of public sector institutions. *Innovation: Management. Policy & Practice*, *12*(2), 138–153. doi:10.5172/impp.12.2.138

Mack, W. R., Green, D., & Vedlitz, A. (2008). Innovation and implementation in the public sector: An examination of public entrepreneurship. *Review of Policy Research*, *25*(3), 233–252. doi:10.1111/j.1541-1338.2008.00325.x

Massa, L., & Tucci, L. C. (2013). Business Model Innovation. In M. Dodgson, D. M. Gann, & N. Phillips (Eds.), *The Oxford Handbook of Innovation Management* (pp. 420–441). Oxford, UK: Oxford University Press.

Meynhardt, T., & Diefenbach, F. E. (2012). What Drives Entrepreneurial Orientation in the Public Sector? Evidence from Germanys Federal Labor Agency. *Journal of Public Administration: Research and Theory*, *22*(4), 761–792. doi:10.1093/jopart/mus013

Morris, M. H., & Jones, F. F. (1999). Entrepreneurship in established organizations: The case of the public sector. *Entrepreneurship Theory and Practice*, *24*(1), 73–93.

Oborn, E., Barrett, M., & Exworthy, M. (2011). Policy entrepreneurship in the development of public sector strategy: The case of London health reform. *Public Administration*, *89*(2), 325–344. doi:10.1111/j.1467-9299.2010.01889.x PMID:22069793

Osborne, S., Radnor, Z., Vidal, I., & Kinder, T. (2014). A sustainable business model for public service organizations? *Public Management Review*, *16*(2), 165–172. doi:10.1080/14719037.2013.872435

Osterwalder, A., & Pigneur, Y. (2010). *Business Model Generation: A Handbook for Visionaries, Game Changers, and Challengers*. Hoboken, NJ: John Wiley and Sons.

Ostrom, E. (1965). *Public Entrepreneurship: A Case Study in Ground Water Basin Management* (Doctoral dissertation). University of California. Retrieved from http://dlc.dlib.indiana.edu/dlc/handle/10535/3581

Ramamurti, R. (1986). Public Entrepreneurs: Who They Are and How They Operate. *California Management Review*, *28*(3), 142–158. doi:10.2307/41165207

Sørensen, E., & Torfing, J. (2012). Collaborative Innovation in the Public Sector. *The Innovation Journal*, *17*(1), 1–14.

Van Dooren, W., Bouckaert, G., & Halligan, J. (2015). *Performance Management in the Public Sector* (2nd ed.). London: Routledge.

Windrum, P. (2008). Innovation and entrepreneurship in public services. In P. Windrum & P. Koch (Eds.), *Innovation in Public Sector Services: Entrepreneurship, Creativity and Management* (pp. 3–20). Northampton, MA: Edward Elgar Publishing.

Yeazdanshenas, M. (2014). Designing a Conceptual Framework for Organizational Entrepreneurship in the Public Sector in Iran. *Iranian Journal of Management Studies*, *7*(2), 351–376.

Zerbinati, S., & Souitaris, V. (2005). Entrepreneurship in the public sector: A framework of analysis in European local governments. *Entrepreneurship & Regional Development*, *17*(1), 43–64. doi:10.1080/0898562042000310723

Zott, C., Amit, R., & Massa, L. (2011). The Business Model: Recent Developments and Future Research. *Journal of Management*, *37*(4), 1019–1042. doi:10.1177/0149206311406265

Acknowledgment

The editors would like to acknowledge the help of all the people involved in this project and, more specifically, to the authors and reviewers that took part in the review process. Without their support, this book would not have become a reality.

First, the editors would like to thank each one of the authors for their contributions. Our sincere gratitude goes to the chapter's authors who contributed their time and expertise to this book.

Second, the editors wish to acknowledge the valuable contributions of the reviewers regarding the improvement of quality, coherence, and content presentation of chapters. Most of the authors also served as referees; we highly appreciate their double task.

Mateusz Lewandowski
Jagiellonian University in Kraków, Poland

Barbara Kożuch
Jagiellonian University in Kraków, Poland

Chapter 1
Between Davos and Porto Alegre:
Democratic Entrepreneurship as Crowdsourcing for Ideas

Slawomir Jan Magala
Erasmus University, The Netherlands

ABSTRACT

Surfing contemporary media communications allows us to notice the names of a Swiss town of Davos and a Brazilian city of Porto Alegre. They stand for two different strategies of pursuing knowledge, allocating it to social institutions and organizing social learning processes. The first is for the rich and powerful. The second is for the poor and powerless. Both symbolize global class struggles, also in producing and applying knowledge, which is the fastest growing human resource. What if we apply the idea of a democratic entrepreneurship and explore contemporary possibilities of information and communication technologies from the point of social learning processes? Can improved learning be imagined as the only sustainable guideline in the longer run? Can primitive accumulation of knowledge give way prevent costly cognitive failures?

MOTTO

Very much like the US Americans, we believe in God. But as opposed to them, we don't trust Him. - Laybach Group, Slovenia

Is it OK to be a Luddite? - Thomas Pynchon

DOI: 10.4018/978-1-5225-2215-7.ch001

INTRODUCTION AND BACKGROUND

Ideological Bias

Political processes in complex societies flow in the shadow of two dangers looming large over frequently declining public attention and diminishing civic participation. One of them is the power of ideological hegemony exercised by networks of influence in multi-mediated social communications, which prevent targeted citizens from unbiased self-reflection on their authentic interests and possible futures. This is why poor US voters in Wyoming identified with George W. Bush in spite of the fact that his policies harmed them more than any other group of the US citizens, and why voters who are most likely to suffer from policies announced by Donald Trump in his race for the presidential nomination on behalf of the Republican party also tend to support him - rather than choosing Bernie Sanders, who is the strongest advocate of social security in the US political arena. Nor is the problem limited to established democracies – Russian voters also tend to support the absolute rule of the current president - Putin (at least in so far as one can judge from biased reporting in censored media and lack of mass protests against the invasion of Ukraine, Crimea or Syria). In all three cases mentioned above an astonishing display of counterproductive preferences seems to be anchored in ideological preferences and unchecked beliefs, which can be manipulated in multi-mediated ecology of mass media (which manipulate bias and install frames of ideological reference) and social media (which cater to a private world of personal relationships and regulate the level of ideological contamination). Can we design an effective social clearing house for ideological contamination – following the lead of anti-Milosevic opposition in Serbia, which had subverted ideological domination of the ruling elite with subtle irony and succeeded in undermining of the coordinated media campaigns?

Another shadow looming large over public participation of informed citizens in democratic decision-making is disguised as the iron necessity of anonymous forces of progress and technological change. The emergent supercomputerized hyper-intelligent robot-like HAL9000 (to use the name of the rebellious supercomputer from Stanley Kubrick's "2001: A Space Odyssey") embodies the threat of unexpected consequences of technological change enclosing individuals in a seamless web of predetermined technological choices, manipulated by military-industrial complexes and other uncontrolled networks of power and wealth. The impact of Information and Communication Technologies, and of searchable large data bases, upon surveillance of citizens - partly revealed by the Wikileaks and their counterparts in numerous other countries (the Polish "Sowa" restaurant tapes with politicians' private conversations are a case in point) clearly demonstrate a threat posed by the growth of intelligent computers, profiling techniques and the experiments of Google, Facebook and Microsoft. This threat is as real as a large and growing family of big brothers watching most of us most of the time. Does this mean that the future of most professions as we know it will be undermined by computers as much as the weavers' profession had been undermined by a steam-powered spinning wheel? Should we have been more alert and smash the Commodore desk computer before it grew into a mobile phone and portable iPad? No, not really, really not. We should invent more democratic forms of deciding about the future technologies, not close technologies down hoping that the earlier ones will be easier to control. Social learning does not have to involve throwing out these parts of a curriculum, which are hardest for the student to master.

PROBLEMS OF SOCIAL LEARNING

When people think and talk, they exchange views, ideas, words, concepts and predictions about future, which is designed and negotiated before it becomes a cultural fact (cf. Appadurai, 2013). These dialogues and "multilogues" are usually taking place in institutionally defined frames. Public discussions are printed (which means that newspapers and books can circulate and broad audiences can read), broadcast (which means that networks of waves are available as well as individual receivers) or digital (wires, glass fibers and computerized centers enable us to enjoy a webcam a website or a webinar, which are cases of gradual merger of the platforms of social communication and new flexible and open learning plat-forms). Learning is a very complex and multilayers process and it is an egalitarian pursuit followed at different speeds by individuals - egalitarian in allowing increasing numbers of increasingly better educated individuals to converge and participate in parliaments without walls, musea without walls or concert halls without walls. More human individuals currently alive have access to more discussion platforms and to the exchange of ideas or to the sharing of experiences than ever before.

But learning is also oligarchic in organizing and focusing attention of individuals and organizations on meritocracy and by managing the instrumental assessment of learning results, or of the ranking of teaching institutions. Egalitarians profit from information and communication technologies, but so do the "elitarians". Oligarchies assemble think tanks to imagine and design futures before the powerful wave of those, who are heading down the emergent social hierarchies, organize a protest movement, orchestrate an institutional pressure and wage a competitive class power struggle. When new inequalities emerge, the question about the speed of learning becomes crucial. Think tanks begin to matter. A think tank assembles individuals, who are defined as experts, who have thought more than others about a given subject and who had demonstrated their knowledge of a given domain of academically defined inquiry. A think tank does not only mobilize the academic analysts, with knowledge of the matter, but also media professionals, who know how to make knowledge matter and PR persuaders, who control spans of attention. A media professional in PR and marketing knows that an Einstein needs a Beyonce filmed in a refugee camp to make an attractive package and sell a political message with a designed ideological prompting. A serious TV program targeting concerned citizens assembles celebrities, whose fame may rest to a certain extent on expertise, even though selection itself does not have to correspond to the actual depth of learning. Things do not differ very much inside academia. Any contemporary university assembles individuals, who had been trained in certain ways of learning and who became fluent in coaching, assisting the learning of others, but who differ from the point of their readiness to share or hoard knowledge they are co-creating. Increasingly, contemporary universities face the problem of two tracks of knowledge development. One is the cloning of intellectual aristocracy, which publishes in top academic journals strictly separated from broader public, another is assembling the most varied and multidisciplinary teams capable of conducting a public debate, especially in case of difficult and fuzzy problems. To the intellectual aristocracy gate-keeping is essential and learning has clear milestones – PhD's or Nobel prizes. Institutes of Advanced Studies are the breeding grounds of "intellectual excellence". To the more populist and egalitarian academics, milestones of learning are to be found in hearts and minds of learning masses, not in the enclaves of the teaching professionals. One never graduates, nor should one believe that graduation from life-long learning is either possible or desirable.

SOLUTIONS IN CLASS-BIASED EDUCATION

One of the least researched dimensions of those differences in social learning trajectories is the dimension of a class struggle. Some academic professionals share their knowledge with the academic experts, who are interested in helping the oppressed and exploited masses – a world-system analysis school exemplified by Immanuel Wallerstein (cf. Wallerstein, 2004) is the case in point. Some academics share with the institutional centers of establishment's power – International Monetary Fund or the World Bank experts (for instance, Kenneth Rogoff, Joseph Stiglitz) are cases in point. Some, John Kenneth Galbraith or Amartya Sen, are consulted by the United Nations Organization and occasionally mentioned in the media, but fighting world poverty in addressing the aid for third world countries or comparing birth control campaigns in China and India are issues which, if tackled by the UNO, are less easily assigned a political or ideological color.

Let us compare some of the diagnoses and prognoses generated by two learning events, which had emerged from the mainstream of social learning and teaching trajectories in contemporary societies. They are a reflection of a class struggle in social learning processes organized, controlled and managed by contemporary societies. Two large and media-reported learning events have already assumed a role of MOOC's (Massive Open Online Colleges) even before some universities had decided to offer them on a global scale: meetings of selected elite individuals in Davos and assemblies of rank and file citizens in Porto Alegre. The composition of the participants in annual events organized in Davos, reads like a "who-is-who" list of a snobbish club. Davos itself is a place rich in meaningful associations because of the symbolic location of the action of the novel "The Magic Mountain" by Thomas Mann and because of the ambiguous role played by the Swiss banks in offering safe haven for non-democratic elites. The mix of invited Davos guests differs from a composition of participants who had been meeting since 1989 in Porto Alegre in Brazil in order to participate in budgetary decision-making within a globally followed democratic teach-in (in 2001 more than 16.000 individuals assembled in various locations around the city in order to discuss the most desirable ways of spending the next year's public budgets, cf. Smith, 2016).

Let us compare these two events as attempts to re-invent democracy, generate institutional entrepreneurship and develop online events as learning experience for civic societies and for the elites. Can academic networks and institutional clusters of knowledge-generating, cooperative links become the lynch-pins in democratic entrepreneurship based on more universal access to knowledge and on the re-design of both Google corporation and Google search engines? Are we moving the first generation of giga-data navigators towards the United Wikileaks Organization as a free knowledge-handling platform of mankind?

Well, we certainly are learning that we should learn rather for the sake of learning than learn in order to meet the expectations defined by teachers in diplomas and ritualized in graduation ceremonies. Learn, young and old man and woman, learn, but do not believe that learning ends when graduation has taken place. "Non scholae sed vitae discimus" (in a free translation - we do not learn to please schools, but to improve lives). This relative lack of respect for the graduating ceremonies and props does suggest a preference of an ideal learner to lean towards the egalitarian rather than "elitarian" end of the political spectrum. Let us note that egalitarian does not mean "left" as in "communist" or "socialist", though it may mean "anarchist" or "anarcho-capitalist" (as in the Spanish school represented by Madrid economists, for instance Philipp Bagus). It means a focus on the world-system approach of Wallerstein, Braudel and the Polish historians of the manorial-serf economy (which had emerged at the eastern European periphery of the emergent capitalist world-system in the XVIth and XVIIth century, parallel to the re-emergence of slavery in the Americas) – Malowist, Rutkowski, Kula, Topolski. Without the analyses offered by these sociologists and historians one can understand neither the recent migration crisis nor growing difficulties in containing it within new geographies of power and unequal exchange. "Histories create geographies, and not vice versa" as Appadurai had observed (Appadurai, 2013,66). Marx, Simmel, Wallerstein did reinforce social learning processes facilitating learning about systemic power of capitalist world-system of exchange. But learners did not graduate with the Russian (1917) or Chinese (1949) communist revolutions, nor did social learning processes come to a standstill with the reversal of the economic policies of the Soviet's Russia or People's China. Learning how to cope with unpredictable breakdowns of daily routines, learners had recently discovered that religion ceased to be the opium for the masses and became the social bank of values, providing loans for social organizers and entrepreneurs of all ideological, political and economic colors, including the violent ones (e.g. The Islamic State). Democratic entrepreneurship does not grow only out of a barrel of a gun - it can also be sparked by a sermon. Governments are not the only social carriers of political, collective actions. Churches, believers, religion in general can play a vital role in reducing complexities of social imaginaries (cf. Taylor, 2003), which is part and parcel of social learning and in charting the course of action for emergent social actors and agencies trying to understand and govern processes of learning and doing.

SOLUTIONS AND RECOMMENDATIONS

Creativity by Trial and Error

Political parties competing for government positions claim that that they will increase their constituencies' creative inputs by stimulating innovativeness. Less waste, less human hardship, less energy consumption, etc. Business corporations announce new Human Resource policies based on a hunt for the most talented employees, who will presumably guarantee a high level of creative inventions and innovations. The best and the brightest graduates of top educational institutions are employed by the top research centers run by the networks of states and markets. Start-ups are breeding at an accelerated rate not only in specially designed high-tech zones (usually based on an idealized image of a silicon Valley where teenagers make history concocting new supercomputers in their parents' garage) but also in specially designed incubators, which are supposed to stimulate creativity, increase inventiveness, improve productivity and competitiveness – and in the longer run bring about the new post-industrial and post-digital revolution (more than 70% of Stanford University students, one of the richest and most prestigious universities in the world, declare a wish to create a start-up). It might seem that learning has become the main target of all individuals in all walks and talks of life and that the rise of creative class (Florida, 2005) in complex cities is the main game in global town, with major battles being fought not for territory or glory, but for attention (Franck, 1998, 2005). However, social learning does not happen in a classless vacuum; governing knowledge and learning requires a continuous oscillation between Davos and Porto Alegre, between truth and democracy:

The fundamental problems facing any democracy – such as our solution to the economic crisis – are not problems of political will as much as they are cognitive failures that must be resolved with both a greater knowledge of the complex realities over which we govern and a fine tuning of the tools of governance. (Innerarity, 2013,ix)

The language, in which the author of "The Democracy of Knowledge" and "The Future and Its Enemies" has phrased his views, includes the term "cognitive failure". This presupposes a metaphoric view of history as a long learning trajectory of human individuals and collectives. But mandarins learn differently from peasants, and Davos speakers talk differently from Porto Alegre presenters. For our purposes we may accept this metaphor, provided we can account for the complexities of learning processes caused by class struggle. Classes learn and compete with what they had learned – with different, usually unexpected results.

Innerarity does not explicitly say so, but we might consider two world wars of the past century as the most serious, grave and tragic consequences of the most important cognitive failures of mankind, namely communism and fascism, which emerged out of class struggles as illusory alternatives to capitalist inequalities and democratic deficits. Illusory but ideologically promoted as viable and robust – as robust as bodies of young Germans or Russians presented *en masse* to Hitler and Stalin during festive parades. From the point of view, which is focused on avoiding comparable consequences, the period 1945-2015 can be considered a very successful one. In a sense, mankind, class struggle or not, learned to avoid world wars, and classes settled for strikes, parliaments, negotiations and a relatively peaceful coexistence after Hitler was removed and Stalin died. When the Soviet communist leader embraced the US American fridges, cars and better homes and gardens exposed on a consumer fair in Moscow in mid-1950ies, after the post-stalinist "thaw", he had signaled the recognition of the successful learning process – capitalist market was recognized as superior to the state-regulated economy. The Chinese had followed after the death of Mao Tse Tung with "to be rich is glorious" campaign of the capitalist reorganization of state owned enterprises. It might seem that "a lesson learned" was "do not turn back to communism".

However, there are limits to the learning metaphor as an explanation reducing social complexities and facilitating new interpretations of historical sequences of events. Considering class struggle, we should not forget that learning was not limited to the ruling classes – the masses had also learned and their "American dream" of a better home with garden, had basically been recognized and implemented after steam, electricity and oil pushed the second industrial revolution of late 19th - early 20th century forward. Ironically, the country, in which trade unions and strikes were crushed and socialism never flourished, made life of the working class much better than countries with strong social-democratic or communist parties. Paradoxically, WWII also had an egalitarian influence upon the American society in the US, first by a universal draft and second by the Marshall's 'GI Bill" after demobilization. Reading contemporary debates among economists, we learn that in the years between the 1927 financial breakdown and 1973 OPEC policies of raising the oil prizes (the US historians speak of long-term cycles and select the dates 1920 to 1970) the rise of the living standards worked towards diminishing of inequality. Quality of daily life of the masses increased, hardships of domestic chores had been greatly reduced, urban and national transportation became cheaper and safer. This has not continued. In the period between 1970 and 2016 inequality grew, while economic growth stagnated and the quality of life of the masses did not improve as dramatically as in the years 1920-1970 (cf. Gordon, 2016). According to some historians of economic growth and economists studying history of economic cycles, comparing economic growth in the years 1920-1970 to the period of 1970-2016 we get a rather contrasting set of social realities. Growth has slowed down or even stopped altogether. Apple iPad or iPhone are brilliant gadgets, but their influence upon daily life of the masses is less beneficial and egalitarian than was the influence of a washing machine, indoor plumbing or safe processed food. Neither creativity of Apple designers and engineers, nor ingenuity of Uber's Travis Kalanick can enable our economies and societies to match the past rate of growth and improvement of the quality of life for the masses in the USA.

Current situation looks differently if we do not look at the US but at China or India – there growth of the middle class continues at an impressive rate. The rise in middle class living standards is frequently mentioned as a visible result of the growth of economy and of the improvement of living standards of hundreds of millions of individuals. It continues, or rather picks up while the US living standards stagnate. A cautious conclusion might be that the US Americans stumbled upon the right mixes of market growth and egalitarian influences of public authorities by coincidence, by trial and error after the Great Depression of 1927-1929, while the Chinese and the Indians have designed theirs at the beginning of the last two decades of the 20th century as an imitation of the past US policies within the local contexts. Learning by trial, error and design?

FUTURE RESEARCH AND CONCLUSION

Is Crowdsourcing an Option?

Some of the problems facing us today are the results of successful solutions of yesterday. Car transportation has become accessible to everyone, but traffic congestion, pollution and parking difficulties prompt us to look for cars and bikes, which can be rented for a ride, not owned for life. Uber-like solutions to the problem of unused car capacities are probably a premonition of things to come, because we learn fast to do more with less and to create and run project networks and platforms without re-creating the pyramids of power and ownership. Crowdsourcing can become a way of reducing the rate of cognitive failures, by making it possible to accept creative inputs from Porto Alegre, not only from Davos. If the first great irony of fate happened after the 1927-1929 crisis, when the inequalities of the "gilded age" were significantly reduced and lower and middle classes could increase productivity and the quality of their lives at the same time, then maybe the second great irony of historical fate is unfolding now, when millions of migrating individuals head towards the few richest EU countries and the USA, guided by the image of their past as "middle class paradises". When they arrive, if they are not kept in the twilight zones in Turkey or Greece, or the US-Mexico border, they find the hostile populists who exploit the fears of the local masses and the ambivalent politicians who exploit political correctness instead of building a more sustainable solidarity. Ability to successfully perform crowdsourcing is thus not only a matter of avoiding cognitive failures in our increasingly complex and extensive knowledge farming, but also a contribution towards increasing the vigor of democracy as a way of life and an egalitarian matrix of sustainable social forms. Crowdsourcing, in other words, could bring us closer to the new arrangement of existing knowledge and a knowledge-to-come. Perhaps it could help us end the long process of the primitive accumulation of knowledge and enter the period in which our iPhones and plug-in sockets at home will give all individuals as easy and universal access to information as they already have to electricity. The possibility is already explored by a few, hopefully it will be successfully turned into reality by many:

As the twentieth century outgrew its atomic zeitgeist, information became the new ace in everyone's hand - Hidalgo, 2015, p. 13

REFERENCES

Appadurai, A. (2013). *The Future as Cultural Fact*. London: Verso.

Brown, W. (2015). *Undoing the Demos. Neoliberalism's Stealth Revolution*. Brooklyn, NY: Zone Books.

Florida, R. (2005). *Cities and the Creative Class*. New York: Routledge.

Franck, G. (1998). *Oekonomie der Aufmerksamheit; ein Entwurf*. Munich: Carl Hanser Verlag.

Franck, G. (2005). *Mentaler Kapitalismus*. Munich: Carl Hanser Verlag.

Gordon, R. J. (2016). *The Rise and Fall of American Growth. The U.S. Standard of living Since the Civil War*. Princeton, NJ: Princeton University Press. doi:10.1515/9781400873302

Hidalgo, C. (2015). *Why Information Grows. The Evolution of Order from Atoms to Economies*. New York: Basic Books.

Innerarity, D. (2013). *The Democracy of Knowledge*. New York: Bloomsbury.

Magala, S. (2009). *The Management of Meaning in Organizations*. Basingstoke, UK: Palgrave. doi:10.1057/9780230236691

Magala, S. (2016). From Linnaeus to Darwin. From Cultural grid to "Going Native". In M. Rozkwitalska, L. Sulkowski, & S. Magala (Eds.), *Intercultural Interactions in the Multicultural Workplace. Traditional and Positive Organizational Scholarship*. Springer International Publishing AG.

Memmi, A. (2006). *Decolonization and the Decolonized*. Minneapolis, MN: University of Minnesota Press.

Smith, G. (2016). *2009), Democratic innovations. Designing Institutions for Citizen Participation* (1st ed.). Cambridge, UK: Cambridge University Press.

Taylor, C. (2003). *Modern Social Imaginaries*. Duke University Press. doi:10.1215/9780822385806

Wallerstein, I. (2004). *World-systems Analysis. An Introduction*. Duke University Press.

Webster, J. G. (2014). *The Marketplace of Attention. Howe Audiences Take Shape in a Digital Age*. Cambridge, MA: The MIT Press.

KEY TERMS AND DEFINITIONS

Cognitive Failures: Paradigmatic research programs corrected by reality checks and transformed into the academic policies of "scientific revolutions."

Creativity: Ability to design and envision new relations between the past patterns, present improvisations and future outcomes, tenuously linked to start-up incubators.

Crowdsourcing: A technology of re-inventing democracy in economic, political and cultural decision-making tested in online accumulation exercises.

Elitarian Bias: Patterns of social imaginaries about desirable education modelled after elitist institutions of the past beginning with Plato's Academy and ending with Harvard Business School.

Emergent Actors: Crystallizations of pockets of resistance inside the mainstream social learning trajectories leading towards political articulation and ideological virtual representation.

Right Mix: A historical cluster of institutional policies stimulating market growth and egalitarian redistribution (e.g. New Deal in the USA).

Social Learning: A complex process of multiple dialogues cutting through patterned interactions and institutionalized communications and emerging as social learning trajectories.

Chapter 2
Adopting the Concept of Business Models in Public Management

Barbara Kożuch
Jagiellonian University in Kraków, Poland

Adam Jabłoński
University of Dąbrowa Górnicza, Poland

ABSTRACT

The aim of the chapter is to propose the principles of adopting the concept of business models in public management. The scope of the work includes the specific principles of business management and public management, examined in terms of integrating the attributes of public organizations that ensure they achieve appropriate functionality. The result of scientific reflections is an attempt to design the canvas of the public organization business model based on an analogy taken from business management for the conceptualization and operationalization of the specific key attributes of the public organization business model. The justifiability of adopting the concept of a business model in the theory and practice of public organization functioning will be illustrated by the solutions used in local units of public employment services in Poland.

INTRODUCTION

The dynamics of changes in management theory and practice determine new research and implicational perspectives. The emergence of new management concepts and the creation of new space in the conceptualization and operationalization of ontological beings is a challenge for current management sciences. The interpenetration of different solutions in various areas of business, public and social affairs almost becomes normalized. The interpenetration creates a new dimension of the value embedded in the needs of its customers expecting adequate contributions and incentives from the relationship with business, public or social organizations. The expected value can be delivered, *inter alia*, through the ontological being which is the business model, understood as a specific model of an organization demonstrating how

DOI: 10.4018/978-1-5225-2215-7.ch002

the organization performs its genotypic function. The ontological being is reserved not only for business, but can also be used for the design, development and growth of social and public organizations. This may sound like an oxymoron, but if one becomes more familiar with the concept of business models, opportunities and new mechanisms emerge for creating value for the benefit of social and public services through modified solutions of the concepts of business models, adapted to social and public services. It is important to move from business models typical of companies to social and public business models and the business models of other types of organizations. The concept of new public management (NPM) is particularly conducive. It is based on the efficient use of business solutions in public organizations. Other models of public management, however, do not prevent the development of relevant business models. It should be noted that NPM also involves a shift from economic to social and public value. The result is the transition from economic to social and public dividend.

Addressing such complex problems, we chose a component approach popularized by A. Osterwalder. Then we focused our analyses on the significance of distinguished groups of public organizations' stakeholders, on citizen-centric orientation in building the business models of public organizations, and also we looked for linkages between sustainability and public value. As a result, we proposed the internal canvas of a public business model, the external canvas of a public business model, and the extension canvas of a public business model.

BACKGROUND

For many years, the concept of business models has played a key role in the implementation of strategic management mechanisms. This ontological being has become an inherent part of management science in both theory and practice. Nowadays competing by means of business models has become a reality. The peculiar configuration of tangible and intangible resources that captures value from the market through one's own value created and supported with the understandable logic of generating income for the organization determines the development of modern entities. What is amazing is that the concept, due to its universality, can be applied not only in business but also to the rules of the functioning of social and public entities, due to the fact that collaborative advantage is of similar importance to competitive advantage in companies in the latter. In view of the above, it can be stated that the dynamics of using the concept of business models in the theory and practice of modern organizations is highly surprising. In particular, since 1995 there has been an increased emergence, among others, of a number of scientific publications in both global and local scientific journals. An even greater increase took place after 2005.

At the same time, people have always known how to organize numerous individual and collective actions at work, during a fight, in religious practices, and even during games. A well-known example of great achievements in the field of work organization is building a fleet of 120 large rowing warships for the summer campaign with Carthage in the Roman shipyards within a few months in 260 BC. This was possible due to the use of the assembly line. A similar solution was used in the 12th and 13th centuries in English manufactories for the production of arrow points. This invention of the ancient times had the largest application in the 20th century due to its introduction into the automotive industry by H. Ford.

The issue of organized human activities has been reflected on for a long time as well. The first scientific reflection of this kind emerged thousands of years before the discipline called today management sciences was developed.

A need to synchronize, coordinate and integrate activities on a significant scale resulted from the birth of the ancient phalanx that based its efficiency on cooperation and a sense of community. This necessity resulted from the construction of Sumerian irrigation channels, Greek urban complexes, Roman roads and aqueducts etc.

These examples emphasize the fact that the issues related to the organized activities of human teams initially focused on the organization of social and public life. They were the subject of interest of the best ancient minds, including Confucius, Plato, Xenophon and Aristotle. The following examples are from the Renaissance: N. Machiavelli, T. More, and T. Campanella, and in the 19th century these were C.H. Saint-Simon, Ch. Fourier, and R. Owen.

In theories of public administration and management a few models were centered on formal organization. The Weber's model is the most popular. It was focused on searching for a structure of public administration that would make it resist political incursions and was seen as the method of achieving neutral and competent governance. It embraced: hierarchy, division of labor, professional recruitment, promotion based on seniority, tenure, discipline under predictable forms, dismissal by means of court action (Lane, 2009). In this context the words of Weber (1947) are significant: "Bureaucratic administration means fundamentally the exercise of control on the basis of knowledge. This is the feature of it which makes it specifically rational" (p. 339).

A hundred years ago bureaucracy was the answer to search for effectiveness. The criticisms of the bureaucracy model and following search for more effective public sector has led to other models of public management. For example, Peters (1996) distinguished and characterized the four models of public management: Market Government, Participative Government, Flexible Government, Deregulated Government (Guy Peters, 1998).

New Public Management is a model of public administration based on management similar to business management. This model is described also by many other terms, which include: managerialism (Pollitt 1990), market-based public administration (Zhiyong & Rosenbloom, 1992), entrepreneurial government (Osborne & Gaebler, 1992), business–like management, and public managerialism (Kickert, 1997).

The synthetic characteristics of contemporary public management models are illustrated in Table 1.

Table 1. Characteristics of contemporary management models

No.	Models	Theoretical foundations	Point of reference	Shared values	Overall manner by which public interest is accomplished	Regulation mechanism
1.	PA	Administration science; Political sciences	Political system	Ethos of officers and the entire public sector	Formulation of public policies and their implementation	Hierarchy
2.	NPM	Public choice theory; Management sciences – economic aspects	Individual Organizations	Effectiveness of the market mechanism	Management of organizational resources and attainment of planned outcomes	Market and bilateral and tripartite agreements
3.	NPG	Organization theory; coop. and Network theories; Management sciences – humanistic aspects	Organizations together with their environment	Dispersed and questioned	Agreeing on basic values and the meaning of management in organizations, relationship management	Trust-based networks and agreements

Source: adapted from: (Osborne, 2010, p. 10).

The recent comparison of key public management models (Kożuch, Kożuch, & Sienkiewicz-Małyjurek, 2016) reveals fundamental changes of theoretical sources and manners by which problems are addressed which meet the expectations of citizens and other stakeholder groups. Firstly, it was administration and political sciences, then economic aspects of management sciences, and presently humanistic aspects of management sciences. The point of reference evolved from a political system through individual organizations to organizations together with their environment. Overall manner by which public interest is accomplished has changed. That means the shift implementation public policies and attainment of planned outcomes to relation management. This is the transition from the prevalence of legal and administrative aspects, through economic to social and humanistic aspects.

In processes of development models of public management organizational effectiveness of public sector entities and its key criteria play a significant role.

Effectiveness occupies an important place among the principles underpinning good governance as defined by the European Commission (European Governance, 2001), along with openness, accountability, participation, and coherence. This issue may be traced using the general model for organizational effectiveness in public management (Figure 1). It draws on the fundamental forms of organizational performance, i.e. efficiency, effectiveness, and utility and sustainability. Typically, effectiveness is defined as the degree to which we approach the aim, accomplish it, or enable or facilitate its delivery. When assessing effectiveness, the cost is neglected as effectiveness is neutral on efficiency (Lawrence, 2014, p. 21), and from among the ingredients of a useful result, only the predictable effects are taken into consideration.

In this model, the objectives are represented by the anticipated utility, while the results by the achieved utility and sustainability. Indirect results can also be distinguished which are termed as effects and final results as impacts. Meanwhile, the relationship between inputs and outputs emerges as efficiency, namely the form of efficiency as achieved maximum productivity with minimum effort or expense. Attention is

Figure 1. General model for organizational effectiveness in public management
Source: adapted from: (Pollitt & Bouckaert, 2004, p. 106; Kożuch, 2004, p. 78).

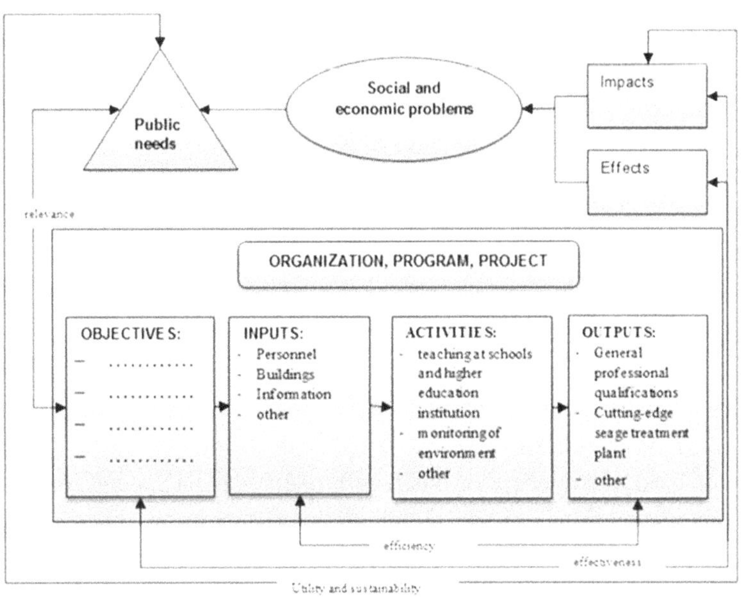

also paid to the category of utility and sustainability which implies to what extent results are attained, that is how their impacts contribute to the fulfilment of public needs.

While referring the concept of organizational effectiveness to public management models, the role of citizen involvement may be discerned (Kożuch & Kożuch, 2016, p. 122-125). Moreover, as is widely accepted, strategic management in public organizations is significantly influenced by socio-economic problems in a given country or region, which give rise to public needs. Decisions pertaining to which of these needs will be met by public authorities are taken within the framework of political processes, leading to the formulation of public policies. As part of social participation, citizens and grassroot organizations may influence political decision-making processes. Participation leads to the involvement of citizens in the formulation of public policies. It is of particular importance at the local level, where citizens jointly decide on the final dimension of public policies defined at the central level in relation to their community. Public participation therefore allows for tailoring the local needs of public policies implemented by public administration and other public organizations. It contributes to the increased effectiveness of public management in such a way that it forms the basis of formulating organizational objectives that meet public needs. Without the involvement of citizens, it would be impossible to identify organizational goals properly, because officials play a dominant role in these processes, and their choices do not have to coincide with the preferences of stakeholders. Therefore, it is of vital importance to create the conditions for including citizens in public management processes.

APPROACHES TO BUSINESS MODELS

The concept of business models is now becoming one of the phenomena of modern management. Competing by means of business models became a reality a long time ago. Companies are convinced of the fact that the market victory can be ensured by an effective and efficient business model, which turns the value offered in the market into profits for the company. Such an interpretation of present reality may attract attention to the evolution of an interest in the concept of business models in management sciences in recent years.

A sudden increase in scientific and non-scientific publications around 1995 (Zott, Amit, & Massa, 2010) may also have resulted from the fact that the concepts of strategy were exhausted in a dynamic increase in organization performance. It turns out that strategy alone is no longer sufficient. There is a need for something new, different, unique and effective. A business model has therefore become an opportunity for many organizations as a concept for creating a new dimension of strategic management, regardless of the type of organization. As a result of such thinking, the concept of business models has often appeared in many contexts.

Such a level of interest in business models has resulted in recent years in the emergence of many definitions, and consequently, great confusion in the strategic management of the organization.

Many perspectives, views and interpretations have made this a ‘hot topic’ throughout the world. However, it has also opened up numerous possibilities for applying business models, especially for social, public and hybrid organizations.

As previously mentioned, the concept of business models plays a special role in the strategic management of both commercial entities and social and public organizations. There are many interpretation approaches to this issue; it is vital, however, to examine business models in terms of components. The component approach is of particular importance in creating the proper configuration of tangible and intangible resources or strategic factors determining the ability of certain entities to create value. The component approach should also be determined taking into account the contextual approach. A context discussed has undoubtedly an important dimension in relation to the environment in which an organization is embedded. This applies both to the environment examined in terms of a system and a network. By analyzing and synthesizing the contextual and component approaches, two tables were developed. Table 2 shows the definitions of business models taking into account: the general definitions of the business model, the definitions of the business model relevant to business activity, and the definitions of the business model relevant to social and public activities. The appendix shows the components of business models in four contexts: the general context, business context, social and public context, and integrated context. A general context refers to the general view of the components of the business model that can be applied to different types of organizations. A business context focuses exclusively on mechanisms relevant to management from the business perspective. A social and public context is typical of social and public organizations. In this context, other attributes are defined, which represent the dimension of a business model that can be applied in the above organizations. An integrated context, as its name implies, makes it possible to combine components for different types of organizations. This also applies to hybrid organizations combining economic and social objectives.

Business models should also be looked at not only through the dimension of the real world but also through the rules of the virtual world. As we live and function both in the real world and the virtual, business models and their components emerge that exist in parallel in these two worlds.

This determines market opportunities in each of these worlds separately and concurrently. We therefore deal with business models and e-business models. Nowadays a scientific debate is taking place as to whether the first business models emerged out of business or e-business. Notwithstanding the above, organizations operating today use both business and e-business models.

Within these two worlds, the new definitions of e-business models have been developed that contributed not only to business but also the social and public sectors. The public sector has also been virtualized through the concept of e-government.

The study of business models includes the methods of their conceptualization and operationalization. Subsequently, the understanding of the concept of business models may result from a number of possible scenarios.

Conceptualization is associated with a design of business models such that allows for their appropriate operationalization, which is focused on the application of optimal management mechanisms to capture the desired value from a business model.

In order to effectively express the concept of the business model, the authors quote the definition by Teece (2010), who believes that a business model defines how the enterprise creates and delivers value to customers and then converts payments received to profits. The following is a synthetic review of the literature on the concept of business models from many perspectives.

The business model approach understood as a type of a market player in the value chain is highlighted, for example, by Obłój (2010): operator, integrator, and conductor; Gołębiowski, Dudzik, Lewandowska and Witek-Hajduk (2008): traditionalist, market player, contractor- specialist, distributor, and integrator. The approach to the e-business model from the perspective of the market player is presented, for example, by Timmers (1999): e-shop, e-procurement, e-mall, e-auction, value chain service providers, virtual business community, and cooperation platform; Rappa (2002): Advertising, Brokerage, Community, Infomediary, Manufacturer, Merchant, Subscription, and Utility; and Applegate (2000): Focused Distributor Models-Retailer, Marketplace, Aggregator, Infomediary, Exchange, Portal Models – Horizontal

Portals, Vertical Portals, Affinity Portals, Producer Models - Manufacturer, Service Provider, Educator, Advisor, Information and news services, Custom Supplier and Infrastructure Provider Models- a number of sub-models, e.g. infrastructure portals.

A business model understood through the prism of company profitability has been presented, among others, by Slywotzky (2000). Together with his team, he described 22 profitable business models based on the experiences of American companies. A cognitive approach to the business model has been presented by Magretta (2002).

The link between the business model and strategy and business processes is highlighted by Osterwalder and Pigneur (2002), and Bossidy and Charan (2010), and Niemczyk (2010). In terms of value creation, the definition of the business model is presented, among others, by: Seddon, Lewis, Freeman and Shanks (2004), and de Witt and Meyer (2002). The following authors focus on studying the business model from the perspective of stakeholders: Hoque (2002), and Voelpel, Leibold, Tekie, and von Krogh (2005), or Jabłoński (2008). The definitions of networked business models are presented, inter alia, by Perechuda (2013) and Jabłoński et al. (2013). The link between the business model and the resource-based view is highlighted by Krzakiewicz and Cyfert (2011). The business model ensuring the stability and continuity of the company is presented, among others, by Demil and Lecocq (2010), Sandberg (2002), Afuah, and Tucci (2003). Nogalski (2009) defines a business model from the development perspective. A hybrid business model is presented by Deodhar, Saxena, Gupta and Ruohonen (2012), and Jabłoński (2015).

The definition of a sustainable business model is presented, among others, by Stubbs and Cocklin (2008), and Boons and Lüdeke-Freund (2013), and Jabłoński (2013), while Neely and Delbridge (2007) focus on a geometric business model. Technological business models are presented by Muegge (2012). Open business models are defined by Chesbrough (2007). A component approach to the business model was presented by Ostervalder (2004).

The above approaches describe the particular complexity of the concept of business models in management science. They indicate a multidimensional look at the business model and create further implications for research. In addition, the interpretation scenarios presented determine directions for the further implementation of the concept of business models. An important factor in determining the issue of how to apply them is to answer the question of whether the business model has to be stable or suitably dynamic.

The selection of this model depends on the area of application. In the changeable market economy it is a dynamic model, while in sectors with low volatility it is a stable model. The key question is which business model best meets the needs of citizens in the area of social and public services?

The following (Table 2) is an overview of the definitions of business models by key strategic management researchers on a global scale, taking into account the general definitions of the business model and the definitions relevant to the business as well as social and public activities.

The above approaches emphasize the particular complexity of the concept of business models in management science. They highlight the multi-dimensional view of the business model and create further implications thereof. They allow one to observe the essential differences in the characteristics of the business models of companies and social and public organizations.

The definitions presented in Table 2 have appeared in the last twenty years. A specific context with a general dimension presents the general definitions of the business model, typical of a standard approach. In this approach it is the structure, key processes, key activities, and customer needs that play a special role. As regards the second approach, the definitions of the business model relevant to business activity were highlighted. Here, the important factors include elements related to the business model intertwined

Table 2. Main definitions of business models

No	General definitions of the business model	Definitions of the business model relevant to business activity	Definitions of the business model relevant to social and public activities
	A business model is a new unit of analysis that is distinct from the product, firm, industry, or network; it is centered on a focal firm, but its boundaries are wider than those of the firm; (2) business models emphasize a system-level, holistic approach to explaining how firms "do business"; (3) the activities of a focal firm and its partners play an important role in the various conceptualizations of business models that have been proposed; and (4) business models seek to explain both economic value creation and value capture. These emerging themes could serve as important catalysts for a more unified study of business models. (Zott, Amit, & Massa, 2011)	A business model articulates the logic and provides data and other evidence that demonstrates how a business creates and delivers value to customers. It also outlines the architecture of revenues, costs, and profits associated with the business enterprise delivering that economic value. (Teece, 2009)	Social business models, which require new value propositions, value constellations and profit equations, and as such, resemble business model innovation. (Yunus, Moingeon, Lehman-Ortega, 2009) The same applies to public activities.
	A business model is based on the cost of innovation, bringing economic benefits for the company and is characterized by the principle of more for less. (Williamson, 2010)	A business model is an organization's approach to generating revenue at a reasonable cost, and incorporates assumptions about how it will both create and capture value (Brandenburger, Harborne, & Stuart, 1996)	A business model refers to the logic of the firm, the way it operates and how it creates value for its stakeholders (Casadesus-Masanell & Ricart, 2010)
	A business model is a mechanism for turning ideas into revenue at a reasonable cost. (Gambardella & Mc Gahan, 2010)	A business model explains how a venture is expected to make money (Fiet & Patel, 2008)	A business model may only be said to be consistent when the choices about its core RCOV components lead to sustainable performance. The open-ended interactions between core components and managers' entrepreneurial initiatives mean business models are always changing. Managers must monitor consistency to ensure sustainable performance. (Demil & Lecocg 2010) A business model explains how a venture is intended to meet justified social or public needs.
	A business model is a crossroads of competence and consumer needs. (Sabatier, Rousselle, & Mangematin, 2010)	A business system used by a company is the configuration of resources (input), actions (processing) and offered products or services (output) directed at creating value for customers. (de Witt & Meyer, 2007)	A business model is a unique blend of three streams that are critical to the business. These include value streams for business partners and the buyers, revenue stream logistical streams. (Mahadevan, 2010)

continued on next page

Table 2. Continued

No	General definitions of the business model	Definitions of the business model relevant to business activity	Definitions of the business model relevant to social and public activities
	A business model includes combined elements that define who, how, when, why and for how much a company provides customers with products and services. (Mitchel & Coles, 2004)	BM is a formal description representing the components and functions of the business implemented by a company to generate revenue and make profit from a sequence of activities. (Ceravolo, Damiani, Fasoli & Gianini, 2010	The business model is a conceptual and architectural implementation of business strategy and the basis for the introduction of business processes. It is a description of the value a company offers to one or several segments of customers and of the architecture of the firm and its network of partners for creating, marketing, and delivering this value relationship capital, to generate profitable and sustainable revenue streams. (Osterwalder & Pigneur 2002)
	A business model is a description of the complexity of the business, which enables the study of its structure, the relationship between the structural elements and how a company- will react in the real world. The structure of the business model is based on the organization's industrial logic (Applegate, 2000, 2001)	The BM is the heuristic logic connecting technical potential with the realization of economic value. (Chaesbrough & Rosenbloom, 2002)	A business model outlines the essential details of the firm's value proposition for its various stakeholders and activity systems the firm uses to create and deliver value to its customers... (Seddon, Lewis, Freeman, & Shanks 2004)
	An ideal business model: • draws a boundary around what is done, • outlines how it is done, explains how to connect internal processes with customers' external requirements and strategic objectives. (Neely & Delbridge 2007)	A BM is venture strategy, architecture and economics, which aim to create sustainable competitive ad-vantage in target markets. BM includes value proposition, customer, internal processes and capabilities, external positioning, economic model and person-al/investor factors. (Morris et al., 2005)	A business model is a concise representation of how an interrelated set of decision variables in the areas of venture strategy, architecture and economics are addressed to create sustainable competitive advantage in defined markets. (Morris, Schindehutte, & Allen 2005)
	A business model describes a holistic, robust and reality-based process for thinking about the specifics of the business. It shows how to connect the financial goals necessary to meet the external conditions in which the firm operates, as well as internal processes such as strategy development, operational tactics and staff selection and development (Bossidy & Charan, 2010)		
	"...Working business models aren't embedded in organizational structures – rather they're organizational mind-sets." (Linder & Cantrell, 2002)		
	"Business Models, though, are anything but arcane. They are, at heart, stories - stories that explain how enterprises work." (Magretta, 2002)		

Source: Own study.

with the elements of strategy. Then we can talk about so-called clean business models and hybrid business models. The third approach focuses on the definitions of the business model relevant to social and public activities. In particular, this approach can be applied to adapting the concept of business models to public management due to strategic factors and elements such as new value propositions, value for stakeholders and value relationship capital.

In the processes of development models of public management organizational effectiveness of public sector entities and its key criteria play a significant role. The outlined specificity of public management and approaches to business models underlie the complexity of public sector business models.

COMPLEXITY OF THE CONCEPT OF PUBLIC SECTOR BUSINESS MODELS

A key element of the discussion on business models is to define their specific components, determining the appropriate configuration. This configuration affects the achievement of the assumed results expressed in economic, social or public terms.

In the approach presented an attempt is made to show the important components of business models in context proposed in the relevant literature. These components are analyzed in four contexts: general, business, public and social, and integrated one (Appendix).

As regards our research, general and social contexts are the most important. The business context allows for showing significant differences between the concepts of business models. In turn, the integrated context is a reference point of future research.

The contexts analyzed associated with the component approach to building a business model have an interdisciplinary character. Each context presented and specific components have a different dimension as each author presents a different approach to configuring business models. Thus, in this perspective, the most common components include value propositions, a value network, a revenue model, cost, customers, profits, and product. In the business context, the following can be distinguished: a position of value, the external environment, financial objectives, competitive strategy factors, sources of revenue, value proposition, and products / services. As regards the social context, specific components include: relationships with customers, abilities/competencies, revenue/price, value chain, revenue mechanisms, resources assessment, stakeholder network, a value model, a resources model, corporate governance, business units, services, and units. The integrated context combines all of the approaches presented. The added value of such a presentation is an in-depth analysis of the selection of the individual components of a business model. This is particularly important for the social and public context, where the components presented are different from other contexts. At the same time, these components may be combined in any way by configuring the social and business contexts in order to create a kind of a strategic hybrid. A multi-dimensional approach to this interpretation is contained in the appendix. It deserves further research.

USING A CONCEPT OF STAKEHOLDERS IN ADOPTING BUSINESS MODELS IN PUBLIC MANAGEMENT

A key area of scientific discussions on the adaptation of business solutions for public management is defining entities, the receivers of value created in the place where business activities are conducted. Using the concept of multiparadigmness in management, two key paradigms may be defined that determine the development of management science, namely a system paradigm and a network paradigm. Entities focused on value in the sector will be stakeholders or network actors.

It can be assumed that the concept of stakeholders as groups interested in the so-called stake offered by the sector will be embedded in the system paradigm and the concept of network actors will be embedded in the network paradigm. Focusing on the concept of stakeholders in the context of public organizations, attention should be paid to the key interpretations of the concept of stakeholders.

The concept of stakeholders has become an inherent part of the mechanisms of modern management. Although this concept has been embedded in the theory and practice of management for many years, linking it with other management concepts still opens new perspectives.

The origins of the stakeholder theory date back to the 1960s. The term 'stakeholder theory' was first used by Ansoff and his co-researchers in the analysis of planning processes in the company in the early 1960s. The issue was, however, not taking into account benefits or value creation for stakeholders, but rather the prevention of conflicts between them (Ansoff, 1985). As a result of the activities of the Stanford Research Institute (SRI), a memorandum was drawn up, in which, according to R.E. Freeman, the concept of a stakeholder appeared for the first time. This organization highlighted the fact that managers should pay special attention to and seek the support of various social groups, as only this can ensure the long-term success of the organization, shaping the relationships between them. However, it was not until 20 years later in 1984, thanks to E. Freeman, that this concept was widely recognized among management concepts and methods (Freeman, 1984). In his opinion, a stakeholder is any group or individual who can affect or is affected by the achievement of the organization's objectives (Freeman, 1984). Since that time stakeholders have been classified into different groups and new definitions thereof have been defined. In 1994, Clarkson, among others, divided stakeholders into active and passive and he defines them as "risk-bearers." An active stakeholder bears some form of risk as a result of having invested some form of capital, human or financial, something of value, in a firm. On the other hand, a passive stakeholder is placed at risk as a result of the firm's activities (Clarkson, 1994). In 1994, Langtry also pointed out that stakeholders are responsible for the well-being of the company (Langtry, 1994). Harrison refers the concept of stakeholders to rights and interests that are the result of transactions, agreements and other legal actions, of moral, individual and group nature (Harrison, 1997). According to Mitchell (1997), stakeholders are groups on which the survival of the organization depends. Freeman and Reed (1998) argue that stakeholders are individuals or groups that can affect the organization's activities or are affected by the organization's activities. This means that the establishment of relationships with stakeholders has the market dimension and is included in the concept of a resource-based view and market orientation. Relationships with stakeholders are a resource that, when skillfully used in terms of a wide market orientation most relevant to the assumptions for the stakeholder theory, can lead to the achievement of significant competitive advantage. As a consequence, through appropriate value drivers, the value of a socially responsible company will be created. Stakeholder collaboration was highlighted in the model by Svendsen (1998) (it is worth noting that the concept of stakeholder management is not used here, only stakeholder collaboration as a space for dialogue). The dynamics of strategic changes and the strength to compete in the free market forces companies to create new sources of competitive advantages, often oscillating over the search for creative forms of market expansion. The search for value in resource potential is becoming a factor of the essence of the struggle in a turbulent environment. Considering the core principles of the resource-based view associated with the construction and configuration of unique, distinctive resources and skills, companies should pay attention not only to the objective nature of resources related to the tangible and intangible assets of the company, but also to the subjective nature of resources, inherent in stakeholders gathered around the company. Links and relationships with stakeholders are a kind of resource that, when efficiently and effectively used, can lead to the company gaining significant competitive advantage. These actions will result in building a socially responsible company. Resources in the form of stakeholders affect both the financial and competitive strength of the company, and they also depend on sectoral conditions that determine the stability, attractiveness and maturity of the sector. They ensure the development and growth of the company and are a source of their renewal, thus maintaining the strategic continuity of a company.

This comes down, among other factors, to building a system of stakeholder management. Building such a system together with shareholders provides a logical platform including the redefinition of the place and role of stakeholders and their relationships with the business by determining the influence of relationships and the correlation between the organization's activities with value for stakeholders and determining the influence of stakeholders on value for shareholders. The achievement in practice of the full model of the abovementioned approach requires managers to have interdisciplinary skills and can be implemented only when managers are not only aware of their responsibility to the company, but also of the stakeholders' requirements and exposure to risk. In such an approach, a business model – stakeholders relationship is particularly important. A business model, on the one hand, is the source of value creation for stakeholders; while on the other hand, stakeholders are the recipients of this value.

When undertaking an in-depth reflection on the concept of stakeholders in relation to public management, the different types of stakeholders have been reviewed in order to select such stakeholder groups that are optimal for a public organization.

The issue is, in particular, the subjective character of the business model, which, for a public organization, contains the configuration of different types of stakeholders contributing to the organization by the value of a higher order it offers.

Table 3 shows the typology of stakeholders in terms of the business model of the individual organization.

Table 3. Typology of stakeholders in terms of the business model of the individual organization

No	Criteria	Typology of stakeholders
1.	Focus on the social and public sector	1) Social stakeholders, Public stakeholders
		2) Non-stakeholders
2.	Location to the organization	3) Insiders
		4) Outsiders
3.	Role in strategic management	5) Primary stakeholders
		6) Secondary stakeholders
		7) Key stakeholders
4.	Criterion of the concluded cooperation agreements	8) Consubstancial stakeholders
		9) Contractual stakeholders
		10) Contextual stakeholders
5.	Market orientation	11) Non-market stakeholders
		12) Market stakeholders

continued on next page

Table 3. Continued

No	Criteria	Typology of stakeholders
6.	Degree of relationship dynamics	13) Silent stakeholders
		14) Real stakeholders
7.	The degree of influence	15) Global stakeholders
		16) Regional stakeholders
		17) Local stakeholders
		18) Sectoral stakeholders
8.	The degree of mutual relationship	19) Universal stakeholders
		20) Specific stakeholders
9.	The degree of mutual relationship	21) Primary stakeholder
		22) Secondary stakeholders
		63) Third-degree stakeholders
10.	The degree of influence strength	24) Majority stakeholders
		25) Minority stakeholders
11.	The degree of influence strength	26) Dominant stakeholders
		27) Dangerous stakeholders
		28) Dormant stakeholders
		29) Decisive stakeholders
		30) Demanding stakeholders
		31) Flexible stakeholders
12.	The degree of impact	32) Dependent stakeholders
13.	The degree of impact	33) Pressure groups
14.	The degree of mutual relationship	34) Stakewatches
15.	The degree of mutual relationship	35) Stakekeepers
16.	The degree of mutual relationship	36) Multi-stakeholders networks
17.	Type of influence	37) Business units
18.	Type of influence	38) Civil society
19.	Type of influence	39) Governmental institutions
20.	Type of influence	40) Public-policy network

Source: Own study.

Based on the 40 types of stakeholders identified, a nine-element group of stakeholders contributing the business model for public organizations was further defined. Consubstantial stakeholders are all those who co-create the company by investing their work, competence, skills, knowledge, and equity. These are the stakeholders who are within the organization: employees, shareholders, and owners. Social stakeholders generate social value and are also the recipients of value from the organization. Demanding stakeholders are stakeholders who expect more from the organization, thus creating higher value becomes a priority for them. Flexible stakeholders can provide specific support for a social organization through their dynamism for its benefit. Global stakeholders are those who have a global impact, who may affect a public organization through social and political influence. A public policy network contributes to the understanding of the mutual relationship and the opening to bonds in the network. Pressure groups, through emphasis, not only significantly affect but also try to change the functioning of public organizations, while stakekeepers uphold the social expectations of all key impact groups.

When examining the concepts of stakeholders in relation to the principles of strategic management, this issue should be observed from the multi-dimensional perspective. If we adopt an angle typical of companies operating in business, we could observe that this typology is not adequate to the specificity of stakeholders of public and social organizations of stakeholders will be different from the typical approach to social and public organizations. Thus, the proposed typology of stakeholders of public and social organizations can make it easier for managers to adopt an appropriate methodology and use it skillfully in a decision-making systems aimed at a particular type of stakeholders. It must be emphasized that, on the one hand, it is a complex process, and on the other hand, it is extremely important for adopting business models to public management.

The following is a proposal of a list of nine key types of stakeholders appropriate for a public organization.

An attempt to select the types of stakeholders allows for a broader view of the public organization, whose aim is to meet the expectations of citizens and other stakeholders, while achieving the expected results from a social, public and also economic perspective. Stakeholders therefore are an important component of a public business model that generates expectations fulfilled publicly by its business model.

THE CITIZEN-CENTRIC ORIENTATION IN BUILDING A BUSINESS MODEL OF PUBLIC ORGANIZATIONS

The definitions of a business model from the perspective of stakeholders focus on four dimensions: value creation, value capture, value networks and strategic choices.

Figure 2. A list of nine key types of stakeholders appropriate for a public organization
Source: Own study.

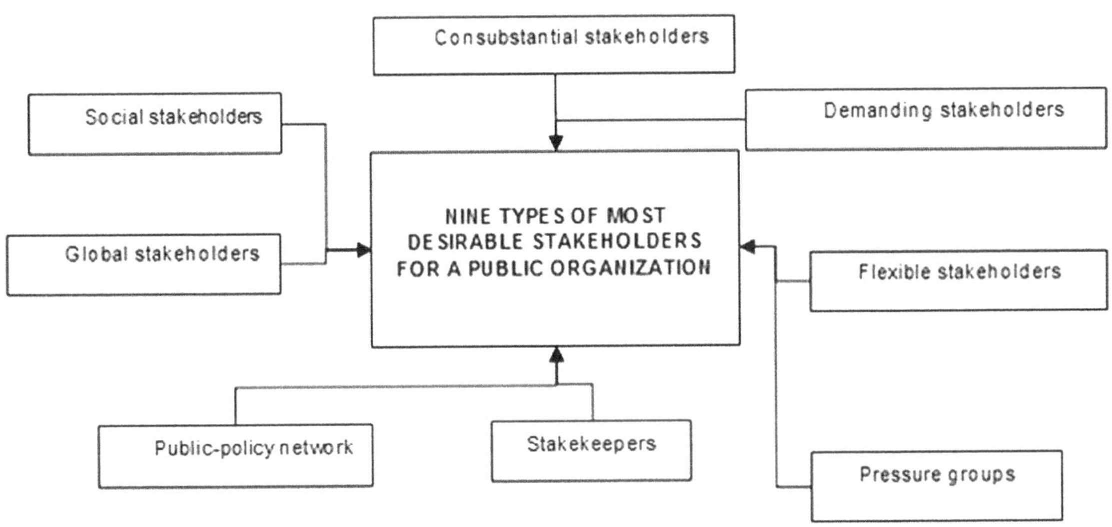

For example, Chesbrough and Rosenbloom (2002) opine that a business model has two important functions. It must create value within the value chain and it must capture a piece of value for the focal firm in that chain (Chesbrough & Rosenbloom, 2002). Zott and Amit (2010) argue that a business model is geared toward total value creation for all parties. The greater the total, the greater the focal firm's bargaining power, and the greater the amount of value it can appropriate (Zott & Amit, 2010). Casadesus-Masanell and Ricart (2009) claim that, put succinctly, a business model refers to the logic of the firm, the way it operates and how it creates value for its stakeholders. Strategy refers to the choice of business model through which the firm will compete in the marketplace. Tactics refer to the residual choices open to a firm by virtue of the business model that it employs (Casadesus-Masanell & Ricart, 2009). A business model is also a structural template of how a focal firm transacts with customers, partners, and vendors. It captures the pattern of the firm's boundary, spanning connections with factor and product markets (Zott & Amit, 2008).

These few abovementioned definitions indicate that the concept of business models is embedded in the process of value-based management. This issue can be approached in terms of results. Then the focus is on the search for organizational high performance with an emphasis on its internal needs. The examination of the organization through the prism of external expectations will direct mental processes to a wide range of stakeholders who are the recipients of many values created not only within the organization but above all during interaction with its environment. Then, the use of the concept of customer-centric orientation becomes meaningful. The result of such thinking is the customer-centric value proposition. Expanding this concept by means of stakeholder-centric orientation is the natural behavior of today's organizations. As regards public organization, a citizen-centric orientation approach can be proposed.

The definition of Citizen Centric Solutions (CCS) implies the centrality of consumption in determining sustainability outcomes. Therefore, we explore the relationship between consumption and the three dimensions of the CCS performance metric. Consumption is not only a basic necessity for survival, it is also critical to our personal, social and economic well-being. In fact, consumption is one of the most commonly used measures of welfare-the higher the level of consumption, the higher the expected quality of life. Consumption, especially in competitive market economies, is also the principal driver of economic growth, which is regarded as the key to a prosperous society.

Booz, Allen, and Hamilton propose a transition from product orientation to customer-centric orientation. In their concept, the following are of particular importance: customer life cycle orientation, building customer understanding at each interaction, working as an insider, and a high degree of organizational trust. Hämäläinen (2014) combines the concept of value-based management with the concept of the business model through the customer-centric concept. Others pay attention to the effect of Customer-Centric Structure on Long-Term Financial Performance (Lee, et al., 2014).

By adapting the abovementioned approaches to the business models of public organizations, it can be stated that it is crucial for public organizations to combine the mechanisms of creating higher order values, as well as to place a citizen at the center of attention as a key recipient of that value through the application of the concept of customer-centric orientation.

Thus, public management aims to create higher order value and it is the citizen who is the primary recipient of the value offered by the organization, such that he or she is in the center of the organizational orientation. Public interest, understood as serving higher goals, those which are most important from the point of view of the whole community of citizens and reflected in shared values, harmonized with those purposes (Kożuch & Kożuch, 2014), becomes a noticeable factor in the creation and management of value.

At the same time, values shared by citizens and the satisfaction of the public interest by the organization may become a key foundation for the conceptualization and operationalization of its business model.

Table 4 shows that the distinguishing features of public business models may include citizens as main stakeholders, government society as a recipient of value, social and public value instead of economic value, collaborating as a dominating activity, and also inter-organizational coordination. These features justify the selection of a citizen-centric approach to public business models.

SUSTAINABILITY AND ITS LINKAGE WITH PUBLIC VALUE IN THE FRAMEWORK OF ADOPTING BUSINESS MODELS IN PUBLIC MANAGEMENT

Sustainable Business Models and Social Business Models are a starting point when considering public business models. Lüdeke-Freund (2010, p. 17) points out that the competitive advantage of the eco-innovation business is created through "superior customer value and contributes to a sustainable development of the company and society where sufficiency, efficiency and consistency can be strategic and normative orientations for innovation" (Lüdeke-Freund, 2010). Dyllick and Hockerts (2002) present a model based on the concept of corporate sustainability (balancing and integrating the company's activities) mapped in the form of a triangle. In the three corners of the triangle there are: the business case, the natural case and the societal case (Dyllick & Hockerts, 2002). Bocken, Short, Rana, and Evans (2013) argue that sustainable business models (SBM) incorporate a triple bottom line approach and consider a wide range of stakeholder interests, including environment and society. They are important in driving and implementing corporate innovation for sustainability, can help embed sustainability into business purpose and processes, and serve as a key driver of competitive advantage (Bocken et al., 2013).

According to Boons and Lüdeke-Freund (2012), sustainable business models enable social entrepreneurs to create social value and maximize social profit; of significance is the business models' ability to act as a market device that helps in creating and further developing markets for innovations with a social purpose. Epstein and Rejc Buhovac (2010) argue that sustainable business models are better understood through comprehension of: role of different drivers (inputs and processes) in balancing, the causal relationship in respect to various actions to be taken, the impact of these activities on sustainable results, and the potential and actual impact on financial performance. McDonough and Braungart (1998)

Table 4. Features differentiating business models and public business models

Attribute	Business model	Public business model
Relations	Shareholders and Stakeholders	Main Stakeholders - Citizens
Recipient of value	Shareholders	Government Society
Value	economic	Social and public
High value	economic	Public
Paradigm	System and Network	System and Network
Dominating character of activities	Competing	Collaborating
Type of coordination	Mainly internal	Mainly inter-organizational

Source: Own study.

present a model of corporate sustainability in the form of a fractal triangle, the corners of which include: ecology - ecology, equity - equity, and economy - economy.

An interesting model of sustainable business based on the original concept of SMART (sustainability, modeling and reporting system) has been developed by Ahmed and Sundaram (2012). In this model, they have defined the sustainable business transformation roadmap, whose key elements are: design, transformation, monitoring and control, discovery and learning, and strategy.

Social entrepreneurship designs these new business models to "create social value relying on individuals skilled at gathering and mobilizing human, financial, and political resources" (Seelos & Mair, 2005).

Yunus, Moingeon, and Lehmann-Ortega (2010) define the concept of a social business model. They have developed five principles of building a social business model consisting of two areas: also common to innovative models, and areas specific to social models. The similarities with conventional and innovative business models include: the challenges of conventional wisdom and fundamental assumptions, the discovery of complementary business partners, and endeavors towards improving process experiments. Specific objectives for social business models include: creating favorable conditions for social orientation in terms of profit by the shareholders, and clear, specific objectives for profit for society.

A social business model is adopted by a social company. Kotler, Kartajaya and Setiawan (2010) define three success measures of a social company that will indicate whether the company will be able to strengthen the economic foundations of society. Using these measures, it is easy to say which company is a social company, and which is not. First of all, such a company attains disposable income. Secondly, it extends this income. And thirdly, it increases it (Kotler, Kartajaya, & Setiawan, 2010). On the basis of these assumptions, they have built a four-component, social business model.

Smith, Binns, and Tushman (2010) define a business model as "the design by which an organization converts a given set of strategic choices about markets, customers and value propositions into value, and uses a particular organizational architecture of people, competencies, processes, culture and measurement systems, in order to create and capture this value" (p. 450). Increasingly, public-sector organizations are adopting new business models to better engage with private-sector partners and thus alleviate barriers to innovation. New business models can encourage greater emphasis on tangible benefits and improvements for citizens and shift the balance of power to the front line, where services are provided (Micheli et al., 2012).

On the basis of the above discussion, therefore, and drawing upon recent theoretical developments in the discipline of public management, we argue here for seven propositions to form the basis of a sustainable business model both for public services in general and for PSOs in particular in the twenty-first century. These propositions are that:

- Public services are systems and not just organizations, or even inter-organizational networks, and need to be governed as such, embracing all of their elements (Osborne et al. 2014),
- Individual PSOs need to be sustainable in their own right in the short term – but this is a necessary and not a sufficient condition for the long-term sustainability of PSOs and of public service systems (Bozeman, 2002; Grindle & Hilderbrand 1995),
- Consequently, such internal efficiency is necessary for individual PSOs but will not produce sustainable public service systems; rather PSOs need to be outwardly focused on external effectiveness for service users and on creating sustainable public value for local communities (Radnor and Osborne 2013; Vidal 2013),

- The contemporary key resource and route to effectiveness for PSOs is knowledge (both of professionals and of services users) and the key tools for its transformation into successful public services are relational rather than discrete and transactional – this transformation is currently emphasized and supported by the information generation, sharing, and utilization possibilities offered by social media and digital technology (Bekkers et al., 2011; Margetts, 2009),
- Sustainable PSOs are dependent upon building long-term relationships across service systems rather than seeking short-term discrete and transactional value (McGuire 2012; McLaughlin et al., 2009),
- Co-production is at the heart of public service delivery and is the source both of effective performance and of innovation in public services (Osborne and Strokosch 2013; Sienkiewicz-Małyjurek 2016), and
- Public service systems also need to embrace environmental sustainability, if they are to be truly sustainable into the second half of the twenty-first century and beyond (Osborne et al. 2010)
- Specificity of PSOs includes the motives of rendering public services and delivering public goods, a system of public values, criteria of public rationality, and patterns of attitudes and behaviors of participants in organized activities for *publico bono* (Kożuch, 2015).

The processes of creating public values depend on models of public management. For example, in the New Public Management model, public interest is articulated by citizens with the help of public managers. It encompasses shared values and the aggregated interest of society members. Operations of public organizations are targeted at recipients of public goods and services. Entrepreneurial spirit is a value which permeates public organizations, as well as conviction on the need for the "reduction" of the state. Management of public affairs has a form of steering that unleashes opportunities inherent to the market mechanism. Goals are largely accomplished through creating mechanisms and structures even going beyond the borders of the public sector. A clearly defined external responsibility to recipients of goods and services or to citizens exists, and specific results come under assessment. Organizations enjoy relatively large levels of autonomy in delivering entrepreneurial targets. In contrast, in the New Public Governance model, an emphasis is placed on constructing networks of public interrelations and managing them. The network comprises central authorities, regional and local authorities, social and political groups and interest groups, as well as social and business organizations. Public interest is created with the help of public managers in the process of governance and public values are shared by network participants, and the conditions for its realization are negotiated with partners and groups of interest (Kożuch & Dobrowolski, 2014, pp. 221-222).

The propositions presented highlight a system of public values, criteria of public organizations' rationality, and patterns of attitudes and behaviors of participants in organized activities.

It is important to try to define the concept of a business model for public organizations while presenting scientific arguments in relation to it.

The authors define a public business model as the configuration of external and internal factors constituting a key tool for strategic management to ensure the provision of public services in view of the rising expectations of citizens, taking into account the needs of other stakeholders, which has resulted in the social and public rate of return. Social rate of return is understood as a specific value generated by the organization, in the form of non-material resources, the recipients of which are company stakeholders.

This may be the security and life stability of organization employees that provides them with a sustainable ability to function in the labor market, employees, company suppliers and all entities throughout the value chain associated with the organization. It may be the increased value of the region where the organizational activity is conducted, due to a reduced unemployment rate. The quality of life in the region is improving owing to the fact that the organization operates in the region. It might be a source of human capital development. Employees with high competence that are employed do better in society and they are less predisposed to criminal and pathological behavior. They are an important educational factor and authority for new generations. The social rate of return may also be related to shaping the image of the organization as socially responsible. By creating social value through legal, ethical, economic and philanthropic behavior, such an organization reveals itself to be a tool for creating social capital as a source of establishing pro-social relationships, a factor stimulating life activity and a tool for improving the quality of life of its stakeholders.

In this sense, the social rate of return becomes an element that attracts stakeholders to the organization. It should be noted, however, that it arises most often in organizations that want to create value in the long term. In contrast, the public rate of return is understood as a specific, higher-order value generated by the organization in the form of non-material resources, the recipients of which are citizens. This may be the result of professional service that fulfills citizens' expectations. This may be the level of emotional commitment of those who provide services to citizens. Finally, it may be the result of the fulfillment of the public service mission for the benefit of citizens. It might also be a kind of cooperation between a public organization and citizens, and the source of establishing further dialogue justifying mutual relationships. The public rate of return, therefore, is an element in establishing interrelationships towards ethical, fair and transparent functioning on the basis of strong mutual ties.

The idea of 'Public Value' has been championed through the work of the Kennedy School's Mark Moore (Moore, 2000). Moore's original work emphasized three aspects of performance for public agencies: delivering actual services; achieving social outcomes; and maintaining trust and the legitimacy of the agency.

Two ways of creating value in the sense of satisfying the desires of citizens and clients are also underlined. The first one is deploying the money and authority entrusted to public managers to produce things of value to particular individuals. In other words, public values are created through public sector production, which are not always physical products or services consumed by individual beneficiaries. The second way is meeting citizens' desires for properly managed and productive public organizations through establishing a mechanism of public accountability. These processes must economize on the use of financial means as well as on the use of public authority. One of the main responsibilities of public managers is creating public value (Moore, 2000, pp. 52-57).

The processes of creating public values depend on models of public management. For example, in the New Public Management model, public interest is articulated by citizens with the help of public managers. It encompasses shared values and the aggregated interest of society members. Operations of public organizations are targeted at recipients of public goods and services. Entrepreneurial spirit is a value which permeates public organizations, as well as conviction on the need for the "reduction" of the state. Management of public affairs has a form of steering that unleashes opportunities inherent to the market mechanism. Goals are largely accomplished through creating mechanisms and structures even going beyond the borders of the public sector. A clearly defined external responsibility to recipients of goods and services or to citizens exists, and specific results come under assessment. Organizations enjoy relatively large levels of autonomy in delivering entrepreneurial targets. In contrast, in the New Public Governance model, an emphasis is placed on constructing networks of public interrelations and managing them. The network comprises central authorities, regional and local authorities, social and political groups and interest groups, as well as social and business organizations. Public interest is created with the help of public managers in the process of governance and public values are shared by network participants, and the conditions for its realization are negotiated with partners and groups of interest (Kożuch & Dobrowolski, 2014, pp. 221-222)

The literature of public management offers the main definitions of public values. Generally, references to public interest are based on shared values and the long-term and common interests of the people. These are justice, fairness and equity, community, accountability, integrity, fair procedures, sound decisions, and policy achievements. (Denhardt & Denhardt, 2003, pp. 78; Pollitt & Bouckaert 2004, p. 9; Dibben, James, Roper, & Wood, 2007, pp. 22-25; Horton & Farnham, 2002, p. 78, 86; Kettl, 2012, p. 472).

In the cases considered here, where the innovations seem to relocate important is relocation either responsibilities for producing publicly valued results, or the right to decide what constitutes publicly valued results, or some combination of the two, one is forced to evaluate the innovations not only in terms of efficiency and cost effectiveness, but also in terms of what might be considered the 'right' relationships in society – some notion of justice and fairness (Moore & Hartley, 2008).

A model of public value creation has been developed in a transparent way by John M. Bryson, among others. In terms of considering collaboration as a public value, the existence of multiple stakeholders and conflicting values within the public sector means that collaboration is interlinked with many other public values over both service outcomes and the processes by which they are achieved.

Based on the analyzes public value in a given society can be defined as a shared set of general values such as respect for all people, democracy, integrity, and sustainability that are interlinked with specific values such as serving the public interest, shared responsibility, public responsiveness, public participation, democratic morality, public trust, and openness to collaboration. All these values are subject to operationalization according to different situations and citizens' changing expectations, and also the technical possibilities of public organizations. However, in general it is the balance between various stakeholders, citizen-centric orientation and sustainability that delineates the components of public sector business model, and allows to conceptualize its framework, hence to better understand the phenomenon.

SOLUTIONS AND RECOMMENDATIONS

When conducting a multivariate analysis of the concept of business models from the perspective of possible applications for public organizations, it is essential to build a business model canvas. The canvas may be based on internal and external factors.

Internal factors will determine the extent to which the organization's resources are used to meet the expectations of citizens and other stakeholders. The proposed canvas consists of nine elements forming a coherent whole. Its key component is value for network stakeholders / actors, which is achieved by successive components. Processes, key resources, and the internal structure affect the external environment. A collaborative factor as another component along with public government affects the relationship with citizens and other stakeholders. The result of using a business model is, however, the social rate of return and the public rate of return.

The proposal of a nine-element public business canvas based on internal factors, prepared according to the interpretation approach, is shown in the following Figure 3.

Figure 3. Nine-element public business model canvas based on internal factors prepared in the interpretation approach focused on value
Source: Own study.

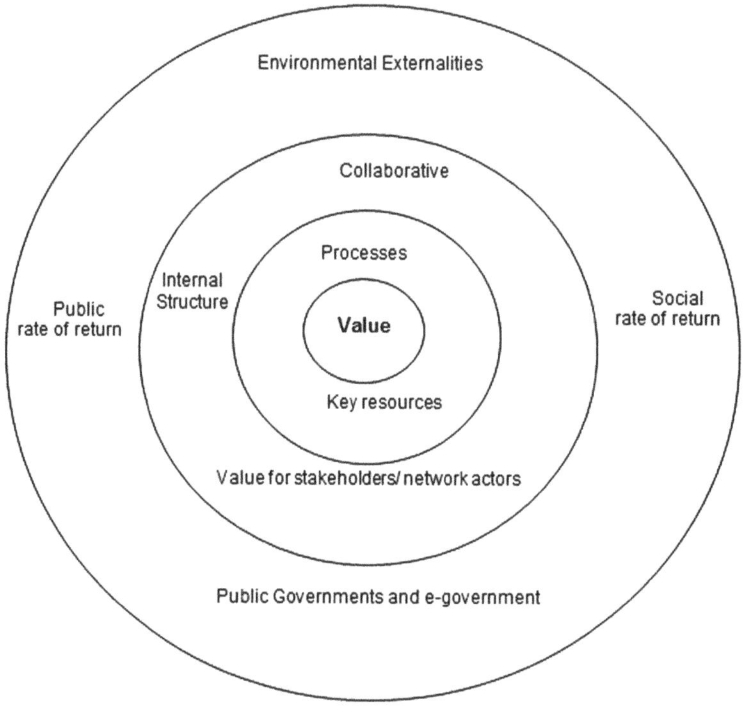

The above business model canvas has been expressed in circles, which contain individual components of the value presented, which is the core of the business model canvas. This model refers to the internal elements presented in a hierarchy.

Another element in the discussion on the range of business models is to think how a business model canvas based on external factors can be described. In this analysis, a key factor is the dialog supported by confidence. It ensures an adequate level of cooperation and relationship and the ability to operate in the network. In addition, the canvas includes norms and balance in relation to resources and between stakeholders as well. The result of this process is to provide common goods and legitimacy.

The proposal of a nine-element public business canvas based on external factors, prepared according to the interpretation approach, is shown in the following Figure 4.

A public business model canvas is also expressed in circles, which contain the individual components of the value presented, which is the core of the business model canvas. This model refers to the external elements presented in a hierarchy.

A further stage of the authors' discussion is an attempt to define a holistic public business model that is an inspiration for the creation of value for citizens and other stakeholders through a grassroots organization. Thus, based on a broad review of the literature, a 36-element canvas of a public business model for a public organization has been proposed.

The logic presented indicates that each public organization, via 36 components proposed, selects those components that will determine its success. This selection will depend on the specific conditions typical of a public organization and its configuration will depend on the skills of those who manage this business model. The following is an extensive, 36-element business model canvas for public organizations.

Figure 4. Public business model canvas based on external factors focused on value
Source: Own study.

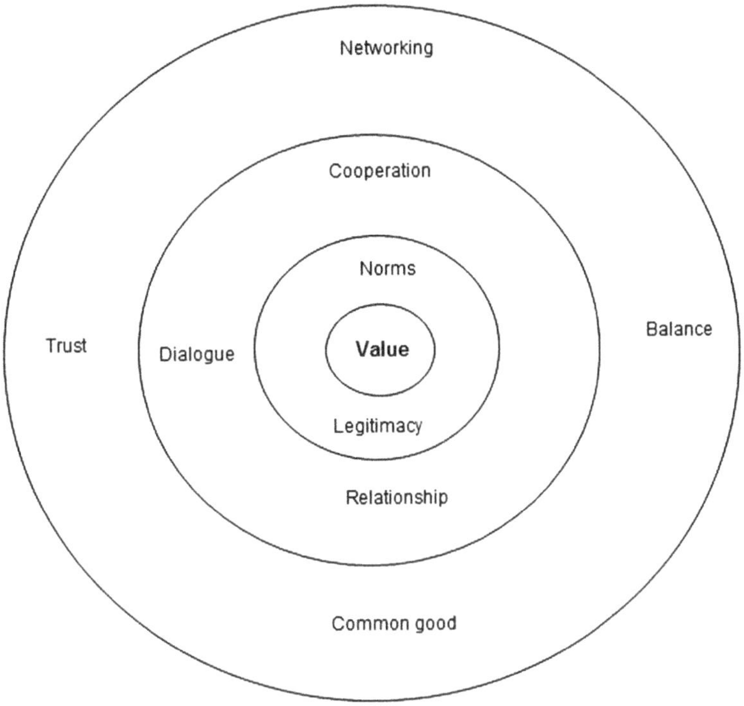

Figure 5. Extensive, 36-element business model canvas for public organizations focused on value
Source: Own study.

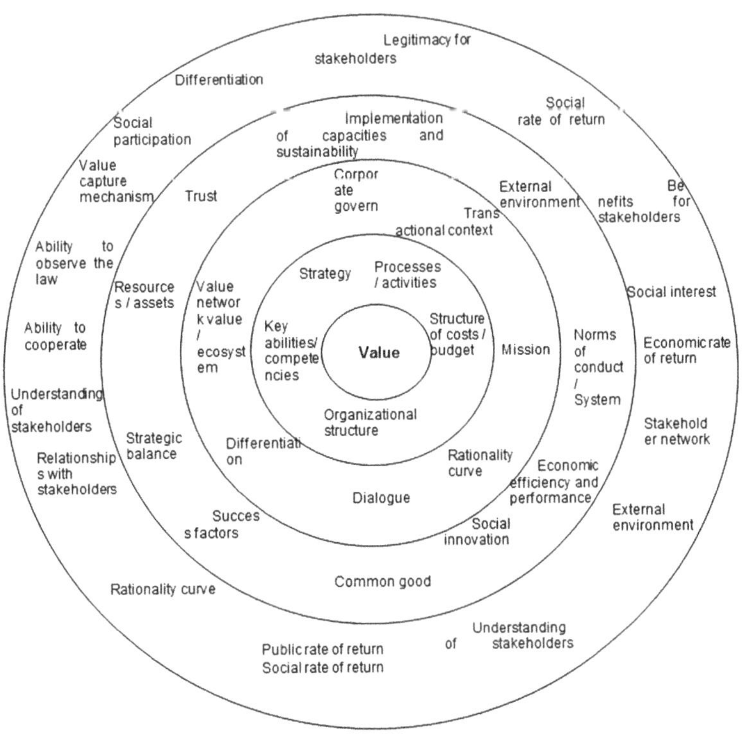

A public business model canvas is also expressed in circles, which include 36 elements centered around value, which is the core of the business model canvas. This model refers to external and internal elements presented in a hierarchy. Individual circles represent the following areas: organizational structure, costs/budget structure, the ability to cooperate, economic effectiveness and efficacy.

Referring to public business models, it should be emphasized that openness and attitude to external participation are conducive to organizational participation, which is an important incentive for citizens - both as voters and consumers of public services - to establish relationships with public organizations, especially at the level of basic territorial structures.

Researchers usually pay attention to social or political participation, without recognizing the fact that any political activity has an organizational dimension. A positive attitude on the part of citizens largely depends on what conditions for becoming involved in public affairs are created by public organizations. Every experience allows citizens as voters to gain knowledge of the conditions of observing the standards of a democratic state and the use of the possibilities inherent in a democratic system.

The cooperation of citizens in the process of public services is beneficial in terms of acquiring knowledge of the functioning of public organizations, as well as establishing personal relationships. Taking part in the processes of organizational participation creates favorable conditions for improving the effectiveness of public management. It is based on knowledge acquired during political and organizational processes as well as personal relationships between the representatives of public organizations and citizens themselves and representatives of civic organizations.

Assuming the relative sustainability of participation, it must be emphasized that the growing knowledge of the democratic system directly contributes to the involvement of citizens in the processes of developing public policies. Knowledge of the functioning of public organizations and experience in conducting joint activities are equally important.

A discussion on the concept of business models for public organizations is a very important element of scientific reflection. The use of the concept of the business model for public organizations may be problematic for some researchers. The authors are aware of this process. Nevertheless, the concepts of NPM as firmly embedded in the literature indicate the possibility of adapting different management methods and tools (including strategic management) to other types of organizations. It is also worth noting that the development of the types of public organizations is today nothing short of remarkable. Thus, there is a natural need to seek solutions conducive to performance. Undoubtedly, the concept of business models is a challenge for many public organizations as it may be a platform for further evolution to achieve the previously described social and public rates of return. Thus a citizen, as a key recipient of the value created by the public organization, is central to the interest of the public organization. A business model examined according to the system approach ensures the ability of the organization to fulfill its strategic obligations towards citizens, creating a kind of higher order value.

Creating this value helps produce specific, mutual relationships based on dialogue and trust.

FUTURE RESEARCH DIRECTIONS

Theoretical and research limitations resulting from the above reflections include: the relatively small amount of scientific research on the business models of public organizations, the high complexity of the issue that may be seen from many separate approaches, and the relatively small number of interpretations of the concepts of business models for public organizations in both theory and practice. Hence, some recommended directions for further research would comprise:

1. Specific components may be a key attribute of public business model design and operationalization.
2. Due to public business model components may be obtained from the network.
3. Key components may develop new sources of innovation, being an accelerator of growth of public organizations efficiency in the network.
4. Public business models may be a source of faster development and growth of public organizations value.
5. The role of the network environment to design and operationalization of public business models.
6. The role of the public business models in improving business efficiency.

CONCLUSION

Bearing the scientific discussion in mind, the authors pay attention to the complexity of the issue presented and its different dimension of interpretation. However, the ability to adapt the concept of business models to be used for public organizations is crucial. Such an attempt made in this chapter shows the new areas of discussion, determining the success of public organizations not only through the strategies which they apply but primarily through innovative business models. This is a new way of searching for the effectiveness of public organizations.

A business model becomes a tool for searching for higher order values and its operationalization to the benefit of citizens and other stakeholders. As key recipients of value together with a public organization, citizens start a dialogue with each other and become a platform for mutual relationships. This results in a constant increase of social and public rates of return.

To sum up the theses contained in the paper, the core conclusions that are the basis for further scientific discussion should be defined.

- A business model of a public organization can be primarily oriented towards higher order values and the concept of customer-centric orientation;
- The effect of conceptualization and operationalization of a public business model is collaborative advantage;
- The key recipients of the public business model are citizens and other stakeholders;
- The business model canvas of a public organization may be focused on internal and external factors;
- The result of a public business model may be a social and public rate of return;
- Each public organization can build its own configuration of a public business model based on its own public model;
- A nine-element public business model will be useful in determining the general direction of changes;
- A 36-element public business model has been proposed to build detailed solutions.

REFERENCES

Afuah, A., & Tucci, C. (2003). *Biznes internetowy strategie i modele*. Kraków: Oficyna Ekonomiczna.

Afuah, A., & Tucci, C. L. (2001). *Internet business models*. New York: McGraw-Hill/ Irwin.

Alt, R., & Zimmerman, H. D. (2001). Introduction to special section on business models. *Electronic Markets*, *11*(1), 3–9.

Amit, R., & Zott, C. (2001). Value creation in e-business. *Strategic Management Journal*, *22*(2), 493–520. doi:10.1002/smj.187

Ansoff, H. I. (1985). *Zarządzanie strategiczne*. Warszawa: PWE.

Applegate, L. M. (2000). E-business Models: Making sense of the internet business landscape. In G. Dickson & G. Desanctis (Eds.), Information Technology and the Future Enterprise, New Models for Managers. Prentice-Hall.

Applegate, L. M. (2001). Emerging e-business models. *Harvard Business Review*, *79*(1), 79–87. PMID:11189465

Applegate, L. M. (2001). *Emerging e-business models: Lessons from the field*. HBS No.9-801-172. Harvard Business School.

Bekkers, V., Edwards, A., Moody, R., & Beunders, H. (2011). Cauht by surprise? Micro-mobilization, new media and management of strategic surprises. *Public Management Review*, *13*(7), 1003–1022. doi:10.1080/14719037.2011.589615

Betz, F. (2002). Strategic business models. *Engineering Management Journal*, *14*(1), 21–27. doi:10.1080/10429247.2002.11415145

Bocken, N. M. P., Short, S. W., Rana, P., & Evans, S. (2014). A literature and practice review to develop sustainable business model archetypes. *Journal of Cleaner Production*, *65*, 42–56. doi:10.1016/j.jclepro.2013.11.039

Boons, F., & Lüdeke-Freund, F. (2013). Business models for sustainable innovation: State-of-the-art and steps towards a research agenda. *Journal of Cleaner Production*, *45*, 9–19. doi:10.1016/j.jclepro.2012.07.007

Booz, A. (n.d.). *Hamilton*. Retrieved from http://www.strategyand.pwc.com/media/file/The_Customer_Centric_Organization.pdf

Bossidy, L., & Charan, R. (2002). *Execution: The Discipline of Getting Things Done*. Crown Business.

Bossidy, L., & Charan, R. (2010). *Szósty zmysł w zarządzaniu firmą. Tworzenie wykonalnych planów i modeli biznesowych*. Warszawa: MT Biznes.

Bozeman B. (2002), Public-Value Failure: When Efficient Markets May Not Do. *Public Administration Review*. DOI: 10.1111/0033-3352.00165

Brandenburger, A., & Harborne Stuart, S. (1996). Value-based business strategy. *Journal of Economics & Management Strategy*, *5*(1), 5–25. doi:10.1111/j.1430-9134.1996.00005.x

Casadesus-Masanell, R., & Ricart, J. E. (2010). From Strategy to Business Models and onto Tactics. *Long Range Planning*, *43*(2-3), 195–215. doi:10.1016/j.lrp.2010.01.004

Chesbrough, H., & Rosenbloom, R. (2002). The Role of Business Model in Capturing Value from Innovations: Evidence from Xerox Corporations Technology Spin-off Companies. *Industrial and Corporate Change*, *11*(3), 529–555. doi:10.1093/icc/11.3.529

Chesbrough, H. W. (2007). Why Companies Should Have Open Business Models. *MIT Sloan Management Review*, *48*(2), 22–28.

Chesbrough, H.W., & Appleyard, M.M. (2007). *Open Innovation and Strategy*. Portland State University PDXScholar.

Chodyński, A. (2011). *Odpowiedzialność ekologiczna w proaktywnym rozwoju przedsiębiorstw. Krakowska Akademia im.* Kraków: Jana Frycza Modrzejewskiego.

Clarkson, M. (1994). A risk Based Model of Stakeholder Theory. In *Proceedings of the Second Toronto Conference of Stakeholder Theory*. Centre for Corporate Social Performance and Ethics, University of Toronto.

Cyfert, S., & Krzakiewicz, K. (2011). Wykorzystanie koncepcji modeli biznesu w zasobowej teorii firmy. In Rozwój szkoły zasobowej zarządzania strategicznego. Prace Naukowe Wałbrzyskiej Wyższej Szkoły Zarządzania i Przedsiębiorczości.

Daud Ahmed, M., & Sundaram, D. (2005). Sustainability modeling and reporting: From roadmap to implementation. *Decision Support Systems*, *53*, 2012.

de Witt, B., & Meyer, R. (2007). *Synteza strategii. Tworzenie strategii konkurencyjnej przez tworzenie paradoksów*. Warszawa: PWE.

Demil, B., & Lecocq, X. (2010). Business Model Evolution: In Search of Dynamic Consistency. *Long Range Planning*, *43*(2-3), 227–246. doi:10.1016/j.lrp.2010.02.004

Denhardt, J. V., & Denhardt, R. B. (2003). *New Public Service. Serving, not steering*. Armonk: M.E. Sharpe.

Denhardt, R. B. (2011). *Theories of Public Organization*. Boston: Wadsworth Cengage Learning.

Deodhar, S.J., Saxena, K., Gupta, R.K., Ruohonen, M. (2012). Strategies for software-based hybrid business models. *Journal of Strategic Information Systems*, (4), 274–294.

Dibben, P., James, P., Roper, I., & Wood, G. (Eds.). (2007). Modernising Work in Public Services. Redefining Roles and Relationships in Britain's Changing Workshops. Palgrave Macmillan.

Dubosson-Torbay, M., Osterwalder, A., & Pigneur, Y. (2001). E-business model design, classification and measurements. *Thunderbird International Business Review*, *44*(1), 5–23. doi:10.1002/tie.1036

Dyllick, T., & Hockerts, K. (2002). Beyond the business case for corporate sustainability. *Business Strategy and the Environment*, *11*(2), 2. doi:10.1002/bse.323

Epstein, M. J., & Rejc Buhovac, A. (2010). Solving the sustainability implementation challenge. *Organizational Dynamics*, 39.

Fiet, J.O., & Patel, P.C. (2008). *Forgiving Business Models for New Ventures*. Doi: 10.1111/j.1540-6520.2008.00252.x

Freeman, E., & Reed, D. (1998). Stockholders and Stakeholders: A New Perspective on Corporate Governance. In B. De Wit & R. Meyer (Eds.), Strategy. Process, Conten. Context.

Freeman, R. E. (1984). *Strategic Management: A Stakeholder Approach*. Boston: Pitman.

Gambardella, A., & McGahan, A. M. (2010). Business-Model Innovation: General Purpose Technologies and their Implications for Industry Structure. *Long Range Planning*, *43*(2-3), 262–271. doi:10.1016/j.lrp.2009.07.009

Gartner. (2003). Available at: http://www3.gartner.com

Gołębiowski, T., Dudzik, T. M., Lewandowska, M., & Witek-Hajduk, M. (2008). *Modele biznesu polskich przedsiębiorstwa*. Warszawa: Szkoła Główna Handlowa.

Grindle, M. S., & Hilderbrand, M. E. (1995). Building sustainable capacity in the public sector: What can be done? *Public Administration and Development*, *15*(5), 441–463. doi:10.1002/pad.4230150502

Guy, P.B. (1998). Governance Without Government? Rethinking Public Administration. *Journal of Public Administration Research and Theory*, *2*, 223-243.

Hämäläinen, M. (2014). *Customer-centric and value-based business model design Impacts of the additive manufacturing technology on firm's business model*. Jyväskylän yliopisto Tietojenkäsittelytieteiden laitos.

Hamel, G. (2001). *Leading the revolution*. Boston, MA: Harvard Business School Press.

Harrison, J. G. (1994). *Strategic Management of Organisation and Stakeholder*. St. Paul, MN: West.

Hedman, J., & Kalling, T. (2003). The business model concept: Theoretical underpinnings and empirical illustrations. *European Journal of Information Systems*, *12*(1), 49–59. doi:10.1057/palgrave.ejis.3000446

Hoque, F. (2002). *The Alignment Effect: How to Get Real Business Value Out of Technology*. Financial Times Prentice Hall.

Horowitz, A. S. (1996). The real value of VARS: Resellers lead a movement to a new service and support. *Mark Comput*, *16*(4), 31–36.

Horton, S., & Farnham, D. (Eds.). (2002). Public Management in Britain. Palgrave.

Jabłoński, A. (2008). *Modele biznesu w sektorach pojawiających się i schyłkowych. Tworzenie przewagi konkurencyjnej przedsiębiorstwa opartej na jakości i kryteriach ekologicznych*. Wydawnictwo Wyższej Szkoły Biznesu w Dąbrowie Górniczej, Dąbrowa Górnicza.

Jabłoński, A. (2013). *Modele zrównoważonego biznesu w budowie długoterminowej wartości przedsiębiorstw z uwzględnieniem ich społecznej odpowiedzialności*. Warszawa: Difin.

Jabłoński, A. (2015). *Spójność hybrydy strategicznej w środowisku sieciowym*. Warszawa: Difin.

Jabłoński, A., & Jabłoński, M. (2013). Projektowanie sieciowych modeli biznesu. *Ekonomika i Organizacja Przedsiębiorstwa*, *12*, 30.

Janson, C. (1981). Some Problems of Longitudinal Research in the Social Sciences. In F. Schulsinger, S. Mednick, & J. Knop (Eds.), *Longitudinal Research: Methods and Uses in Behavioral Science*. Boston: Martinus Nijhoff Publishing. doi:10.1007/978-94-009-8147-8_2

Johnson, M.W., Christensen, C.M., & Kagermann, H. (2009). Jak fundamentalnie zmienić model biznesowy firmy. *Harvard Business Review Polska*, 160–161.

Jourdan, Z., Rainer, K., & Marshall, T. (2008). Business Intelligence: An Analysis of the Literature. *Information Systems Management*, 25.

Ju-Yeon, L., Shrihari, S., Henderson, M. C., Palmatier, R.W. (2014). Effect of Customer-Centric Structure on Long- Term Financial Performance. *Marketing Science*. .10.1287/mksc.2014.0878

Kettl, D. F. (2012). *The Politics of the Administrative Process*. Washington, DC: CQ Press.

Kickert Walter, J.M. (2002). *Public Governance in the Netherlands: An Alternative to Anglo-American 'Managerialism'*. doi: 10.1111/1467-9299.00084

Kimberly, J. R. (1976). Issues in the Design of Longitudinal Organizational Research. *Sociological Methods & Research*, 4.

Kindström, D. (2010). Towards a service-based business model – Key aspects for future competitive advantage. *European Management Journal*, *28*(6), 479–490. doi:10.1016/j.emj.2010.07.002

Kotler, P., Kartajaya, H., & Setiawan, I. (2010). Marketing 3.0, Dobry produkt? Zadowolony klient? Spełniony Człowiek! MT Biznes Sp. zo.o.

Kożuch, B. (2015). Zarządzanie publiczne jako teoria i praktyka zarządzania. In B. Kożuch & Ł. Sułkowski (Eds.), *Instrumentarium zarządzania publicznego*. Warszawa: Difin.

Kożuch, B. (2015). Zarządzanie publiczne jako teoria i praktyka zarządzania. In B. Kożuch & Ł. Sułkowski (Eds.), *Instrumentarium zarządzania publicznego*. Warszawa: Difin.

Kożuch, B., & Dobrowolski, Z. (2014). *Public Trust. Organisational Perspective*. Frankfurt am Main: Peter Lang.

Kożuch, B., Kożuch, A. (2014). Responsywność w zarządzaniu usługami publicznymi. *Prace Naukowe Uniwersytetu Ekonomicznego we Wrocławiu*, *355*.

Kożuch, B., Kożuch, A. (2016). Partycypacja publiczna a skuteczność zarządzania publicznego. *Przedsiębiorczość i Zarządzanie*, *16*(12).

Kożuch, B., Kożuch, A., & Sienkiewicz-Małyjurek, K. (2016). Contemporary faces of Public Management. *Przegląd Organizacj,* 3.

Lane, J. E. (2009). *State Management. An enquiry into models of public administration and management.* London: Routledge Taylor and Francis Group.

Langtry, B., (1994). Stakeholders and Moral Responsibilities of Business. *Business Ethics Quarterly, 4.*

Lee, I. W., Feiock, R. C., & Lee, Y. (2012). Competitors and Cooperators: A Micro-Level Analysis of Regional Economic Development Collaboration Networks. *Public Administration Review*, *72*(2), 253–262. doi:10.1111/j.1540-6210.2011.02501.x

Linder, J. C., & Cantrell, S. (2001). *Changing Business Models: Surveying the Landscape*. Working Paper. Institute for Strategic Change, Accenture.

Linder, J. C., & Cantrell, S. (2002). It's All in the Mind(set). *Across the Board*, *39*(3).

Lisein, O., Pichault, F., & Desmecht, J. (2009). Les business models des sociétés de service actives dans le secteur Open Source. Systémes d'Information et Management, 14(2).

Lüdeke-Freund, F. (2010). *Towards a conceptual framework of business models for sustainability.* Lüneburg: Centre for Sustainability Management.

Magretta, J. (2002). Why Business Models Matter. *Harvard Business Review*, *80*(5), 86–92. PMID:12024761

Mahadevan, B. (2000). Business Models for Internet based E-Commerce, An Anatomy. California Management Review, 42(4).

Margetts, H. (2009). The internet and public policy. *Policy and Internet*, *1*(1), 1–21. doi:10.2202/1944-2866.1029

Massaro, M., Dumay, J., Guthrie, J. (2015). On the Shoulders of Giants: Undertaking a Structured Literature Review in Accounting, Accounting. *Audit. Account. J.*, 267 – 284.

Mc Donough, W., & Braugnart M. (1998, October). The next industrial revolution. *The Atlantic Monthly.*

McGuire, L. (2012). Slippery concepts in context. Relationship marketing and public services. *Public Management Review*, *14*(4), 541–555. doi:10.1080/14719037.2011.649975

McLaughlin, K., Osbourne, S., & Chew, C. (2009). Developing the marketing function in UK public service organizations: The contribution of theory practice. *Public Money and Management*, *29*(1), 35–42. doi:10.1080/09540960802617343

Micheli, P., Schoeman, M., Baxter, D., & Goffin, K. (2012). *New Business Models for Public-Sector Innovation Successful Technological Innovation for Government.* Research-Technology Management. doi:10.1007/978-3-642-32150-4

Mitchell, D. W., & Coles, C. B. (2004). Business model innovation breakthrough moves. *The Journal of Business Strategy*, *25*(1), 16–26. doi:10.1108/02756660410515976

Mitchell, R. K. (1997). Toward A Theory of Stakeholder. Identification and Salience: Defining the Principle of Who and What Really Counts. *Academy of Management Review.*

Moore, M., & Hartley, J. (2008). Innovations in Governance. *Public Management Review*, *10*(1), 3-20.

Moore, M. H. (1995). *Creating public value: Strategic management in government.* Cambridge, MA: Harvard University Press.

Moore, M. H. (2000). *Creating Public Value. Strategic Management i Government.* Cambridge, MA: Harvard University Press.

Morris, M., Schindehutte, M., & Allen, J. (2005). The entrepreneurs business model: Toward a unified perspective. *Journal of Business Research, 58*(6), 726–735. doi:10.1016/j.jbusres.2003.11.001

Muegge, S. (2012). *Business Model Discovery by Technology Entrepreneurs.* Technology Innovation Management Review.

Neely, A., & Delbridge, R. (2007). *Effective Business Models: What Do They Mean for Whitehall.* Retrieved from www.nationalschool.gov.uk/sunningdaleiinstitute

Niemczyk, J. (2010). *Modele biznesowe.* Warszawa: C.H. Beck.

Nogalski, B. (2009). Modele biznesu jako narzędzie reorientacji strategicznej przedsiębiorstw. *Master of Business Administrati*on, *2*.

Obłój, K. (2010). *Pasja i dyscyplina strategii, Jak z marzeń i decyzji zbudować sukces firmy*. Poltext.

Osborne, D., & Gaebler, T. (1992). *Reinventing Government: How the Entrepreneurial Spirit Is Transforming the Public Sector.* Addison-Wesley Publishing, Company, Inc.

Osborne, S. (2010). *The New Public Governance? Emerging Perspectives on the Theory and Practice of Public Governance.* London: Routledge.

Osborne, S. P., Radnor, Z., Vidal, I., & Kinder, T. (2014). A sustainable business model for public service organizations? *Public Management Review, 16*(2), 165–172. doi:10.1080/14719037.2013.872435

Osborne, S.P., & Strokosch, K. (2013). It takes two to tango? Understanding the co-production of public services by integrating the services management and public administration perspectives. *British Journal of Management, 24*(SI), S31-S47.

Osterwalder, A. (2004). The business model ontology: a proposition in a design science approach. InBusiness Model Generation, A Handbook for Visionaries, Game Changers, and Challengers. John Wiley & Sons, Inc.

Osterwalder, A., & Pigneur, Y. (2002). *An e-Business Model Ontology for Modeling e-Business.* 15th Bled Electronic Commerce Conference e-Reality: Constructing the e-Economy, Bled, Slovenia.

Perechuda, K. (2013). Dyfuzja wiedzy w przedsiębiorstwie sieciowym, wizualizacja i kompozycja. Wydawnictwo Uniwersytetu Ekonomicznego we Wrocławiu.

Petrovic, O., Kittl, C., & Teksten, R. D. (2001). Developing business models for e-business.*International Electronic Commerce Conference.*

Pollitt, C. (1990). *Managerialism and the Public Services: The Anglo-American Experience.* Basil Blackwell.

Pollitt, C., & Bouckaert, G. (2004). *Public Management Reform. A Comparative Analysis*. Oxford, UK: Oxford University Press.

Rappa, M. (2002). *Managing the digital enterprise: Business models on the web*. Retrieved from http://digitalenterprise.org/models/models.html

Rayport, J. F., & Jaworski, B. J. (2001). *E-commerce*. New York: McGraw-Hill/Irwin.

Sabatier, V., Rousselle, T., & Mangematin, V. (2010). From recipe to dinner: Business model portfolio in the European biopharmaceutical industry. *Long Range Planning*, *43*(2/3), 431–447. doi:10.1016/j.lrp.2010.02.001

Sandberg, K. D. (2002). Is It Time to Trade. In *Your Business Model?* Harvard University Update.

Scott, M., Shafera, H., Smitha, J., & Linder, J. C. (2005). The power of business models. *Business Horizons*, *48*(3), 199–207. doi:10.1016/j.bushor.2004.10.014

Seddon, P. B., Lewis, G. P., & Freeman, P. (2004). The Case for Viewing Business Models as Abstractions of Strategy. Communications of the Association for Information Systems, 13.

Seelos, C., & Mair, J. (2005). Social entrepreneurship: Creating new business models to serve the poor. *Business Horizons*, *48*(3), 241–246. doi:10.1016/j.bushor.2004.11.006

Sienkiewicz-Małyjurek, K. (2016). Innowacyjność koprodukcji w zarządzaniu publicznym na poziomie lokalnym. *Zeszyty Naukowe Politechniki Śląskiej, Seria: Organizacja i Zarządzanie, 89*.

Slyvotzky, A.J., Morrison, D.J., & Andelman, B. (2000). *Strefa zysku*. PWE.

Smith, W., Binns, A., & Tushman, M. (2010). Complex Business Models: Managing Strategic Paradoxes Simultaneously. *Long Range Planning*, *43*(2).

Stubbs, W., & Cocklin, C. (2008). Conceptualizing a „sustainability business model. Organization & Environment, 21(2), 103–127.

Svendsen, A. (1998). *The Stakeholder Strategy. Profiting from Collaborative Business Relationships*. San Francisco: Berret-Koehler Publishers, Inc.

Teece, D. J. (2010). Business Models, Business Strategy and Innovation. *Long Range Planning*, *43*(2-3), 172–194. doi:10.1016/j.lrp.2009.07.003

Timmers, P. (1999). *Electronic Commerce: Strategies and Models from business-to-business trading*. Chichester, UK: Wiley & Sons.

Venkatesh, A., & Vitalari, N. P. (1991). Longitudinal Surveys in Information Systems Research: An Examination of Issues, Methods, and Applications. In *The Information Systems Challenge: Survey Research Methods*. Harvard University Press.

Vidal, I. (2013). Governance of social enterprises as producers of public services. In P. Valkama, S. J. Bailey, & A.-V. Anttiroiko (Eds.), *Organisational Innovation in Public Services: Forms and Governance*. Basingstoke, UK: Palgrave. doi:10.1057/9781137011848_10

Viscio, A. J., & Pasternack, B. A. (1996). Toward a new business model. *Strategic Business*, *2*(1), 125–134.

Voelpel, S., Leibold, M., Tekie, E., & von Krogh, G. (2005). Escaping the Red Queen Effect in Competitive Strategy: Sense-testing Business Models. *European Management Journal*, *23*(1), 37–49. doi:10.1016/j.emj.2004.12.008

Weber, M. (1947). *The theory of Social and Economic Organization*. New York: Free Press.

Williamson, P.J. (2010). Cost Innovation: Preparing for a 'Value-for-Money' Revolution. *Long Range Planning*, *43*(2-3), 343–353. 10.1016/j.lrp.2009.07.008

Yip, G. S. (2000). Global Stategy in the Internet Ea. *Business Strategy Review*, *11*(4), 1–14. doi:10.1111/1467-8616.00152

Yunus, M., Moingeon, B., & Lehman-Ortega, L. (2010). Biulding Social Business Models: Lessons from the Grameen Experience. *Long Range Planning*, *43*(2-3), 308–325. doi:10.1016/j.lrp.2009.12.005

Zhiyong, L., & Rosenbloom, D. H. (1992).. . *Public Administration Review*, *52*(6), 535–537. doi:10.2307/977163

Zott, Ch., & Amit, R. (2008). The fit between product market strategy and business model: Implications for firm performance. *Strategic Management Journal*, *29*(1), 1–26. doi:10.1002/smj.642

Zott, C., & Amit, R. (2010). Business model design: An activity system perspective. *Long Range Planning*, *43*(2–3).

Zott, C., Amit, R., & Massa, L. (2010). *The Business Model: Theoretical Roots, Recent Developments, and Future Research*. IESE. Retrieved from http://www.iese.edu/research/pdfs/DI-0862-E.pdf

Zott, Ch., Amit, R., Massa L. (2011). The Business Model: Recent Developments and Future Research. *Journal of Management*. doi:10.1177/0149206311406265

KEY TERMS AND DEFINITIONS

Business Model: Implementation logic of the social mission, which fulfillment of requires the application of appropriate resource configuration and how you use them to achieve the objectives of a public organization.

Collaborative Dividend: Division of value generated by public organizations for the benefit of its stakeholders.

Component-Based Approach: A set of components creating the configuration of the business model, the structure of which should serve efficiency, effectiveness the mission of a public organization.

Public Management: In the context of business models, public management can be considered as a set of measures aimed at social good while maintaining organizational efficiency.

Public Trust: One of the key components of the business model of public organization.

Public Value: Benefits of a social nature accepted in accordance with the mission of a public organization.

Social Dividend, Public Dividend: Consumption of value generated by public organization for the benefit of its stakeholders.

APPENDIX

Table 5. Components of business models in contexts

Business model components	General context	Business context	Social and public context	Integrated context
	1	2	3	4
The dimensions of business models: ■ Performance ■ Uniqueness ■ Adaptation ■ Revenue accelerators Business model components: ■ The logic of the customer, ■ Customer service, ■ Information and overtaking, ■ Rational dynamics ■ Price structure, ■ Purpose of the strategy, ■ Products and market segments ■ Differentiation from competitors in terms of products and services. ■ Resources and skills, ■ Strategic resources, Methodology of manufacturing processes, Networks of suppliers, partners, and alliances. (Hamel, 2001)	x			x
■ The concept of business ■ Abilities ■ Position of value (Applegate, 2001)	x	x		
Four main pillars: ■ product innovation ■ relationships with customers, ■ infrastructure management, ■ financial aspects Ontological components: ■ products and services ■ infrastructure and the network of partners ■ capital of relationships with customers, financial aspects (revenue model and cost structure of Osterwalder and Pigneur (2010)			x	x
■ Market ■ Value proposition ■ Value chain ■ Costs and profits ■ Value network ■ Competitive strategy (Chesbrough & Rosenbaum, 2000)	x		x	x
■ Type of competitive advantage ■ Main resources and skills ■ Configuration of the value chain ■ Resources ■ Skills Configuration of links - the value chain (Obłój, 2010)				

contined on next page

Table 5. Continued

Business model components	General context	Business context	Social and public context	Integrated context
	1	2	3	4
■ Value proposition ■ Nature of inputs ■ Understanding how to transform these inputs (including technology) ■ Nature of outputs ■ Vertical and horizontal scope ■ Geographic scope ■ Nature of customers ■ How to organize ■ (Yip, 2004)			x	x
■ External environment ■ Financial objectives ■ Internal capabilities (the appropriateness of internal processes and structures) (Bossidy & Charan, 2002)		x		
■ Value proposition for a customer ■ Customer ■ Source of competence ■ Competitive strategy factors ■ Sources of revenue ■ Growth factors (speed, range, and size of ambition) (Morris, Schindehutte, & Allen, 2005).		x		
Four main categories of components: Strategic choices, Value creation, Value network, Value capture Strategic choices include: - Customer (the target market) - Value proposition - Abilities/competencies - Revenue /price - Competition - Product offer - Strategy - Trademarks - Diversification - Mission Value creation includes: ■ Resources/assets ■ Processes / activities - Value network includes: - Costs - Financial aspects Profit (Shafer, Smith, & Linder, 2005)		x	x	x
■ Value proposition ■ Target market ■ Value chain ■ Revenue mechanisms ■ Value network or ecosystem ■ Competitive strategy (Chesbrough & Appleyard, 2007)			x	x
■ Infrastructure: the configuration of values, core capabilities, network partners ■ Offer: value proposition ■ Customer: target customers, distribution channels, customer relationships ■ Finance: revenue model, cost structure (Neely & Delbridge, 2007)			x	x

contined on next page

Table 5. Continued

Business model components	General context	Business context	Social and public context	Integrated context
	1	2	3	4
■ Value proposition for the customer ■ Profit formula (includes a revenue model, cost structure, margin model and resource velocity) (Johnson, Christensen, & Kagermann, 2009)		x		
■ Customers ■ Products / services ■ Distribution channels (Lisein, Pichault & Desmecht, 2009)		x		
■ Value proposition for customers ■ Core resources and activities ■ Relationships with partners (Nogalski, 2009)		x	x	x
■ Value proposition ■ Revenue mechanisms ■ Value chain ■ Value network ■ Competitive strategy ■ Target market (Kindström, 2010)		x	x	x
■ Segments of customers ■ Value proposition ■ Value delivery channels ■ Relationships with customers ■ Revenue streams ■ Key resources ■ Key activities ■ Key partners ■ Cost structure (Osterwalder & Pigneur, 2002)		x	x	x
RCOV structure: ■ Resources and competencies ■ Organizational structure ■ Value proposition (Demil & Lecocq, 2010)		x	x	x
■ Target market segment ■ Customer benefits (used products and services) ■ Technology and features of products / services ■ Mechanisms to capture value ■ Maintaining earned revenue streams (Teece, 2010)		x		
■ Price, product, ■ distribution, ■ organization & technology characteristics (Horowitz, 1996)	x			
■ Global skills, ■ corporate governance, ■ business units, ■ services, ■ units' connections. (Viscio & Pasternak 1996)	x	x	x	
■ Product/service, ■ information flow, ■ business actors and their roles, ■ advantages for actors, ■ sources of income, ■ marketing strategy. (Timmers, 1998)			x	x

contined on next page

Table 5. Continued

Business model components	General context	Business context	Social and public context	Integrated context
	1	2	3	4
▪ Product innovations, ▪ relations with customers, ▪ infrastructure management, ▪ financial aspects. (Markides, 1999)	x	x		
▪ Understanding of customer, ▪ marketing strategy, ▪ corporate governance, ▪ Internet & Intranet capacities. (Donath, 1999)			x	
▪ Actors, ▪ market segments, ▪ value offer, ▪ value activity, ▪ stakeholder network, ▪ value interfaces, ▪ value changes. (Gordijn et. al., 2001)			x	x
▪ Price model, ▪ revenue model, ▪ business processes model, ▪ e-trade opportunities, ▪ organization forms, ▪ value proposals. (Lindner & Cantrell 2001)		x	x	
▪ Value proposals, ▪ market goals, ▪ internal value chain structure, ▪ costs structure, ▪ profitability model, ▪ value network, ▪ competitiveness strategy. (Chesbrough & Rosenbaum, 2000)		x	x	x
▪ Market offer, ▪ skills, ▪ investing in key technologies, ▪ annual net profit/loss (bottom line). (Gartner, 2003)	x		x	
▪ Key strategy, ▪ strategic resources, ▪ value network, ▪ links with customers. (Hamel, 2001)		x	x	
▪ Value model, ▪ resources model, ▪ production model, ▪ model of customer relationships, ▪ revenues model, ▪ capital model, ▪ market model. (Petrovic et. al., 2001)		x		
▪ Products, ▪ relationships with customers, ▪ infrastructure and partner network, financial aspects. (Dubosson – Torbay et. al., 2001)			x	x

contined on next page

Table 5. Continued

Business model components	General context	Business context	Social and public context	Integrated context
	1	2	3	4
■ Value for customer, ■ scope, ■ price, ■ revenues, ■ relations between activities, implementation, capacities and sustainability. (Afuah & Tucci 2001)		x	x	x
■ Concept of capacities and values. (Applegate, 2001)	x	x	x	
■ Transaction content & structure, ■ corporate transactions. (Amit & Zott, 2001)			x	x
■ Mission, ■ structure, ■ processes, ■ revenues, ■ legal conformity, ■ technologies. (Alt & Zimmerman, 2001)	x			
■ Value cluster, ■ market offer scope, ■ resources system, ■ financial model. (Rayport & Jaworski, 2001)	x		x	x
■ Resources, ■ sales, profits and capital. (Betz, 2002)		x		
■ Market goals, ■ value proposal, ■ costs, ■ profits, ■ advantages for customers. (Magretta, 2002)	x	x		
■ Suppliers' value network, ■ value proposal, ■ skills and competences, ■ support, ■ infrastructure, ■ specific characteristics. (Von der Horst et. al., 2002)		x		x
■ Suppliers' value network, ■ resources assessment, ■ value proposal, ■ skills and competences, ■ processes, ■ competitiveness, ■ initial offer. (Hedman & Kalling 2003)	x	x	x	

Source: own development

Chapter 3
Public Organizations and Business Model Innovation:
The Role of Public Service Design

Mateusz Lewandowski
Jagiellonian University in Kraków, Poland

ABSTRACT

Improvement of public services and raising the citizens' quality of life is one of the biggest concerns of public sector. Changing the way public organizations operate could support such an improvement. In this regard, business model framework is an emerging field of research that could shed some new light. Yet, in case of public organizations, business model innovation remains heavily underexplored. The purpose of this chapter is threefold, to indicate and discuss the opportunities of adapting the business model innovation theory to public organizations, to conceptualize its theoretical framework, and to explore the role of public service design in the process of innovating public sector business models.

INTRODUCTION

Governments, authorities and all other public sector organizations pursue the ideas how to improve public services and raise the citizens quality of life. In this regard the researchers and practitioner of public management recognized innovation as significant driver to improve performance of delivering public services (Hollanders et al., 2013; Hughes, Moore, & Kataria, 2011; Innobarometer 2010 Analytical Report Innovation in Public Administration, 2011). Although, there are various types of innovation in the public sector (de Lancer Julnes, 2016; Lewandowski, 2015), one of the most recent research area is innovation pertaining to the business model (Zott, Amit, & Massa, 2011a). Yet, it remains underexplored not only in the business sector organizations, (Zott et al., 2011a), but, as Julnes, Gibson, and Park (2016) implicitly predict, also in the public organizations. Thus, there is a need to investigate business model innovation in the public sector.

The main aims of the chapter are: (1) to indicate and discuss the opportunities of adapting the business model innovation theory to public organizations, (2) to conceptualize the theoretical framework of

DOI: 10.4018/978-1-5225-2215-7.ch003

business model innovation for public organizations, and (3) the role of public service design in this the process of innovating public sector business models.

The chapter is structured as follows. The background section outlines the specificity of business model innovation as a research field located on the overlapping area of its two baseline theories: innovation theory, and business model theory. Then, according to the same scheme, the third section investigates the construct of public sector innovation, and public sector business model. Its main output – the General Framework of Public Sector Business Models - is then used to conceptualize how public sector business models may be innovated. The final point is made in the fourth section, where Public Service Design is presented through the lens of public sector business model as an important strategy to increase citizens quality of life. In the end conclusions summarize the main findings.

BACKGROUND

Innovation

Innovation has been deeply explored over the last couple of decades in several different fields of research, such as business and management, economics, organization studies, innovation and entrepreneurship, technology, science and engineering, knowledge management and marketing (Baregheh, Rowley, & Sambrook, 2009; Cooper, 1998). Despite the variety of research disciplines, the classic Schumpeterian definition of innovation says that it is an introduction of a new production method, product or its quality, the opening up of a new market or a new source for raw materials or semi-manufactures, or the creation of a new organizational structure in industry (Schumpeter, 1934, p. 66). According to Damanpour (1996) innovations encompass new products or services, new process technology, new organization structure or administrative systems, or new plans or programs pertaining to organization members. In turn, in the Oslo Manual (OECD/Eurostat, 2005) innovation is conceived as "the implementation of a new or significantly improved product (good or service), or process, a new marketing method, or a new organizational method in business practices, workplace organization or external relations" (p.46). Those recognized definitions only implicitly refer to the components of a business model framework, but not acknowledge business model innovation as such. Even more recent and comprehensive approaches not always do that (Baregheh et al., 2009; Cooper, 1998).

Cooper (1998) claims that every innovation is defined at the same time by three dichotomous dimensions, encompassing product versus process, radical versus incremental, and technological versus administrative. Baregheh et al. (2009) examined 60 definitions from aforementioned fields, and synthesized the six attributes of the innovation process:

1. Stages of innovation: creation, generation, implementation, development, adoption;
2. Social context: organizations, firms, customers, social systems, employees, developers;
3. Means of innovation: technology, ideas, inventions, creativity, market;
4. Nature of innovation: New, improve, change;
5. Type of innovation: Product, service, process, technical;
6. Aim of innovation: succeed, differentiate, compete.

In turn, Crossan & Apaydin (2010) presented the complex multi-dimensional framework of innovation, based on a very broad literature studies. Their model distinguishes three groups of determinants and two dimensions of innovations. According to this innovations are determined on three levels:

1. Individual and group level focused on leadership, encompassing: Chief Executive, Officer's, Top Management Team's and Board of Directors' ability and motivation to innovate;
2. Organizational level focused on managerial levers, embracing: mission, goals and strategy, structure and systems, resource allocation, organizational learning and knowledge management, and organizational culture;
3. Process level focused on business processes, including: initiation and decision-making, portfolio management, development and implementation, project management, commercialization;

Two dimensions of innovations point out their ambivalent nature. On the one hand, innovation as a process is associated with: an individual, group or firm level, a driver, such as resources or market opportunity, a top-down or bottom-up direction, a source, such as invention or adoption, a locus – firm or network, and a tacit or explicit nature. On the other hand, innovation is also an outcome, described by: a form, such as product, service, process or business model, an incremental or radical magnitude, a referent, such as firm, market or industry, and an administrative or technical type (Crossan & Apaydin, 2010). This complex and comprehensive approach acknowledges business model as a form of innovation, in contrast to previous cited works.

Business Model

In general, business model depicts the rationale of how an organization creates, delivers, and captures value (economic, social, or other) in relationship with a network of exchange partners (Massa & Tucci, 2013). Others define it as a unit of analysis to describe how the business of a firm works (Frankenberger, Weiblen, Csik, & Gassmann, 2013), or as a system describing how the pieces of a business fit together (Magretta, 2002). Over a decade ago, Pateli & Giaglis (2004) identified eight sub-domains of research in the area of business models, such as: definitions, components, taxonomies, conceptual models, design methods and tools, evaluation models, adoption factors and change methodologies.

Researchers made quite many attempts to conceptualize the business models framework (Afuah & Tucci, 2000; Al-debei, El-Haddadeh, & Avison, 2008; Chesbrough & Rosenbloom, 2002; Frankenberger et al., 2013; Linder & Cantrell, 2000; Mahadevan, 2000; Papakiriakopoulos, Poylumenakou, & Doukidis, 2001). But two approaches gained more recognition – one tries to provide most simple framework (Frankenberger et al., 2013; Johnson, Christensen, & Kagermann, 2008), and the other complex one (Osterwalder & Pigneur, 2010; Osterwalder, Pigneur, Bernarda, & Smith, 2014; Wirtz, 2011).

Johnson's et al. (2008) study outlines four elements of successful business model: customer value proposition, profit formula, key resources, and key processes. Customer value proposition encompasses identification of the target customers, their problems and needs related to particular jobs they do, and offering addressed to satisfy the problem or fulfil the needs. Profit formula, in turn, is associated with a revenue model, cost structure, margin model and resource velocity. Key resources are those, which are necessary to deliver customer value proposition profitably, and may include, among others, people, technology, partnerships, brand. Those resources are transformed in key processes, consisting of the processes, metrics, and norms (Johnson et al., 2008). Frankenberger (et al., 2013) proposed to turn similar four major dimensions of business model architecture into questions: Who is the customer? What is offered to the target customer (what the customer values)? How to build and distribute the value proposition? Why the business model is financially viable?

Wirtz (2011) made a systematic overview of the business model concept, and proposed an integrated business model consisting of nine partial models divided into three main components—strategic, customer and market, value creation. The strategic component comprises three models regarding the strategy (mission, strategic positions and development paths, value proposition), resources (core competencies and assets), and network (business model networks and partners). The customer and market components consist of customer model (customer relationships/target group, channel configuration, customer touchpoint), market offer model (competitors, market structure, value offering/products and services), and revenue model (revenue streams and revenue differentiation). The value creation component encompasses production of goods and services (manufacturing model and value generation), procurement model (resource acquisition and information), and financial model (financing model, capital model and cost structure model).

A more recognized and applied complex business model framework also distinguishes nine building blocks (Osterwalder, Pigneur, & Tucci, 2005), but is conceptualized as the business model canvas, showing how the components fit together (Osterwalder & Pigneur, 2010). Table 1 contains a comparison of the components of chosen simple and complex business model frameworks.

Most recently customer value proposition design has been developed, and comparing to Johnson's et al. (2008) proposition, comprises six building blocks, which are a detailed description of the value propositions and customer segments, two components of Business Model Canvas (Osterwalder et al., 2014). Value proposition is composed of the products and services offered to the customer, the relievers

Table 1. Simple and complex view of business model components

Complex (according to Business Model Canvas) (Osterwalder & Pigneur, 2010)	Simple (Johnson et al., 2008)	Simple (Frankenberger et al., 2013)
Customer segments that an organization serves	Customer value proposition	Who is the customer?
Value propositions that seek to solve customers' problems and satisfy their needs		What is offered to the target customer (what the customer values)?
Channels which an organization uses to deliver, communicate and sell value propositions	Key resources	How to build and distribute the value proposition?
Customer relationships which an organization builds and maintains with each customer segment		
Key resources as the assets required to offer and deliver the aforementioned elements		
Key partnerships being a network of suppliers and partners that support the business model execution by providing some resources and performing some activities		
Key activities which are performed to offered and deliver the aforementioned elements	Key processes	
Revenue streams resulting from value propositions successfully offered to customers	Profit formula	Why the business model is financially viable?
Cost structure comprising all the costs incurred when operating a business model		

of customers pains, and the creators of customer gains pertaining to the tasks and jobs he or she needs to accomplish with the assistance of the offered product or service. Thus, on the customer's side are the jobs, pains and gains related to doing the jobs.

Especially adoption factors and change methodologies of business models (Pateli & Giaglis, 2004; Scott, 2015) justified the need and pointed some directions very relevant for further exploration, such as, among others, sustainable and circular business models (Jong, Engelaer, & Mendoza, 2015; Joustra, de Jong, & Engelaer, 2013; Roome & Louche, 2015; Stubbs & Cocklin, 2008) and business model innovation (Amit & Zott, 2012; Gambardella & McGahan, 2010; Gauthier & Gilomen, 2015; Malhotra, 2000; Mentink, 2014).

Business Model Innovation

Business models foster innovation in two ways. On the one hand they are considered as vehicles of innovative products to appropriate markets, as companies commercialize innovative ideas and technologies through their business models. On the other hand, they are a new subject of innovation, distinct, albeit complementary, to traditional types of innovation such as product, process or organizational (Massa & Tucci, 2013; Mitchell & Coles, 2003; Zott et al., 2011a).

Massa and Tucci (2013) propose that BMI may refer to the design of novel BMs for newly formed organization, or the reconfiguration of existing BMs. Business model design (BMD) refers to the entrepreneurial activity of creating, implementing and validating a BM for a newly formed organization. Business model reconfiguration (BMR) means reconfiguration of organizational resources (and acquiring new ones) to change an existing BM. The output of design and/or reconfiguration must be novel or unique at least to some extent. Although, BMI is a set of business model design and/or reconfiguration, both of them imply different activities and conditions. Conditions of BMD are typical for new organizations, like lack of resources or legitimacy for instance. BMR is accompanied by problems typical for existing, more mature organizations, like for example management process, models of organizational learning and change (Massa & Tucci, 2013). It is strictly related to the concept of entrepreneurship, which Sharma & Chrisman (1999), define as the "acts of organizational creation, renewal, or innovation that occur within or outside an existing organization" (p.18).

Giesen, Berman, Bell, and Blitz (2007) identified three types of BMI:

1. Industry model innovation, consisting of innovation to the industry value chain through moving into new industries, redefining existing ones or creating new industries;
2. Revenue model innovation, pertaining to the new way revenues are generated;
3. Enterprise model innovation, changing the role of a firm in a value chain.

Enkel and Mezger (2013) investigated imitation of business models across industry boundaries, and showed that companies can facilitate cross-industry innovation on business model level through a process of abstraction, analogy identification and adaptation.

From a managerial point of view BMI encompasses innovation the content (nature of activities), the structure (how activities are linked and their sequence), and the governance (how activities are controlled and how responsibility for them is distributed) (Zott & Amit, 2010). This taxonomy has been somewhat simplified, and business models may innovate through adding new activities, linking activities in novel ways, and changing which parties perform an activity (Amit & Zott, 2012).

One of the emerging directions on BMI research focuses on the impact of BMI on sustainability (Massa & Tucci, 2013; Stubbs & Cocklin, 2008) and circular economy (Lewandowski, 2016; Mentink, 2014). The other, already much more recognized pertains to open innovation (Chesbrough, 2013), which requires the adoption of new, open business models (Aranha, Abudd, Garcia, & Corrêa, 2015; Chesbrough, 2007, 2010). Moreover, such models may impact business model innovation in complementary markets, due to the reconfiguration of downstream activities and capabilities (Gambardella & McGahan, 2010; Zott, Amit, & Massa, 2011b). Business model configurations depend on the way the activities of external innovators are organized, usually either as a collaborative community or as a market (Boudreau & Lakhani, 2009; Zott et al., 2011b).

Business model management turns out to be challenging in several ways. Many authors emphasize the need and importance of the langue issues and explanation of a business model (Euchner, 2014; Massa & Tucci, 2013; Spieth, Schneckenberg, & Ricart, 2014). There are also problems related to managing multiple business models (Markides & Charitou, 2004; Massa & Tucci, 2013) or pertaining to the fact, that a business model itself can become part of intellectual property (Rivette, Nothhaft, & Kline, 2000).

Many barriers to innovate business models exist, but may be overcome or reduced by various facilitators (Zott et al., 2011b). Spieth, Schneckenberg, and Matzler (2016) reviewed current perspectives on business model innovation and proposed a role-based approach as an alternative view to conceptualize business models and to get better understanding of the processes of business model innovation. They also showed, that such processes need much wider and detailed attention. Nevertheless, one of the key factors is to determine if the company needs to alter its business model (Johnson et al., 2008).

There are several arguments for the yes answer, to the question if business model innovation is also applicable to the public sector. Firstly, every organization has a business model as long as it creates, delivers and captures value (Kaplan, 2011). Secondly, some previous studies have already confirmed usability of business model framework for public organizations, however they applied strictly private sector business model framework (Coblence, Normandin, & Poisson-de Haro, 2014) or preceded business model theory (Alford, 1993). Thus, adaptation of a business model framework to public organizations, and conceptualization of such a model, remains in its very early stage and needs further elaboration. Because business model innovation is an overlapping field of two major theories, such as business model theory and innovation theory, in the next section of the chapter both of them will be outlined to propose the conceptualization of public sector business model innovation.

TOWARDS THE CONCEPTUAL FRAMEWORK OF PUBLIC SECTOR BUSINESS MODEL INNOVATION

Public Sector Innovation

Public sector innovation has been claimed an important instrument for improving public sector performance (Hollanders et al., 2013; Hughes et al., 2011). It remains acknowledged and recently relatively well cognized as for rather new area. Some main directions encompassed three perspectives: internal, associated with employees and managers, comparative, focused on good practices and benchmarking, and civic, pertaining to citizens and their perception of innovation (Vigoda-Gadot, Shoham, Schwabsky, & Ruvio, 2008).

One of the early definitions of innovation in the public sector considers it as "the conception and implementation of significant new services, ideas or ways of doing things as government policy in order to improve or reform them" (Glor, 2000, p.4). Other, perceives innovation in governance as a "creative idea which is successfully implemented to solve a pressing public problem" and public management innovation as "development of new policy designs and new standard operating procedures by public organizations to address public policy problems" (Alberti & Bertucci, 2006).

There are different types of innovations in public administration. Institutional innovations focus on the creation of new institutions or the renewal of established ones. In turn, organizational innovation is an introduction of new working procedures or management techniques in public administration. Improvement of the quality of public service delivery remains in the scope of process innovation. Conceptual innovation introduces new forms of governance, like interactive policy-making, engaged governance, people's budget reforms, horizontal networks, for instance (Alberti & Bertucci, 2006). Although, public sector innovation is much more varied and many other typologies exist, both for public sector in general and for particular sub-sectors (de Lancer Julnes, 2016; Lewandowski, 2015; Pollock, 2008). General typologies comprise, for instance, nine categories of Innovation, such as: agencies, partnerships, horizontal integration, good fiscal management, public service revitalization, devolution and decentralization, service improvement, systems and process improvements, and regulatory change (Armstrong & Ford, 2001). Borins (2002) distinguished bottom-up innovation, politically-led responses to crisis, and organizational turnarounds engineered by newly appointed agency heads. In turn, sub-sector-specific innovations pertain, among others, to pedagogy and education (Pollock, 2008), culture (Lewandowski, 2015; Zolberg, 1980), health care (Omachonu & Einspruch, 2010), or state-level laws (Colvin, 2006). The variety of typologies is just one of many differences between innovation in public and private sector organizations (Boyne, Gould-Williams, Law, & Walker, 2005; Kożuch, 2009; Ross, Kleingeld, & Lorenzen, 2004), although it well depicts the diversified specificity of public sector innovation.

Despite the differences, the research on the antecedents (enablers and barriers in particular) of public sector innovations (e.g. Borins, 2002; Ross et al., 2004) implicitly supports the argument, that business model could be considers as analytical framework for innovation in the public sector.

Specificity of Public Sector Business Model

There are different approaches to the components and framework of a business model (Johnson et al., 2008; Osterwalder & Pigneur, 2010; Wirtz, 2011). They vary from the most simple – consisting of four pillars (Frankenberger et al., 2013; Johnson et al., 2008) to those encompassing nine or even more components (Lewandowski, 2016; Osterwalder & Pigneur, 2010; Wirtz, 2011). The multi-component ones more are useful for practical, applicative purposes and for advanced studies. Four-component BM are more feasible for early conceptualization of the differences between business models for private profit-oriented organizations and public organizations, and have been already applied as analytical framework for the latter (Coblence et al., 2014).

Key differences between business model and public sector business models, summarized in Table 2, pertain to two out of four components outlined by Johnson et al. (2008): profit formula, and customer value proposition.

Table 2. Key differences between BM and PSBM

BM Component	Business For-Profit Organizations		Public Organizations	
	Specificity	Example	Specificity	Examples
Profit formula	• Always highly related to customer value • Dominant logic is exchange logic • Focused on economic capital (value) • Requires customer engagement and participation	• Buying an • ice-cream in a shopping center	• Not always highly related to customer value, • depends on type of public organization • dominant logic is duty logic • hidden profit formula: exchange logic concerns conversion of capitals (economic, symbolic, power, social etc) and game theory and exchange logic may be considered as corruption	Paying taxes to the municipality •
Customer value proposition	• Usually tailored to target customer needs • Usually one value proposition for one target group • Offering value is not obligatory	• Luxury branded vanilla ice-cream covered with milk chocolate and almonds	• Usually tailored to general customer needs • Usually many value propositions for many target groups • Offering value is obligatory, • Compulsion-based services	Municipal road maintenance, acquiring a building permission

Profit Formula

In case of business organizations profit formula concerns the revenue model, cost structure, margin model and resource velocity (Johnson et al., 2008). This implies that exchange of product or service for money is the driving logic of a business model. In this framework revenue is always highly related to customer value (Osterwalder et al., 2014). In case of public organizations, the scope of activities is more diversified. They are supposed to secure access to public goods, assure that all state duties indicated in the legislation are provided etc. Thus, profit formula is rather a financing formula. Goods and services may be considered as "free", as financing them is often indirect, through redistributed taxes. Redistribution is typical for public sector, and its dilemmas may be well and briefly depicted by three equity standards in public sector performance evaluation, which suggest that resources may be distributed equally, according to the needs, or input (equal opportunities, compensatory equity, and market equity respectively) (Lamb, 1987). Thus, the type of public organization impacts the profit formula applied in a business model. It may be profit oriented, however the cost reduction was and still is the biggest problem (Hildreth & Hildreth, 1989; Wilson, 1887). Despite profit or cost orientation, it often remains directly unrelated to offered customer value proposition (goods or services). This implies broadening the notion of exchange (profit formula) (Alford, 2002). Exchange logic, typical for profit-oriented business organizations, is different, more duty oriented. Duties are usually derived from legislation and state responsibilities.

However, profit formula in its classic business context applies to public organizations in one more way. Wide understanding of exchange and conversion of capitals (Alford, 1993, 2002) applied to the business model framework may explain organizational pathologies, like nepotism or corruption (Kożuch & Dobrowolski, 2014), from a new angle.

Customer Value Proposition

In case of business organizations customer value is tailored to the needs of target customer groups (Osterwalder & Pigneur, 2010). Value proposition design encompasses finding a perfect match between customers jobs, and pains and gains related to it, and offered products and services (Osterwalder et al., 2014). It implies that one value proposition is created for and offered to one target group of customers. However, in certain circumstances two business models may be applied to one market (Markides & Charitou, 2004), or multiple business model management is necessary (Louis, Blumenthal, Gluck, & Stoto, 1989; Massa & Tucci, 2013). In contrast, public organizations, used to offer products and services tailored to general customer needs, although it is changing (Alford, 1998; Osborne & McLaughlin, 2005). Some of public sector institutions, like municipality for instance, offer multiple values for various target groups, and may not resign from offering "customer value". Possibility to change value proposition may also be strictly limited. Eventually, there is a group of compulsion-based services, which citizens may not want, but have to "use", associated with police interventions, imprisonment, audits etc. In such cases "customer value" may be difficult to adjust to expectations. In this regards Lamb (1987) notices that public organizations must often select apathetic, disinterested and opposed targets, and are pressured or required to adopt undifferentiated strategies, and Alford (2002) distinguishes three types of "customers" of public organizations, like paying customers, beneficiaries, and obligatees.

This specificity highlights two major differences between BM and PSBM, indicating two practical questions:

1. How profit formula should be related to customer value proposition?
2. How customer value proposition should be tailored to customers' (citizens') expectations?

The first one embraces many relevant debates, among them on public and private goods (Jun, 1997), or limits of market economy and level of social protection (Warner & Clifton, 2014). As Laing (2003) puts it: "not all public services lend themselves to the application of such a user payment orientation due to both the existence of multiple indirect beneficiaries and the ongoing primary emphasis on social justice within many public services" *(p. 438)*. It is a matter of a key principle that precedes and conditions different ways of organizing public service delivery. The second question is related to the debates on, among others, public service quality (Hsieh, Chou, & Chen, 2002; Redman, Mathews, Wilkinson, & Snape, 1995), and public sector marketing (Laing, 2003; Lamb, 1987; Lee & Kotler, 2007). Discussion mentioned and not mention debates is beyond the aim and scope of this chapter. Nevertheless, they outline on a meta-level, a wider framework for public sector business models. It is depicted as a matrix with two dimensions reflecting both questions, and therefore four options resulting from the interaction of the possible answers (simplified and radicalized for the purpose of this conceptualization) (Table 3).

First one (I), is classic public administration which offers goods and services tailored to expectations of general customers, which are not directly related to the profit formula. This model has evolved through the decades, from *minimal state*, through *unequal partnership*, to *welfare state,* although variously around the world (Osborne & McLaughlin, 2005; Ostrowska, Frączkiewicz-Wronka, & Bratnicki, 2013). In Bovaird's (2007) classification it comprises Traditional professional service provision. Other three options pertain to *the plural state,* encompassing marketization, New Public Management and community governance and co-production.

Table 3. General framework of public sector business models

Profit Formula	**Related to Customer Value**	**(II)** • **Public organizations marketize services** • **Public organizations operate on business basis**	**(III)** • **Business organizations taking over public services and tasks**
	Not Related to Customer Value	**(I)** • **Classic public administration**	**(IV)** • **Public organizations improve services through community governance or community co-production**
		Tailored to expectations of general customers	Tailored to expectations of particular target customers
		Customer Value Proposition	

Second and third option (II & III) pertain to the various attempts to implement marketization and New Public Management, which in general tried to introduce to the public sector managerialism focused upon improved performance of public service creation and delivery (Osborne & McLaughlin, 2005). It encompassed wide range of instruments, like privatization, contracting out, mixed market solutions, cooperative contracting, market testing etc., and took different directions either from cooperation to competition or the other way around (Reichard, 2002; Warner & Clifton, 2014). Some of the reforms, like managerialism, contractualism and the call for customer focus, tend to be in kind of 'one size fits all' (Alford, 1998, p. 129). Nevertheless, in general, in this models public organizations started to operate on business basis or business organizations become responsible for public services and tasks. There are two models of public and private production process, that apply to this context (Alford, 1993), and which may be considered as the early frameworks of business models for the private and public sector.

Fourth option (IV) encompasses community governance, co-production, and other types of civic movements, which are alternative to the failures and shortages of marketization and New Public Management, like for example unsustainable strain on social cohesion or loosing social protection (Bovaird, 2007; Osborne & McLaughlin, 2005; Warner & Clifton, 2014). Importantly, many forms have their history, and have been already applied interchangeably, like co-production which had its decline and renewal state (Alford, 1998). But the core idea behind co-production, community governance etc. is "tailoring the services to the needs of individual person, with the possibility of choice" (Koch & Hauknes, 2005, p. 29).

Bovaird (2007) developed a conceptual framework that allows a more detailed characterization of the relationships between users and communities and professionalized public services. He distinguished following seven types. Traditional professional service provision with user is when services are delivered by professionals, but users and community members are closely involved in the planning and design stages. User co-delivery of professionally designed services happens when users and community members deliver the service which is designed and planned by professionals. In turn, full user – professional coproduction is associated with fully shared task of planning and designing the service, and delivering it, equally involving both the users and professionals. User – community co-delivery of services with professionals, without formal planning or design processes, is a relationship in which users and community groups take responsibility for undertaking activities, and call on professional service expertise when they need it. User – community sole delivery of professionally planned services is when users and other community members take responsibility for delivering services planned by professionals. User/

community sole delivery of co-planned or co-designed services encompasses situations when users or other community members deliver services that they partly also plan and design. In contrast, traditional self-organized community provision is when professionals do not participate in any form in planning, designing and delivering services (Bovaird, 2007). There is also a framework for citizensourcing including three dimensions, such as citizen ideation and innovation, collaborative administration, and collaborative democracy (Hilgers & Ihl, 2010).

Outlined specificity of public sector business model framework allows to conceptualize how it incorporates innovation.

Public Sector Business Model Innovation

The approach to BMI presented by Massa & Tucci (2013) emphasizes in fact its institutionalization, and from this point of view applies to the public sector. Following this argument to a more radical extend could imply the view of public organization as an institutionalized business model. If institutionalization becomes more important than the business model itself, public organization turns into an "empty shell", because it usually lasts and awaits for renewal (Downs, 1964; McCurdy, 1991), instead of achieving the last phase of organizational life cycle (Adizes, 2004). Hence, public sector business models not always expire, and public organization life cycle is not always terminated, like in case of a municipality. This is why the renewal of public organization is so important, and the perspective of business model innovation provides new lens to look at public organizations and their performance. One way to approach public sector business model innovation is to find analogies to the private sector.

Imitation of business model across industry boundaries described by Enkel and Mezger (2013) could be considered for adaptation to the public sector. They claim, that analyzing structural similarities between industries on the level of components of the business model results in developing the systematic process model for business model innovation. Maybe such a model could be developed for the public sector as well. For instance some forms of marketization may be applied to different service sectors (Reichard, 2002). The taxonomy of BMI encompassing new activities to address customers' needs, linking activities in novel ways, or new governance arrangements pertaining to the parties performing activities (Amit & Zott, 2012), is universal and could be directly applied to the public sector. In turn, open innovation in public administration can offer new ways of citizen integration and participation, enhance public value creation and political decision-making process (Hilgers & Ihl, 2010). It is also associated with collaborative innovation (Sørensen & Torfing, 2012). Hence, the concept of open innovation is highly relevant for considering open business model innovation in the public sector.

Very important and emerging trend to change the economy is related to Circular Economy (Ellen MacArthur Foundation, 2013a, 2013b; Joustra et al., 2013; Renswoude, Wolde, & Joustra, 2015). Every organization optimizes its processes, may virtualize its products or processes and uses resources from material loops, thus every business model is both linear and circular to some extent (Lewandowski, 2016; Mentink, 2014; Renswoude et al., 2015). Thus, also public organization may apply circular economy principles to their operating schemes, and turn their business models into more circular ones. For example public hospitals could adopt performance models in procurement, and become leaders in recycling and waste reduction (Ellen MacArthur Foundation, 2015a), and e-government is a good example of virtualization (Ellen MacArthur Foundation, 2013b, 2015b; Moon, 2002).

Table 4 summarizes the attempt to derive the types of Public Sector Business Model Innovation from its business sector origin.

Table 4. Types of public sector business model innovation

Types of Business Model Innovation	Types of Public Sector Business Model Innovation	Examples of PSBMI
Design of novel BM for newly formed organization (Massa & Tucci, 2013)	Creating a new public organization, or a private organization or NGO from a public organization	Spin-off (Shane, 2004)
Innovation to the industry value chain (Industry model innovation) (Giesen et al., 2007)	Changing the way public organizations cooperate in a value chain	Smart city (Nam & Pardo, 2011; Su, Li, & Fu, 2011)
New way revenues are generated (revenue model innovation) (Giesen et al., 2007)	New way of financing	Crowdfunding (Baeck & Collins, 2015; Rosenman, 2007)
New role of a firm in a value chain (enterprise model innovation) (Giesen et al., 2007)	Changing the main role of public organization	Forms of academic entrepreneurship (Klofsten & Jones-evans, 2000; Louis et al., 1989)
New activities to address customers' needs (Zott & Amit, 2010)	Adding new tasks or public services	E-government (Moon, 2002)
Novel ways of linking activities addressing customers' needs/ Reconfiguration of existing BM (Amit & Zott, 2012; Massa & Tucci, 2013)	Link activities in novel ways	Patchwork App - innovative application for social workers (Design Commission, 2013); Cooperative contracting (Reichard, 2002)
Changing which parties perform an activity (governance) (Amit & Zott, 2012)	Changing which parties perform an activity	Contracting out; Privatization (Reichard, 2002)
Cross-industry adaptation a business model (Enkel & Mezger, 2013)	Applying business model from one type of PO to another, or from industry to PO	Forms of marketization may be applied to different service sectors (Reichard, 2002)
Open business models (Aranha et al., 2015; Chesbrough, 2007)	Allowing citizens participation in public service creation and delivery or performing public tasks	Participatory budget in Porto Alegre, or public-private partnership like Villa Family (Bovaird, 2007)
Circular business model innovation (Lewandowski, 2016; Mentink, 2014)	Applying circular economy principles to public organization	Recycling and waste reduction in Dutch hospitals (Ellen MacArthur Foundation, 2015a)

Types of BMI and general framework for PSBM, when combined together, may shed some light on what PSBMI could look like. The general framework of public sector business model indicates four general types of such models, depending on the relation between profit formula and customer value proposition, and to what extend customer value proposition is general or tailored to costumers needs. Hence, two general levels on innovating public sector business model exist. First level pertains to a situation when changing a business model of public organization changes either the relation between profit formula and customer value proposition, or the extent to which it is tailored to costumers needs, or both. It also changes the main tenets of business model framework. Second level encompasses a situation, when the business model of public organization is innovated within unchanged scope of BM tenets. Both options have been illustrated on the Figure 1.

An example of the 1st level PSBMI is a public university when it starts to create spinoffs, the new companies "founded to exploit a piece of intellectual property created in an academic institution" (Shane, 2004, p. 4). Forms of academic entrepreneurship in general, and spin-offs in particular, from the perspective of the traditional public university change its business model tenets into triple helix model (Leydesdorff & Etzkowitz, 1998). Such a change is very sensitive, as it may lead to organizational dys-functions and pathology in academia, when it corrodes ethos and social values (Sułkowski & Zawadzki,

Figure 1. Levels of public sector business model innovation
Source: Own elaboration.

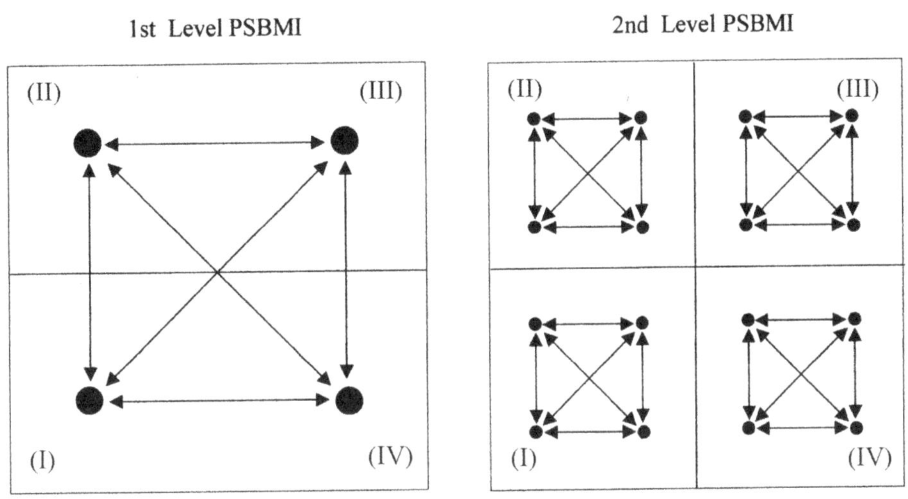

2016), or threatens organizational development (Lewandowski, 2013). The case of using crowdfunding to support public projects also raises several concerns (Rosenman, 2007, 2015).

An example of the 2nd level PSBMI is when a public organization, which already uses contracting out, decides to turn towards cooperative contracting (Reichard, 2002). Such a change of a business model does not impact the main business model tenets, it is a simple shift from one New Public Management solution to the other.

Another way to approach PSBMI is related to the innovation levels (Koch & Hauknes, 2005). In this regard, Enkel and Mezger (2013) attempt to pursue a systematic process model for business model innovation could be taken a step further for the public sector business model. A framework of business model could be elaborated for a city and region to support their sustainable development. For instance, there is the smart city concept already applied by the cities worldwide (Nam & Pardo, 2011). Other possibility to innovate public sector business model is to apply an ideology or methodology to an existing business model, like in case of circular business model (Ellen MacArthur Foundation, 2015a) or design (Barzelay & Thompson, 2010; Radine, 1987).

SOLUTIONS AND RECOMMENDATIONS

Innovating Public Sector Business Models Through the Application of Design Theory to Public Service

The pressure to introduce design into public services, and even to develop rather user-driven services which "involve public service staff and users working together to determine what services are provided and how", is increasing (Donetto, Pierri, Tsianakas, & Robert, 2015; Public Administration Select Committee, 2008, p. 9). At the same time design is an essential component of the public sector innovation theory (Omachonu & Einspruch, 2010), and co-production of public services (Bovaird, 2007), hence also of public sector business model innovation.

In general, service design is an interdisciplinary method for inventing and improving services, helping with (re) designing them from the perspective of the user, based on 'design thinking', true co-creation and collaboration with the user (Thoelen et al. 2015). Another, already recognized in the field, working definition of service design comprises four elements, such as (Saco & Goncalves, 2008, p. 12; Whicher, Swiatek, & Cawood, 2013):

- The aim, which is to create useful, useable, desirable, efficient, and effective services;
- A human-cantered approach pertaining to the customer experience and the quality of service as the focal point and the key value for success;
- A holistic approach integrating design decisions on strategic, system, process, and touch-point level;
- A systematic and iterative process integrating "user-oriented, team-based inter- disciplinary approaches and methods in ever-learning cycles."

Acording to Public Administration Select Committee (2008, p. 9) user-centred public services "actively involve the people using them in service design and delivery" and "entail drawing upon the expertise, views and perspectives of service users to complement the skills and input of service professionals."

Despite still weak conceptualization of what service design really is (Saco & Goncalves, 2008), and methodological challenges how to evaluate its effects (*Ferrari and Manzi (2014)*), several ways how service design may contribute to the public sector business model innovation may be outlined.

Implementation of design to the public sector is not new, as many historical examples exist (Ravneberg, 2009). Although, the argument to use design science or design theory to improve public services and public administration was raised deliberately just a few decades ago (Radine, 1987). For example Radine (1987) suggested to apply four design principles from architecture to the process of administrative law creation (and to statutory law controlling agencies as well), such as:

1. The law should "fit" the organizations it is controlling.
2. Multidisciplinary character in case of administrative law corresponds with social and organizational theories, which should be taken from "outside" to innovative the way law is created.
3. The way of dealing with failures should be based on the comprehensive, systematic feedback about the impacts of law. This would allow to better understand what are the failures in designing law and what are the causes.
4. Paradigm based on body of knowledge and empirical and conceptual model, which is a starting point for every design process, in case of administrative law should encompass a conceptual framework of organizational characteristics and relationships. The most important is the nature of the relationships resulting from the contacts of the agencies with professions, organizations and industries.

Building further on this argument, design theory may and should be considered to apply to all activities performed by every type of public organization. This would mean a big and important shift in the way public sector operates (DESIS Network, 2013; PricewaterhouseCoopers, n.d.). And it has already been recognized. For example, in 2010, for the first time, design was highlighted in the European Commission's Innovation Union (a policy initiative of the Europe 2020 strategy for growth) for its transformative power not only in business, but also in the public sector, and society (Whicher, Swiatek, and Thurston 2016). In the US and UK public policy is already being refocused on establishing design-led government programs (Design Commission, 2013; Public Administration Select Committee, 2008; Yoffee, 2016). Moreover, many methods and tools supporting public service design have already been recognized. They encompass for example Experience-Based Design (Donetto et al., 2015), co-design method called Storytelling Group (Kankainen et al., 2012), Affinity diagram (Beyer and Holtzblatt (1998), behavioral maps (Wang 2014), or emotional maps (Bowen et al., 2013). There is also Do-It-Yourself guide for public organizations how to design public services (Thoelen et al., 2015).

The results of institutional change and application of appropriate design methodology to public services bring, as many cases depict (Table 5), not only positive effects, but also innovation to public sector business models (Bovaird, 2007; Design Commission, 2013; PDR, 2013; Whicher et al., 2013).

Table 5. Innovative changes to PSBM led by public service design

Design-Led Public Services	Innovative Changes to PSBM Components (Based on Johnson et al. 2008)			
	Citizen-User Value Proposition	**Profit Formula**	**Key Resources**	**Key Processes**
GOV.UK (Design Commission, 2013)	A single platform containing all central government websites. It has a unified look and feel and simplifies experience of using the sites. It is a single point of entry for citizens wanting to interact with, or find things out from, central government departments.	Cost reduction due to reducing multiple contracts for 24 government differently specified and procured department websites	New department - Government Digital Service office with skills and top-management allowance to be radical	Ruling design principles were established
London Borough of Barking and Dagenham (Whicher et al., 2013)	The residents involved in the co-design project rated the council more highly in terms of keeping them informed and listening to their needs.	Cost savings being reinvested	- Partnership and knowledge: a design agency trained staff to undertake ethnographic research and led co-design workshops and service prototyping work	- Undertaking ethnographic research, co design, and service prototyping
Patchwork App (Design Commission, 2013)	An app, which allows social workers from multiple agencies to find out quickly and easily who else is working with their family, makes the work simpler and more efficient. It allows for earlier interventions and better outcomes for families.	Time savings, more efficient work	Partnership with a creative agency FutureGov	Performed through Patchwork App

continued on next page

Table 5. Continued

Design-Led Public Services	Innovative Changes to PSBM Components (Based on Johnson et al. 2008)			
	Citizen-User Value Proposition	Profit Formula	Key Resources	Key Processes
Villa Family Project (Bovaird, 2007)	- Elderly people may live in their villages, close to relatives and friends, in a family atmosphere, and they receive professional 24 hour care; - Separate flats in a large house for two families, who each host three elderly people, - With two host families under the same roof, hosts can stand in for each other briefly, such as on annual holidays; - The architecture of the Villa Family is specially designed to overcome typical problems in such arrangements and helps to professionalize the job.	- Both young and elderly benefit from contact with each other; - Elderly people employ the hosts; - Jobs are attractive because of the salary and possibility to bringing up children; - Investor receives the rental income - Municipalities to donate the lease of a plot of land on which to build the Villa Family, and in return, it guarantees that elderly people from the municipality have priority allocation - Free land allows affordable rents to be set, in line with social housing.	Public-private partnership	The county council registers the host's qualifications and the suitability the household and monitors the project
District Labour Office in Żory (PDR, 2013)	- A new system to help clients find their way around the building - An information point in the ground floor lobby as the first point of contact for clients and dedicated information staff to help clients - For parents a children's corner activity area and changing facilities built into the ground floor toilets. - Staff were encouraged to rewrite any documents intended for clients in easy to understand everyday language	- Cost effective improvements - EU funded project covered the costs of external design experts and the workshops	- Minimal investment, mainly by using existing staff and facilities in new ways - Partnership with PDR within the EU funded project	Improved communication Staff Training programs

The Table 5 above may be summarized by a conclusion, that on a very general level public service design may foster innovation in public sector business models in the similar way it does for public service reform, which seem to be the two sides of the same coin. A sentence from the report "Restarting Britain 2: Design and Public Services " illustrates it well: "there are lots of ways to approach public service reform: opening up public services to choice and competition; reducing the public service cost base and improving its overall productivity; and bringing public services into the 21st century through greater use of digital technologies. But to succeed, each of these approaches must incorporate the principles of great design." (Design Commission, 2013, p. 5)

User-centered service design approach may contribute to public sector business model innovation not only in a single public organization, but also on the level of local and central government (Design Commission, 2013), and respond better to the challenges related to multiple-value for multiple stakeholders, as Villa Project shows (Bovaird, 2007).

Although public service design may greatly contribute to citizens quality of life (Ravneberg, 2009), its application is not an easy process. There are several barriers and pitfalls, which may impede PSD (Thoelen et al., 2015), and thus public sector business model innovation. They comprise, among others, several barriers impacting the integrity of business model fit, such as being blind to the "outside world" due to high personal engagement in the service or organization reform, imbalance in listening to the standpoint of the end users and to the needs and wishes of employees, imbalance in focusing on the interaction between the citizens/customers and the employees, and not enough attention for what needs to happen behind the scenes in order to make this a reality, or simply skepticism from stakeholders and being afraid of changes.

FUTURE RESEARCH DIRECTIONS

Some direction of the future research could encompass exploration of public sector value chain, identification of business model types and also business model innovation types typical for the public sector, including similarities and differences across the countries. Moreover, a framework of a business model could be elaborated for a city and region to support their sustainable development. Another interesting point would encompass more detailed comparison of Public Sector Business Model with Social Exchange Theory. Important topic pertains also to the ethical issues resulting from innovating business models of public organizations.

CONCLUSION

This chapter links business model and public management theory in three ways. It identifies the specificity of public sector business model which pertains to profit formula and customer value creation. In the public sector those two main components of a business model are usually indirectly related to each other, and customer value is hardly tailored to target customers' needs. Although, it is changing due to the public sector business model innovation. Such an innovation may exist on two levels. Either it changes the tenets of the general framework of public sector business model, or it happens between the lines drawn by the principles. In this regard, Public Service Design appears to be a strategy to innovate every business model of a public organization, and to deliver much better public services and raise citizens quality of life.

Identified specificity supplements both theories. Business model theory is extended with its public sector context. Public management, in turn, gains a synthetized framework, which has been around for some time in the literature, although not explicitly outlined. Such a perspective provides new lens to analyze how public organizations operate.

REFERENCES

Adizes, I. (2004). *Managing Corporate Lifecycles*. Santa Barbara, CA: Adizes Institute.

Afuah, A., & Tucci, C. L. (2000). *Internet Business Models and Strategies: Text and Cases* (1st ed.). Mcgraw-Hill College.

Al-debei, M. M., El-Haddadeh, R., & Avison, D. (2008). Defining the Business Model in the New World of Digital Business. In *Proceedings of the Fourteenth Americas Conference on Information Systems* (pp. 1–11). Retrieved from http://bura.brunel.ac.uk/bitstream/2438/2887/1/AMCIS2008.pdf

Alberti, A., & Bertucci, G. (2006). Replicating Innovations in Governance: An Overview. In *Innovations in Governance and Public Administration: Replicating what works* (pp. 1–24). New York: United Nations.

Alford, J. (1993). *Towards a New Pubuc Management Model: Beyond "Managerialism" and Its Critics. Australian Journal of Public Administration*. doi:10.1111/j.1467-8500.1993.tb00263.x

Alford, J. (1998). A Public Management Road Less Travelled : Clients as Co-producers of Public Services. *Australian Journal of Public Administration*, *57*(December), 128–137. doi:10.1111/j.1467-8500.1998.tb01568.x

Alford, J. (2002). Defining the Client in the Public Sector: A Social-Exchange Perspective. *Public Administration Review*, *62*(3), 337–346. doi:10.1111/1540-6210.00183

Amit, R., & Zott, C. (2012). Creating Value Through Business Model Innovation. *MIT Sloan Management Review*, *53*(53310), 41–49. doi:10.2139/ssrn.1701660

Aranha, E. A., Abudd, N., Garcia, P., & Corrêa, G. (2015). Open Innovation and Business Model : A Brazilian Company Case Study. *Journal of Technology Management & Innovation*, *10*(4), 91–98. doi:10.4067/S0718-27242015000400010

Baeck, P., & Collins, L. (2015). *Crowdfunding public services - tapping into the crowd to finance public projects*. Retrieved from http://www.nesta.org.uk/blog/crowdfunding-public-services-tapping-crowd-finance-public-projects

Baregheh, A., Rowley, J., & Sambrook, S. (2009). Towards a multidisciplinary definition of innovation. *Management Decision*, *47*(8), 1323–1339. doi:10.1108/00251740910984578

Barzelay, M., & Thompson, F. (2010). Back to the Future: Making Public Administration a Design Science. *Public Administration Review*, 70.

Beyer, H., & Holtzblatt, K. (1998). *Contextual Design: Defining Customer-Centered Systems* (1st ed.). San Francisco, CA: Morgan Kaufmann.

Borins, S. (2002). Leadership and innovation in the public sector. *Leadership and Organization Development Journal*, *23*(8), 467–476. doi:10.1108/01437730210449357

Boudreau, K. J., & Lakhani, K. R. (2009). How to manage outside innovation. *MIT Sloan Management Review, 50*.

Bovaird, T. (2007). Beyond engagement and participation: User and community coproduction of public services. *Public Administration Review*, *67*(5), 846–860. doi:10.1111/j.1540-6210.2007.00773.x

Bowen, S., McSeveny, K., Lockley, E., Wolstenholme, D., Cobb, M., & Dearden, A. (2013). How was it for you? Experiences of participatory design in the UK health service. *Codesign-International Journal of Cocreation in Design and the Arts*, *9*(4), 230–246. doi:10.1080/15710882.2013.846384

Boyne, G. A., Gould-Williams, J. S., Law, J., & Walker, R. M. (2005). Explaining the adoption of innovation: An empirical analysis of public management reform. *Environment and Planning. C, Government & Policy*, *23*(3), 419–435. doi:10.1068/c40m

Chesbrough, H. (2007). Why companies should have open business models. *MIT Sloan Management Review, 48*(2), 22–29. Retrieved from http://doi.org/Article

Chesbrough, H. (2010). Business model innovation: Opportunities and barriers. *Long Range Planning*, *43*(2–3), 354–363. doi:10.1016/j.lrp.2009.07.010

Chesbrough, H. (2013). *Open Innovation. The New Imperative for Creating and Profiting from Technology*. Boston: Harvard Business School Press. http://doi.org/<ALIGNMENT.qj></ALIGNMENT>10.1017/CBO9781107415324.004

Chesbrough, H., & Rosenbloom, R. S. (2002). The role of the business model in capturing value from innovation : Evidence from Xerox Corporation ' s technology spin-off companies. *Industrial and Corporate Change*, *11*(3), 529–555. doi:10.1093/icc/11.3.529

Coblence, E., Normandin, F., & Poisson-de Haro, S. (2014). Sustaining Growth through Business Model Evolution: The Industrialization of the Montreal Museum of Fine Arts (1986–2012). *The Journal of Arts Management, Law, and Society*, *44*(3), 126–144. doi:10.1080/10632921.2014.936077

Colvin, R. A. (2006). Understanding Policy Adoption and Gay Rights: The role of the media and other factors. *The Innovation Journal: The Public Sector Innovation Journal, 11*(2).

Cooper, J. R. (1998). A multidimensional approach to the adoption of innovation. *Management Decision*, *36*(8), 493–502. doi:10.1108/00251749810232565

Crossan, M. M., & Apaydin, M. (2010). A multi-dimensional framework of organizational innovation: A systematic review of the literature. *Journal of Management Studies*, *47*(6), 1154–1191. doi:10.1111/j.1467-6486.2009.00880.x

Damanpour, F. (1996). Organizational complexity and innovation: Developing and testing multiple contingency models. *Management Science*, *42*(5), 693–716. doi:10.1287/mnsc.42.5.693

de Lancer Julnes, P. (2016). The Study of Innovation. State of Art and Framework for Analysis. In Innovation in the Public and Nonprofit Sector (pp. 12–32). New York: Routledge.

Design Commission. (2013). *Restarting Britain 2: Design and Public Services*. Retrieved from files/101/DC_Restarting_Britain_2_report - interactive.pdf

DESIS Network. (2013). *Public and Collaborative: Exploring the intersection of Design, Social Innovation and Public Policy* (E. Manzini & E. Staszowski, Eds.). DESIS Network.

Donetto, S., Pierri, P., Tsianakas, V., & Robert, G. (2015). Experience-based Co-design and Healthcare Improvement: Realizing Participatory Design in the Public Sector. *Design Journal, 18*(2), 227–248. http://doi.org/<ALIGNMENT.qj></ALIGNMENT>10.1111/j.1747-4949.2010.00571.x

Downs, A. (1964). *Inside Bureaucracy*. Rand Corporation Research Study. http://doi.org/<ALIGNMENT.qj></ALIGNMENT>10.2307/447012

Ellen MacArthur Foundation. (2013a). Towards the Circular Economy: Opportunities for the consumer goods sector (vol. 2). Author.

Ellen MacArthur Foundation. (2013b). Towards the Circular Economy: Economic and business rationale for an accelerated transition (Vol. 1). Author.

Ellen MacArthur Foundation. (2015a). *Delivering the circular economy a toolkit for policymakers*. Author.

Ellen MacArthur Foundation. (2015b). Growth within: A circular economy vision for a competitive. Author.

Enkel, E., & Mezger, F. (2013). Imitation processes and their application for business model innovation: An explorative study. *International Journal of Innovation Management*, *17*(1), 1340005. doi:10.1142/S1363919613400057

Euchner, J. (2014). Business Model Innovation in Practice. *Research Technology Management*, (November-December), 33–39. http://doi.org/<ALIGNMENT.qj></ALIGNMENT>10.5437/08956308X5706013

Ferrari, P. A., & Manzi, G. (2014). Citizens evaluate public services: A critical overview of statistical methods for analysing user satisfaction. *Journal of Economic Policy Reform*, *17*(3), 236–252. doi:10.1080/17487870.2014.909313

Frankenberger, K., Weiblen, T., Csik, M., & Gassmann, O. (2013). The 4I-framework of business model innovation: A structured view on process phases and challenges. *International Journal of Product Development*, *18*(3/4), 249–273. doi:10.1504/IJPD.2013.055012

Gambardella, A., & McGahan, A. M. (2010). Business-model innovation: General purpose technologies and their implications for industry structure. *Long Range Planning*, *43*(2–3), 262–271. doi:10.1016/j.lrp.2009.07.009

Gauthier, C., & Gilomen, B. (2015). Business Models for Sustainability: Energy Efficiency in Urban Districts. *Organization & Environment*. doi:10.1177/1086026615592931

Giesen, E., Berman, S. J., Bell, R., & Blitz, A. (2007). Three ways to successfully innovate your business model. *Strategy and Leadership*, *35*(6), 27–33. doi:10.1108/10878570710833732

Glor, E. D. (Ed.). (2002). Is Innovation a Question of Will or Circumstance? An Exploration of the Innovation Process Through the Lens of the Blakeney Government in Saskatchewan. *Innovation*.

Hildreth, W. B., & Hildreth, R. P. (1989). The Business of Public Management. *Public Productivity Review*, *12*(3), 303–321. doi:10.2307/3380120

Hilgers, D., & Ihl, C. (2010). Citizensourcing: Applying the Concept of Open Innovation to the Public Sector. *The International Journal of Public Participation*, *4*(1), 67–88.

Hollanders, H., Arundel, A., Buligescu, B., Peter, V., Roman, L., Simmonds, P., & Es-Sadki, N. (2013). European Public Sector Innovation Scoreboard 2013. A pilot exercise. Belgium: Academic Press.

Hsieh, A.-T., Chou, C.-H., & Chen, C.-M. (2002). Job standardization and service quality: A closer look at the application of total quality management to the public sector. *Total Quality Management*, *13*(7), 899–912. doi:10.1080/0954412022000017012

Hughes, A., Moore, K., & Kataria, N. (2011). *Innovation in Public Sector Organisations: A pilot survey for measuring innovation across the public sector.* London: NESTA. Retrieved from http://scholar.google.com/scholar?hl=en&btnG=Search&q=intitle:Innovation+in+Public+Sector+Organisations+A+pilot+survey+for+measuring+innovation+across+the+public+sector#0

Innobarometer. (2010). *Analytical Report Innovation in Public Administration*. Author.

Johnson, M. W., Christensen, C. M., & Kagermann, H. (2008). Reinventing your business model. *Harvard Business Review*, *86*(12). doi:10.1111/j.0955-6419.2005.00347.x

Jong, E. de, Engelaer, F., & Mendoza, M. (2015). *Realizing opportunities of a circular business model.* Academic Press.

Joustra, D. J., de Jong, E., & Engelaer, F. (2013). *Guided Choices towards a Circular Business Model.* Academic Press.

Julnes, P. de L., Gibson, E., & Park, S. (2016). Refining Our Understanding of the Process of Innovation in Public and Nonprofit Organizations. In P. de L. Julnes & E. Gibson (Eds.), *Innovation in the Public and Nonprofit Sector* (pp. 245–260). New York: Routledge.

Jun, J. S. (1997). Dialectic between the Private Realm and the Public Realm: Renewing Debate on the Public Good. *Administrative Theory & Praxis*, *19*(2), 238–245. doi:10.1097/EDE.0b013e31812e5535

Kankainen, A., Vaajakallio, K., Kantola, V., & Mattelmäki, T. (2012). Storytelling Group – a co-design method for service design. *Behaviour & Information Technology*, *31*(3), 221–230. doi:10.1080/0144929X.2011.563794

Kaplan, S. (2011). Business Models Aren't Just For Business. *Harvard Business Review*, 11.

Klofsten, M., & Jones-evans, D. (2000). Academic Comparing in Europe - Entrepreneurship The Case of Sweden and Ireland. *Small Business Economics*, *14*(4), 299–309. doi:10.1023/A:1008184601282

Koch, P. M., & Hauknes, J. (2005). *On innovation in the public sector – today and beyond.* Publin Report No. D20. Retrieved from http://citeseerx.ist.psu.edu/viewdoc/download?doi=10.1.1.104.3988&rep=rep1&type=pdf

Kożuch, B. (2009). Innowacyjność w sektorze publicznym - bariery i możliwości rozwoju. In B. Kryk & K. Piech (Eds.), *Innowacyjność w skali makro i mikro*. Warszawa: Instytut Wiedzy i Innowacji.

Kożuch, B., & Dobrowolski, Z. (2014). *Creating Public Trust. An Organisational Perspective*. Frankfurt am Main: Peter Lang. doi:10.3726/978-3-653-02970-3

Laing, A. (2003). Marketing in the public sector: Towards a typology of public services. *Marketing Theory*, *3*(4), 427–445. doi:10.1177/1470593103042005

Lamb, C. W. Jr. (1987). Public Sector Marketing is Different. *Business Horizons*, *30*(4), 56–60. doi:10.1016/0007-6813(87)90066-8

Lee, N. R., & Kotler, P. T. (2007). *Marketing in the Public Sector: A Roadmap for Improved Performance*. Pearson Education.

Lewandowski, M. (2013). Introduction to Academic Entrepreneurship. In A. Szopa, W. Karwowski, & P. Ordóñez de Pablos (Eds.), Academic Entrepreneurship and Technological Innovation (pp. 1–28). IGI Global. http://doi.org/ doi:10.4018/978-1-4666-2116-9.ch001

Lewandowski, M. (2015). Types of Innovations in Cultural Organizations. *International Journal of Contemporary Management*, *14*(1), 67–78.

Lewandowski, M. (2016). Designing the Business Models for Circular Economy — Towards the Conceptual Framework. *Sustainability*, *8*(1), 43. doi:10.3390/su8010043

Leydesdorff, L., & Etzkowitz, H. (1998). The Triple Helix as a model for innovation studies. *Science and Public Policy Beech Tree Puhlishing*, *25*(3), 195–203. doi:10.1016/j.vaccine.2010.03.035

Linder, J., & Cantrell, S. (2000). *Changing Business Models: Surveying the Landscape* (Vol. 34). Retrieved from http://scholar.google.com/scholar?hl=en&btnG=Search&q=intitle:Changing+Business+Models:+Surveying+the+Landscape#0

Louis, K. S., Blumenthal, D., Gluck, M. E., & Stoto, M. A. (1989). Entrepreneurs in Academe : An Exploration of Behaviors among Life Scientists. *Administrative Science Quarterly*, *34*(1), 110–131. doi:10.2307/2392988

Magretta, J. (2002). Why Business Models Matter. *Harvard Business Review*. PMID:12024761

Mahadevan, B. (2000). Business Models for Internet-Based E-Commerce: An Anatomy. *California Management Review*, *42*(4), 55–69. doi:10.2307/41166053

Malhotra, Y. (2000). Knowledge management and new organization forms: A framework for business model innovation. *Information Resources Management Journal*, *13*(1), 5–14. doi:10.4018/irmj.2000010101

Markides, C., & Charitou, C. D. (2004). Competing with dual business models: A contingency approach. *The Academy of Management Executive*, *18*(3), 22–36. doi:10.5465/AME.2004.14776164

Massa, L., & Tucci, L. C. (2013). *Business Model Innovation. The Oxford Handbook of Innovation Management*. doi:10.1002/9781118466421.ch4

McCurdy, H. E. (1991). Organizational Decline: NASA and the Life Cycle of Bureaus. *Public Administration Review*, *51*(4), 308–315. doi:10.2307/976746

Mentink, B. (2014). *Circular business model innovation: A process framework and a tool for business model innovation in a circular economy* (MSc Thesis). Delft University of Technology & Leiden University.

Mitchell, D., & Coles, C. (2003). The ultimate competitive advantage of continuing business model innovation. *The Journal of Business Strategy*, *24*(5), 15–21. doi:10.1108/02756660310504924

Moon, M. J. (2002). The Evolution of E-Government among Municipalities: Rhetoric or Reality? *Public Administration Review*, *62*(4), 424–433. doi:10.1111/0033-3352.00196

Nam, T., & Pardo, T. A. (2011). Conceptualizing Smart City with Dimensions of Technology, People, and Institutions. In *Proceedings of the 12th Annual International Conference on Digital Government Research* (pp. 282–291). http://doi.org/ doi:10.1145/2037556.2037602

OECD/Eurostat. (2005). *Oslo manual: Guidelines for collecting and interpreting innovation data.* <ALIGNMENT.qj></ALIGNMENT>10.1787/9789264013100-en

Omachonu, V. K., & Einspruch, N. G. (2010). Innovation in Healthcare Delivery Systems : A Conceptual Framework. *The Innovation Journal: The Public Sector Innovation Journal*, *15*(1), 1–20.

Osborne, S. P., & McLaughlin, K. (2005). The New Public Management in context. In K. McLaughlin, E. Ferlie, & S. Osborne (Eds.), *New Public Management: Current Trends and Future Prospects* (pp. 7–14). London: Routledge.

Osterwalder, A., & Pigneur, Y. (2010). *Business Model Generation: A Handbook for Visionaries, Game Changers, and Challengers*. Hoboken, NJ: John Wiley and Sons.

Osterwalder, A., Pigneur, Y., Bernarda, G., & Smith, A. (2014). *Value Proposition Design: How to Create Products and Services Customers Want*. Hoboken, NJ: John Wiley and Sons.

Osterwalder, A., Pigneur, Y., & Tucci, C. L. (2005). Clarifying Business Models: Origins, Present, and Future of the Concept. *Communications of the Association for Information Systems*, *16*, 1–25. http://doi.org/10.1.1.83.7452

Ostrowska, S., Frączkiewicz-Wronka, A., & Bratnicki, M. (2013). Mission Oriented Scorecard: The Experience of Public Hospitals in Poland. In *7th Conference On Performance Measurement And Management Control*. European Institute for Advanced Studies in Management.

Papakiriakopoulos, D. A., Poylumenakou, A. K., & Doukidis, G. J. (2001). Building e-Business Models: An Analytical Framework and Development Guidelines. In *14th Bled Electronic Commerce Conference* (pp. 446–464). Retrieved from http://domino.fov.uni-mb.si/proceedings.nsf/0/5ba2fc04f2174ec8c1256e9f0030b818/$FILE/28_Papakiri.pdf

Pateli, A. G., & Giaglis, G. M. (2004). A research framework for analysing eBusiness models. *European Journal of Information Systems*, *13*(4), 302–314. doi:10.1057/palgrave.ejis.3000513

PDR. (2013). *Designing Effective Public Services: A practical guide for public service managers*. PDR.

Pollock, K. (2008). The Four Pillars of Innovation: An Elementary School Perspective. *The Innovation Journal: The Public Sector Innovation Journal*, *13*(2), 1–20. Retrieved from http://www.innovation.cc/peer-reviewed/pollack_innovative2.pdf

PricewaterhouseCoopers. (n.d.). *Redefining local government*. Author.

Public Administration Select Committee. (2008). *User Involvement in Public Services*. Sixth Report of Session 2007–08.

Radine, L. B. (1987). Organization Theory in Administrative Law: A Proposal for a Design Science. *The American Sociologist*, *18*(3), 278–283. doi:10.1007/BF02691772

Ravneberg, B. (2009). Identity politics by design: Users, markets and the public service provision for assistive technology in Norway. *Scandinavian Journal of Disability Research*, *11*(2), 101–115. doi:10.1080/15017410902753904

Redman, T., Mathews, B., Wilkinson, A., & Snape, E. (1995). Quality management in services: Is the public sector keeping pace? *International Journal of Public Sector Management*, *8*(7), 21–34. doi:10.1108/09513559510103166

Reichard, C. (2002). Marketisation of Public Services in Germany. *International PublicManagement Review*, *3*(2), 63–80.

Rivette, K. G., Nothhaft, H. R., & Kline, D. (2000, January-February). *Discovering New Value in Intellectual Property. Harvard Business Review*.

Roome, N., & Louche, C. (2015). Journeying Toward Business Models for Sustainability: A Conceptual Model Found Inside the Black Box of Organisational Transformation. *Organization & Environment*, 1–24. doi:10.1177/1086026615595084

Rosenman, M. (2007). Charity Fundraiser-in-Chief: The government's competition with nonprofits. *Stanford Sociar Innovation Review*. Retrieved from http://ssir.org/articles/entry/charity_fundraiser_in_chief#

Rosenman, M. (2015). *Why Crowdfunding Government Is a Bad Idea*. Retrieved June 20, 2016, from http://philanthropynewsdigest.org/commentary-and-opinion/why-crowdfunding-government-is-a-bad-idea

Ross, V. E., Kleingeld, A. W., & Lorenzen, L. (2004). a Topographical Map of the Innovation Landscape. *The Innovation Journal: The Public Sector Innovation Journal*, *9*(2), 1–19. Retrieved from http://linkinghub.elsevier.com/retrieve/pii/S1053810000904586

Saco, R. M., & Goncalves, A. P. (2008). Service Design: An Appraisal. *Design Management Review*, *19*(1), 10–19. doi:10.1111/j.1948-7169.2008.tb00101.x

Schumpeter, J. A. (1934). The Theory of Economic Development: An Inquiry Into Profits, Capital, Credit, Interest, and the Business Cycle. New Brunswick, NJ: Transaction Publishers.

Scott, J. T. (2015). *The Sustainable Business: A Practitioner's Guide to Achieving Long-Term Profitability and Competitiveness* (2nd ed.). Sheffield, UK: Greenleaf Publishing. doi:10.1007/s13398-014-0173-7.2

Shane, S. (2004). *Academic Entrepreneurship: University Spinoffs and Wealth Creation*. Cheltenham, UK: Edward Elgar Publishing. doi:10.4337/9781843769828

Sharma, P., & Chrisman, S. J. J. (1999). Toward a reconciliation of the definitional issues in the field of corporate entrepreneurship. *Entrepreneurship Theory and Practice*, *23*(3), 11–27. doi:10.1007/978-3-540-48543-8_4

Sørensen, E., & Torfing, J. (2012). Collaborative Innovation in the Public Sector. *The Innovation Journal*, *17*(1), 1–14. Retrieved from https://vpn.utm.my/docview/1362242870?accountid=41678

Spieth, P., Schneckenberg, D., & Matzler, K. (2016). Exploring the linkage between business model (&) innovation and the strategy of the firm. *R&D Management*. http://doi.org/<ALIGNMENT.qj></ALIGNMENT>10.1111/radm.12218

Spieth, P., Schneckenberg, D., & Ricart, J. E. (2014). Business model innovation – state of the art and future challenges for the field. *R & D Management*, *44*(3), 237–247. doi:10.1111/radm.12071

Stubbs, W., & Cocklin, C. (2008). Conceptualizing a Sustainability Business Model. *Organization & Environment*, *21*(2), 103–127. doi:10.1177/1086026608318042

Su, K., Li, J., & Fu, H. (2011). Smart city and the applications. In *2011 International Conference on Electronics, Communications and Control, ICECC 2011 - Proceedings* (pp. 1028–1031). http://doi.org/doi:10.1109/ICECC.2011.6066743

Sułkowski, Ł., & Zawadzki, M. (2016). Corporate University: A Critical Approach. *Entrepreneurship and Management, 17*(1).

Thoelen, A., Cleeren, S., Denis, A., Peters, K., Van Ael, K., & Willems, H. (2015). *Public Service Design A guide for the application of service design in public organisations*. Retrieved from http://www.thespiderproject.eu/wp-content/uploads/2015/09/PSD_manual_UK_LR.pdf

van Renswoude, K., ten Wolde, A., & Joustra, D. J. (2015). Circular Business Models. Part 1: An introduction to IMSA's circular business model scan. Amsterdam: Academic Press.

Vigoda-Gadot, E., Shoham, A., Schwabsky, N., & Ruvio, A. (2008). Public sector innovation for Europe: A multinational eight-country exploration of citizens perspectives. *Public Administration*, *86*(2), 307–329. doi:10.1111/j.1467-9299.2008.00731.x

Wang, S.-M. (2014). Public service space remodeling based on service design and behavioral maps. *Journal of Industrial and Production Engineering*, *31*(2), 76–84. doi:10.1080/21681015.2014.887595

Warner, M. E., & Clifton, J. (2014). Marketisation, public services and the city: The potential for Polanyian counter movements. *Cambridge Journal of Regions, Economy and Society*, *7*(1), 45–61. doi:10.1093/cjres/rst028

Whicher, A., Swiatek, P., & Cawood, G. (2013, August). An Overview of Service Design for the Private and Public Sectors. *See Platform*.

Whicher, A., Swiatek, P., & Thurston, P. (2016). Trends in Design and Government in Europe. *Design Management Review, 27*(1), 44–50. http://doi.org/DOI: 10.1111/drev.10348

Wilson, W. (1887). The Study of Administration. *Political Science (Wellington, N.Z.)*, *2*(2), 197–222. doi:10.2307/2139277 PMID:4591257

Wirtz, B. W. (2011). Business Model Management: Design – Instruments – Success Factors (1st ed.). Gabler Verlag.

Yoffee, S. F. (2016). Translating Design Policy into a State Design Commission: Maryland. *Design Management Review, 27*(1), 38–43. http://doi.org/DOI: 10.1111/drev.10347

Zolberg, V. (1980). Displayed Art and Performed Music: Selective Innovation and the Structure of Artistic Media. *The Sociological Quarterly*, *21*(2), 219–231. doi:10.1111/j.1533-8525.1980.tb00606.x

Zott, C., & Amit, R. (2010). Business model design: An activity system perspective. *Long Range Planning*, *43*(2–3), 216–226. doi:10.1016/j.lrp.2009.07.004

Zott, C., Amit, R., & Massa, L. (2011a). The business model: Recent developments and future research. *Journal of Management*, *37*(4), 1019–1042. doi:10.1177/0149206311406265

KEY TERMS AND DEFINITIONS

Business Model: A way of how an organization creates and delivers value to its customers or audience, and how it captures value from them in return.

Business Model Innovation: A new way of creating, delivering and/or capturing value, introduced by an organization purposefully.

Customer Value Proposition: A set of attributes, relevant for a customer or audience, which an offered good or service has.

Design: A plan, drawing, pattern or a way in which parts of a bigger whole are purposeful arranged.

Profit Formula: A way of exchanging customer value proposition (good or service) for other value (usually, but not always, the price of offered good or service).

Public Service: A service highly relevant for citizens quality of life and safety, created and delivered by a public institution or other organization on behalf of a public body.

Service Design: A way to plan, implement and offer a service which would contain the attributes preferred by the customers or audience, based on true and in-depth recognition of those attributes through engagement the customers or audience.

Chapter 4
Options for Applying Functional Overlapping Competing Jurisdictions (FOCJs) for Municipal Cooperation in Russia

Peter Friedrich
University of Tartu, Estonia

Mariia Chebotareva
University of Tartu, Estonia

ABSTRACT

Municipal cooperation is important for transformation countries, like Russia, which have to develop legal, institutional and political environments for public service activities. The authors recommend FOCJs as an instrument for coordinated municipal public service provision. To determine the suitability of FOCJs the analyst has to investigate the relation between FOCJ theory, their financing possibilities, the fiscal effects, and the legal forms in which FOCJs can operate in Russia. The authors define several forms of FOCJs and sort out appropriate public enterprises of private and public law for Russian FOCJ. To analyse the establishment, the operations, and the competition between FOCJs the authors present three types of models. One relates to the establishment of an FOCJ, the second concentrates on financing service activities, and the third model deals with competition among FOCJs and demonstrates the effects of different ways of finance. The article concludes with elaborating recommendations for financing FOCJ under conditions in Russia.

INTRODUCTION[1]

Modern realities and the intensifying crisis in the Russian economy demonstrate a serious need for the juridical, as well as economic and political, transformation of existing Russian public institutions. Following the example of many European countries, Russian governments should modernize the provision of public services and promote public entrepreneurship[2]. Some authors argue for further decentralization

DOI: 10.4018/978-1-5225-2215-7.ch004

(Петров, 2000; Гельман, 2006; Ляпин, 2015) and more cooperation of municipalities in local government service provision (Ирискина, 2010; Ладыгин, 2011).

Special means of cooperation in providing public services – Functional Overlapping Competing Jurisdictions (FOCJs[3]) – are discussed in the Western public finance literature (Frey, 1999, 2000; Frey & Eichenberger, 1995, 1996, 2002, 2006; Friedrich, Ukrainski & Timpmann, 2014). The main fields of discussion concern:

1. The character of an FOCJ as a jurisdiction (Detig, 2004; Friedrich, 2006; Gabbe, 2008; Friedrich & Eckardt, 2014).
2. The character of an FOCJ as an economic subject and political unit (Friedrich, Kaltschütz & Nam, 2004; Friedrich & Fladung, 2008; Friedrich & Eckardt, 2014).
3. The FOCJ as a means of municipal cooperation in relation to other forms of cooperation (Bartholomae & Popescu, 2008, Friedrich, Ukrainski & Timpmann, 2014).
4. The FOCJ as a means to intensify competition in public service provision (Friedrich & Eckardt, 2014).
5. The FOCJ as a reform concept for the European Union (Eckardt & Gritsch, 2016).
6. Cross-border problems with respect to FOCJ (Detig, Feng & Friedrich, 2002; Friedrich & Ukrainski, 2014; Metis, 2014; Eckardt & Gritsch, 2016).
7. Lessons from historical cases of FOCJs (Frey, 2005; Smith, 2011; Fink, 2012; Eckardt & Gritsch, 2016).
8. FOCJs related to several sector activities and problems (Spindler, 1998, 2008a, 2008b; Detig, 2004; Friedrich & Popescu, 2006; Friedrich & Reiljan, 2011).
9. Opportunities for and limits of application of FOCJs (Friedrich & Reiljan, 2011; Friedrich & Eckardt, 2014).

Issues, which are not tackled intensively in literature, are the management of FOCJs (Friedrich & Eckardt, 2014), the problems of entrepreneurship in FOCJs (Friedrich & Ukrainski, 2013), cooperative production in FOCJs, decision-making in FOCJs, application of FOCJs under different law regimes (Leitenstorfer & Török, 2008), particularly that of Russia, and the financing of FOCJs (Fladung & Friedrich, 2008). Therefore, the chapter deals with the principles to finance FOCJs. In addition, whether FOCJs are applicable for Russian municipalities has not been discussed intensively. The authors are especially interested in tackling the last two fields of research. Thus, the research question concerns: whether FOCJ are legally and financially applicable for municipal cooperation in Russia. In particular, the objectives of this chapter are to identify in which legal form they may exist in Russia, to formulate financial models of FOCJ establishment, management and production, and participant attraction, to apply these models to stress the financial and fiscal implications of Russian FOCJs, to bridge the gap between public finance, principles, and FOCJ requirements, and finally to consider the advantages and disadvantages of Russian FOCJs with regards to financial aspects.

The content of section two deals with the definition of FOCJs. The authors tackle the existence of Russian FOCJ legal forms in section three. In the following section the financial models are formulated, and the effects of FOCJs are determined. The implications for financing and municipal cooperation are stressed in the fifth section, leading to the discussion of the advantages and disadvantages of FOCJ application by Russian municipalities. The conclusions comprise the results of the study and point to requirements for future research.

BACKGROUND

Definition and Features of Functional Overlapping Competing Jurisdictions

Public entrepreneurship refers to a public actor who recognizes entrepreneurial opportunities, undertakes investment decisions with a view to innovating, and takes action by using resources to implement a differentiated vision that adds public value (Filion, 2011; Friedrich & Ukrainski, 2013). Public entrepreneurship deals with the coordination of private and public actors to achieve public goals, the formulation of tasks and goals, the design and development of public organisations, the ways of management of resources, and the control of external effects (Klein et al., 2010; Friedrich & Ukrainski, 2013). One can distinguish social entrepreneurship, political entrepreneurship, knowledge entrepreneurship, technology entrepreneurship, and public sector oriented entrepreneurship (Friedrich & Ukrainski, 2013). Functional Overlapping Competing Jurisdictions, as special means of cooperation in providing public services, are an example of the latter.

The term FOCJ consists of four components, referring to:

1. **Functional (F):** An FOCJ as a corporation focuses on one or a few tasks (functions), for example, technology promotion, school services, etc. Its size is therefore adjustable to its functions.
2. **Overlapping (O):** FOCJs are overlapping. Their functions require different expansions considering space and tasks. Often FOCJs overlap if they fulfil the same functions. A transfer of a client or a member to another FOCJ does not mean that he has to accept an entirely different service from the new one.
3. **Competing (C):** FOCJs are competitive. They compete for the clients of their services, e.g. pupils and for FOCJ members, e.g. parents of pupils. The members are allowed to leave and enter the FOCJ. Within the FOCJ democratic decision-making prevails, which is also partly competitive, e.g. taken with majority-based decisions.
4. **Jurisdictions (J):** These are economic entities that, if organized under public law, may possess tax authority and the power of sanctioning. They are able to collect taxes and contributions. For FOCJs under private law the options for contractual penalties, funding obligations, and the fixing fees process must be established in the company statute or contract. FOCJs should complement existing Russian public bodies. They should allow more direct democracy through statutes stipulating the democratic procedures (Frey, 1997; Eichenberger, 1998; de Spindler, 1998; Eichenberger 2002; Friedrich & Eckardt, 2014).

The features of FOCJs are reflected in Table 1, where they determine the columns. In general, four types of FOCJs can be distinguished:

1. FOCJ type I has citizens as members, e.g. municipalities without territories. Citizens living in various municipalities establish FOCJ to provide services.
2. FOCJ type II has governments as members, e.g. municipalities, counties, states, nation states, the European Union, etc.
3. FOCJ type III comprises municipalities, other governments, other subjects of public law and private law (firms, etc.).

4. FOCJ type IV has members who are natural persons, and/or private and public legal persons, e.g. citizens, associations, chambers, churches, municipalities, central government and regions, and private and public firms (Friedrich & Eckardt, 2014).

Successful examples of FOCJs can be found in many European countries, such as Germany (Zweckverband - FOCJ type II),[4] Switzerland (FOCJ type I), the USA (special purpose districts - FOCJ type I), and even cooperation among countries, like the EU today. Some examples are already history.

An open question is how FOCJs are related to municipal cooperation. FOCJs of type I are indirectly linked to coordination of municipalities; rather than municipalities themselves, citizens are members of this kind of FOCJ. The coordination process between municipalities in FOCJ I is conducted through citizens' membership choice. Competition among FOCJ I for citizens as members and for clients is the driving force. "Municipal cooperation is when two or more municipalities agree to work together on any of the tasks assigned to them in order to gain mutual benefits. The term municipal cooperation is a relationship between two or several local authorities (i.e. entities at the first level of territorial administration) having a status of legal persons, endowed with competences, powers, and resources in accordance with the European Charter of Local Self-Government" (IMC Toolkit Manual, 2010, p. 7) .[5] This definition implies the typical features of municipal cooperation, such as: referring to a municipal activity organized at the local level, based on the mutual interest of the municipalities, concerning legal functions of the municipalities that are voluntary or imposed by law. Municipal cooperation involves sharing roles and responsibilities, costs, risks, control, and benefits. It represents a durable, flexible, and dynamic arrangement (IMC Toolkit Manual, 2010). These arrangements can lead to organizational forms like FOCJs, or non-organizational forms, such as short-term and long-term contracts and joint arrangements in Russia (Russian Federation Federal Act Nº 131 "On General Principles of the Organization of Local Self-Government in the Russian Federation", 2016, Article 8). The authors concentrate only on the FOCJ. Sometimes through a FOCJ, higher rank governments may also be involved.

Direct municipal cooperation prevails in types II to IV because municipalities themselves can be members of a FOCJ. They transfer the performance of functions to the organization of FOCJ types I to IV, including enterprises of public and private law and second-level self-government authorities (Wilhelm, 2013). The coordination involves municipal membership competition and service provision competition. Clients are located in the member municipalities and other towns. With both, FOCJ market coordination determines the parameters of the actions of the municipalities. FOCJ coordination differs from municipal coordination by public finance measures and changes through constitutional law or incentive policies by higher rank governments. The authors do not discuss the coordination between municipalities through mutual agreements, informal cooperation, market processes including private actors, such as sub-contracting (Wilhelm, 2013), or through projects in the form of public-private partnership.

Legal Opportunities for FOCJs in Russia

Which types of FOCJ occur in Russia depends primarily on the Russian legal framework. In Russia the division between public and private law is not well-defined. It's doctrinal, but not normative, which means that it's not fixed in law. This differs from divisions, existing in many European countries, like Germany, Poland, and Switzerland. Determining legal forms of companies under public law is rather problematic. Moreover, the term "public forms of property" is not used. Instead, there are two terms, state and municipal forms of property, which are fixed in the Constitution of the Russian Federation

(Винницкий, 2013). All forms of companies (state, municipal, or private) operate under civil law, an aspect of law which belongs to private law. Furthermore, the Russian governments are at an equal level with natural and legal persons (Винницкий, 2010). This can be explained historically. After the fall of the command system, the state was equalized in property relations with natural and legal persons (Винницкий, 2010).

Experts point to the need for increasing the role of public enterprises to form a special public law regime regulating state executive bodies, executive and administrative bodies of local self-government, and public enterprises (Винницкий, 2010). Establishing FOCJs needs a possible legal form, matching the most important features of FOCJs mentioned above.

In Figure 1 the available forms of a legal company under private law are shown. Apart from the sole proprietor (Civil Code of the Russian Federation, 2016, Article 23) the usual commercial organizations in Russia are termed "commercial corporative organizations" (Civil Code of the Russian Federation, 2016). To them belong the general partnership[6] and the partnership where some partners are limited.[7] The corporations comprise limited liability companies[8] and joint stock companies, appearing as non-public joint stock companies[9] and public joint stock companies.[10] Moreover, there are cooperatives for production[11] such as for trade or for activities in several sectors like agriculture or handicraft manufacturing, etc. In Russia a special company form for state and municipal companies exists, called "unitary enterprises".[12] They belong to one public owner only, e.g. the central government, regional government, or municipalities. Companies in Russia may operate for non-commercial purposes under forms of non-commercial corporative and unitary organizations as well. The non-commercial corporative companies comprise other different organizations, such as consumer cooperatives,[13] associations (unions),[14] public organizations[15] like parties, etc. and funds, establishments[16] of sport, medicine, education, and security, etc. (Official website for state and municipal establishments in Russia 2016). Autonomous non-commercial organizations and religious organizations are non-commercial unitary organizations.

The relevant FOCJ characteristics of these company forms depict the columns in the matrix of Table 1. The last column shows for which type of FOCJ the legal form seems appropriate. The rows in the matrix in Table 1 are dedicated to the company forms. Several FOCJ required features are met, such

Figure 1. Legal forms of companies in the Russian Federation
Source: Authors' schematic according to the Russian Civil Code

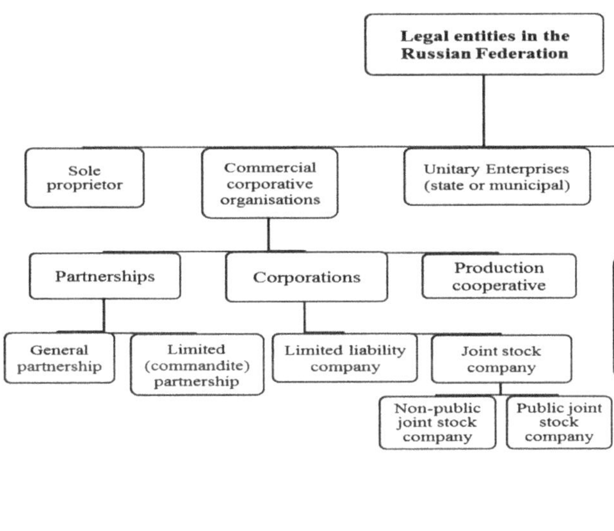

Table 1. Legal form in the Russian Federation under the FOCJ conditions

Legal Forms	FOCJ Conditions												
	Functions	**Private com-panies appli-cable**	**Over-Lapping**	**Compe-tition for mem-bers**	**Kind of Members**	**Member-ship of More Than One**	**Decision-Making Bodies**	**Published Statutes**	**Democratic Procedures**	**Right to Fix Fees and Contri-butions**	**Mem-bers Free Enter and Quit**	**Members Influence on Board of Directors**	**Provision of Services to Members**	**Possible Type of FOCJ**
Sole proprietor	+	-	+	-	Citizens	-	-	-	-	+	-	-	+	No
General partnership	+	+	+	+	Individual entrepreneurs and commercial organisations	+	All members together	Foundation Agreement	+	+ it can be fixed in statute	+	No board of directors	+	FOCJ type III
Limited (commandite) partnership	+	+	+	+	**General members**: individual entrepreneurs and commercial organisations. **Members-investors**: natural persons, legal persons, governmental units	+	All general members	Foundation Agreement	+	+ it can be fixed in statute	+	No board of directors	+	FOCJ type III and IV
Public joint stock company	+ (some activities only with licenses)	+	+	+ (not an aim)	Natural persons, legal persons, governmental units (e.g. municipalities)	+	General meeting of shareholders, board of directors	+	+	+ it can be additionally fixed in statutes	+	+	Not neces-sary	FOCJ type I, II, III, IV
Non-public joint stock company	+ (some activities only with licenses)	+	+	-	Natural persons, legal persons, governmental units (e.g. municipalities)	+	General meeting of shareholders, board of directors	+	+	+ it can be additionally fixed in statutes	-	+	Not necessary	No
Limited liability company	+ (some activities only with licenses)	+	+	+ (not an aim), restricted (mem-bers are not more than 50)	Natural persons, legal persons, governmental units (e.g. municipalities)	+	General meeting of shareholders (board of directors is possible)	+	+	+ it can be additionally fixed in statutes	+	+	Not necessary	FOCJ type I, II, III, IV
Production cooperative	+	+	+	+ max 5 members	Natural persons; natural and legal persons	+	General meeting of members = the chairman and the board of directors	+	+	+ it can be fixed in statute	+	+	+	FOCJ type I and III

continued on next page

Table 1. Continued

Legal Forms	FOCJ Conditions												
	Functions	**Private com-panies appli-cable**	**Over-Lapping**	**Compe-tition for mem-bers**	**Kind of Members**	**Member-ship of More Than One**	**Decision-Making Bodies**	**Published Statutes**	**Democratic Procedures**	**Right to Fix Fees and Contri-butions**	**Mem-bers Free Enter and Quit**	**Members Influence on Board of Directors**	**Provision of Services to Members**	**Possible Type of FOCJ**
Unitary enterprises	+	+	+	-	The Russian Federation, subjects of the Russian Federation or municipalities	-	**Founders** can be the Russian Federation, the subject of the Russian Federation or municipality. **Executive body** is a director	+	-	-	-	-	?	No
Consumer cooperative	+	+	+	+	Natural persons or natural persons and legal persons	+	General meeting of members, Board, Board of directors, Audit commission	+	+	+	+	+	+	FOCJ type I and III
Public organisations	+	+	+	+	Natural persons	+	General meeting of members, Chairman, Board of directors	+	+	+ membership fees	+	+	+	FOCJ type I
Associations (unions)	+	+	+	+	Natural persons and (or) legal persons, municipalities (according to existing practice)	+	General meeting of members, Chairman, Board of directors	+	+	+ membership fees	+	+	+	FOCJ type I, II, III, IV
Non-commercial partnership	+	+	+	+	Natural persons and (or) legal persons	+	General meeting of members, Chairman, Board of directors	+	+	+ membership fees	+	+	+	FOCJ type I, II, III, IV
Self-regulated organizations	+	+	+	+	Individual entrepreneurs and legal persons or natural persons, who carry out professional activity	+ voluntary or obligatory	General meeting of members, governing board, executive body	+	+	+	+	+	+	FOCJ type I, III, IV
Funds	+ different socially useful purposes	+	+	-	Founders Natural persons and (or) legal persons	-	High collegial body, Chairman, Board of directors	+	+	+ it can be fixed in statute	-	+	?	No

continued on next page

Table 1. Continued

Legal Forms	FOCJ Conditions												
	Functions	**Private com-panies appli-cable**	**Over-Lapping**	**Compe-tition for mem-bers**	**Kind of Members**	**Member-ship of More Than One**	**Decision-Making Bodies**	**Published Statutes**	**Democratic Procedures**	**Right to Fix Fees and Contri-butions**	**Mem-bers Free Enter and Quit**	**Members Influence on Board of Directors**	**Provision of Services to Members**	**Possible Type of FOCJ**
Establish-ments	+ non-commercial functions	+	+	-	Natural person, or legal person, or the Russian Federation, subject of the Russian Federation, a municipality - founders	only one founder	Supervisory Board, Chairman, other bodies	+	-	+ it can be fixed in statute	-	+ No Board of Directors	?	No
Autonomous non-commercial organisations	+ different functions in non-commercial spheres	+	+	-	Founders: natural persons and (or) legal per-sons, the Russian Federation, subject of the Russian Federation, a municipality	-	Founders, Chairman (executive body)	+	-	It can be fixed in statute	- founders can quit	- No board of directors	+	FOCJ type I, II, III and IV with difficulties
State corporations	+	+	+	-	Central State	-	Board of Directors, Supervisory Board	- based on special acts	-	-	-	+	-	No
State companies	+	+	+	-	Central State	-	Board of Directors, Supervisory Board	+	-	-	-	+	-	No
Public law companies	+	+	+	-	Central State	-	Supervisory Board, Executive Director, Board of Directors	+	-	-	-	+	-	No

- no possibility
+ there is a possibility
Source: Compiled by authors

as performing functions, overlapping, etc. Many companies compete for market results as well as for members. The decision-making of management is regulated in laws, statues, and company contracts. Democratic decision-making structures exist with some companies. The right to fix fees, prices, and contributions by members is regulated too. Some company forms allow entry of additional members or the leaving of old members. If the FOCJ requirements are met, then the authors depicted a plus; a minus sign appears if the requirements are not fulfilled. Critical for the ability to use FOCJs in a company form is whether a public function can be transferred to and performed by firms of private law. Sometimes this is prohibited by a stipulation in the Constitution, public organizational laws, or sector laws. Another limiting feature refers to the kind and the number of members. For municipal cooperation, the municipalities themselves or their citizens or private firms should be members. Therefore, membership of natural or legal persons must be possible. Public organizations are open for natural persons only. So they are applicable for FOCJ I if natural persons are citizens (e.g., in the form of a public joint stock company or limited liability company). Apart the type I, municipalities can form FOCJ type II (e.g. in the form of a public joint stock company or limited liability companies and associations). Some company forms have only one member, such as unitary enterprises and establishments. They are not suitable for any kind of FOCJ. This is also true for a sole proprietor company. Some companies have, in principle, no members, but founders (e.g. an autonomous non-commercial organization). The founders play a similar role as members, but freely entering and quitting is restricted. Only under special circumstances are these forms appropriate for FOCJ II. The funds have no membership, and the options of fund reorganization are rather limited; they are normally not suitable for FOCJ. FOCJ type I and type III may occur as production cooperatives (with firms and municipalities) and consumer cooperatives (with household citizens and municipalities). Although there is no explicit FOCJ law form in Russia, some private forms of law can be applied to create an FOCJ-like organization.

The law maker in Russia prepares a new form of non-commercial organization. The central state corporations (according to the Act “On non-commercial organisations”, Article 7.1) will be replaced by public-law companies. The eight existing central state corporations of public law were established, which will be changed into the new form. In addition, the so-called state companies (according to the Act “On non-commercial organisations”, Article 7.2) become incorporated into the new form of company as well. Public-law companies fulfil a wide range of public functions, e.g. modernization and innovative development of the economy, implementation of important projects, or socio-economic development of regions. This form of company has already been included in the Russian Civil Code, although the Act “On public-law companies” stipulating the management regulation is still under law-making procedure. These forms show only one owner – the central state. For establishing FOCJ of different types, membership of municipalities must be possible providing significant authoritative, regulatory and supervising functions for inter-municipal cooperation (Проект Федерального закона N° 252441-6 «О публично-правовых компаниях в Российской Федерации и о внесении изменений в отдельные законодательные акты Российской Федерации», 2016).

Public Finance for FOCJs

The public finance literature (Olson, 1969; Frenkel, 1986; Hyman, 1993; Ter-Minassian, 1997; Josselin & Marciano, 2004; Reingewertz, 2014) offers arguments for reasons to finance a jurisdiction, how to finance its establishment, and how to finance its operations. As FOCJs are jurisdictions, these finance principles might be applied to them as well. Some of them originate from the principles of fiscal federalism, which deal with the allocation of tasks and fiscal burdens to jurisdictions and economic units. According to the principle of fiscal equivalence (Olson, 1969) the members of an FOCJ and economic units receiving services from an FOCJ should pay for the FOCJ until the equity between benefit receivers, payers, and deciders is reached. The fiscal equivalence also serves to minimize special external effects, e.g. spillovers (Postlep, 1993; Zimmermann, 1999). Municipalities should be included in the FOCJ until marginal benefits and marginal costs are equal. The FOCJs are allowed to overlap. Those economic units that receive services should pay for them to achieve an equivalence of expenditures, revenues, taxes, and services (Olson, 1969; Frenkel, 1986; Hyman, 1993; Ter-Minassian, 1997).

The financing principle of Tiebout (1956) is also relevant for FOCJ financing. It is called "voting by feet", "voting by exit", and "showing his voice" (Hirschman, 1970; Napp, 1994; Blankart & Borck, 2004). The members contribute to the FOCJ financing and receive benefits from its activities. If the membership promises a positive net benefit, then an economic unit decides to become an FOCJ member. Voting by feet is emphasized in the case of FOCJ competition for members and clients. In a state of equilibrium, no migration of members between the FOCJs will take place and an optimal financial situation for the FOCJ members is achieved (Frey, 1997). In turn, when interpreting a FOCJ as a club, the services of the FOCJ are only available for the club members. The optimal size of the club and the related financial efforts are achieved if the marginal benefits equal the marginal costs of the last-entering club member (Buchanan, 1965). Jurisdictions, such as FOCJs, compete by offering the members a low financial burden when entering the FOCJ. Public finance experts discuss - as an analogy to tax competition (Sinn, 1997, Blankart, 2002) - whether such a financial competition results in a service level that is too low for the FOCJ.

Two further principles refer to allocation of the tasks and the financial burdens in a federation. They can be used for FOCJ financing. The subsidiarity principle (Josselin & Marciano, 2004; Dafflon, 2006) requests that only tasks the municipalities are not able to perform should be assigned to the central state and regions. As a consequence, the municipality is responsible for many tasks. Moreover, those economic agents should decide about services, who benefits from them, and who has to bear the fiscal load. The preferences of citizens are reflected in the organizational mix of the resulting FOCJ. This argument supports FOCJ formation. The connexity principle (Friedrich, Nam & Reiljan, 2009) says that the jurisdiction, which decides about task performance, should finance the respective tasks. That means an FOCJ is responsible for financing its own tasks and has to enforce the provision of financial means from FOCJ members and the service clients. For that purpose, the FOCJ has authority by law or a company contract.

When considering the solidarity principle (Cyr, 2014; OECD, 2016), the distribution problem must be solved. That implies a balancing of the financial load between richer and poorer FOCJ members. Balancing of the service activities takes place by preferring poorer members through special favourable service conditions and introducing appropriate amendments in the FOCJ statute. However, high-income people might concentrate in FOCJs that meet their particular standards. Low-income groups tend to require membership in FOCJs especially oriented to their income level. Higher rank jurisdictions have to ensure inter-FOCJ solidarity (e.g. responsibilities of social insurance companies with respect to membership and service conditions).

Further principles concern the fiscal equalization (Zimmermann, 1999; Blöchliger & Charbit, 2008) between jurisdictions. Equalization refers to vertical fiscal flows and property transfers between jurisdictions of different levels (e.g. central government, regional government, municipalities). Horizontal equalization takes place between jurisdictions of the same rank, e.g., among regions or among municipalities. Incorporated in such fiscal activities are the FOCJs. This occurs vertically through grants or subsidies from the Federation, region, or municipality, and through tax payments to those entities. Assignments of property and real estate, and a supply of free-of-charge services or those that do not cover costs to the FOCJ may take place. Sometimes a higher rank jurisdiction provides credits and warrants for debts. Horizontal equalization between FOCJs is not so often induced by higher rank jurisdictions (e.g., if guarantee funds have to be organized for stabilization of FOCJs). The fiscal equalization principle helps in assigning financial responsibilities within the FOCJ. Moreover, some internal vertical fiscal relations (Köllner, 2001) exist between the members and the FOCJ. Members finance through general contributions (as well as through taxes for FOCJs under public law in some countries), through purpose contributions, through payments for services (through fees and prices), through investments and services in kind, through transfer of property, and through crediting or guarantee for credit. Capacity indicators (Zimmermann, 1999) can express the payment ability of members like in fiscal equalization. They measure contribution capacity, the capacity to provide credits, and the capacity to provide property, etc. in order to collect appropriate sums of financial means. For determining financial requirements, need indicators may express needs for service volumes, investments, internal distributions, or provisions against risks, etc. Such indicators can be determined in advance in the statutes or occasionally by democratic management decisions. The determination of need indicators should allow the necessary amount of financial means to be raised (Neumark, 1970).

Additionally, contribution determining may follow the principles of taxations. The ability to pay principle (Wagner, 1890; Neumark, 1970; Musgrave & Musgrave, 1989) implies that members of the FOCJ of equal financial strength should pay equal contributions. The contribution payments of the members differ according to their ability to pay. If the payment is interpreted as a price for receiving services, the so-called benefit principle (Neumark, 1970; Musgrave & Musgrave, 1989) applies. Benefit principle theories assist in finding optimal contributions within an FOCJ. So-called positive theories explain the existing fees. They refer to models of utility maximizing management, e.g. the directors of an FOCJ. The application of the causation principle (Brümmerhoff & Büttner, 2015) may serve as a yardstick for financing if some members create especially high costs because of capacity extensions, special investments, geographical conditions of supply, etc. In the case of extraordinary member advantages, such as special cost savings, a differentiation of their contributions to the FOCJ seems justified. A simple form of the causation principle governs cost allocations. If members generating costs cannot be identified, common charges covering costs should be paid by all members (common charge principle) (Brümmerhoff & Büttner, 2015).

Those principles point to important features of FOCJ financial management development. However, they are not precise enough to specify a detailed principle for financing the establishment or operation of the FOCJ. The authors try to achieve deeper insights by developing theories of financing for FOCJ establishment and operation and showing the financial implications of membership competition.

MICROECONOMIC MODELLING OF ESTABLISHMENT, FINANCING, AND COMPETITION AMONG FOCJs IN RUSSIA

The Model of Financing FOCJ Establishment

The authors analyse the financial contribution of the members establishing an FOCJ by formulating a founding model that results from the application of an oligopoly model that was developed by Cornes and Harley (2001). For purposes of financial analyses it was adjusted by Detig, Feng, Friedrich (2002) and Friedrich and Eckhardt (2014) in order to show the finance of municipalities' resources dedicated to an FOCJ.This approach can be extended because several different ways of financial participation can be chosen. The authors intend to elaborate propositions of optimal financial behaviour. Out of several financing possibilities they refer to one mode of finance with respect to one municipality. Another attempt concerns the determination of the whole sum of financial means needed. A municipality might participate in fiscal means through tax payment, credit, or resources in kind, e.g. real estate. The authors consider an FOCJ of type II, which has only municipalities as members.

The contribution of a municipality i to the equity capital of the FOCJ is named e_i. It can consist of j types of financing $\sum_{j=1}^{m} e_{ij}$. We consider that the benefits the municipality achieves are related to the participation of the municipality in financing the equity capital which results to $\frac{e_{ij}}{\sum_{j=1}^{n} e_j}$. The sum $\sum_{j=1}^{n} e_j$ is also named E. The financial contribution of the other municipalities equals $E_R = E - e_i$. The benefits of municipality i increase with a higher proportion of that community in FOCJ equity capital, e.g. the voting power of the municipality within the FOCJ might grow and more favourable solutions for FOCJ activities might be achieved. These benefits are illustrated by the parameter b_i. Therefore, the benefits develop as $b_i * \frac{e_{ij}}{E}$. Moreover, there are opportunity costs c_{ij} per financing by e_{ij}, which gives $c_{ij} * e_{ij}$. The utility out of the participation by finance e_{ij} equals

$$U_i = b_i * \frac{\sum_{j=1}^{m} e_{ij}}{E} - \sum_{j=1}^{m} c_{ij} * e_{ij} \tag{1}$$

The differentiation with respect to the financing mode e_{ij} in municipality i yields.

$$\frac{dU_i}{de_{ij}} = b_i * \frac{(E_{R_{ij}} + e_{ij} + \sum_{m-j}^{m} E_j - e_{ij})}{\left(E_{R_{ij}} + e_{ij} + \sum_{m-j}^{m} E_j\right)^2} - c_{ij} = 0 \tag{2}$$

From elaboration of terms one receives:

$$b_i * \frac{(F_{R_{ij}} + \sum_{m\text{-}j}^{m} E_j)}{\left(E_{R_{ij}} + e_{ij} + \sum_{m\text{-}j}^{m} E_j\right)^2} - c_{ij} = 0 \tag{3}$$

Substituting the following expression (4) into (3)

$$E_{R_{ij}} = E - e_{ij} + \sum_{m\text{-}j}^{m} E_j \tag{4}$$

leads to

$$b_i * (E - e_{ij}) = c_{ij} * E^2 \tag{5}$$

and

$$\frac{b_i}{c_{ij}}\left(1 - \frac{e_{ij}}{E}\right) = E \tag{6}$$

Resulting the optimality condition:

$$\frac{e_{ij}}{E} = 1 - \frac{c_{ij}}{b_i} * E \tag{7}$$

is the optimal proportion of one way of financing *j* of municipality *i* in the FOCJ.

With higher benefits b_i of municipality *i*, its share of financing through *j* is increasing. If the opportunity costs of *j* are increasing, the share shrinks. To determine total financial means all shares of finance have to add up to 1.

$$\sum_{i=1}^{n}\sum_{j=1}^{m} \frac{e_{ij}}{E} = \sum_{i=1}^{n}\sum_{m=1}^{m}\left(1 - \frac{c_{ij}}{b_i}\right) E = 1 \tag{8}$$

$$\frac{n-1}{\sum_{i=1}^{n}\sum_{j=1}^{m} c_{ij}/b_i} = E \tag{9}$$

are the total financial means dedicated to equity capital of the FOCJ.

The model illustrates two basic microeconomic finance FOCJ principles:

1. All shares of finance have to add up to one.
2. The relation between the proportion of kind *j* finance by a FOCJ member *i* to the total sum of finance has to equal the following expression. It consists of one minus the product of total finance with the relation between social costs of finance *j* in municipality *i* and social benefits of finance *j* in municipality *i*.

The Model for FOCJ Current Costs Financing

The aim of this section is to elaborate the rules of current financing. The current financing depends on the form of law of the FOCJ; the fiscal situation of the members; subsidization and grant conditions; the capital market; the legal issues concerning the financing possibilities and management conditions; the goals of the directors, council members and delegates of the members; the politics of the supervisory boards; the competition among FOCJs; and the conditions for internal and equity financing, etc. The authors assume that a Russian FOCJ company forms a limited liability company to provide services and has to cover costs. For special services the FOCJ may seek levies. Costs comprise many financial costs such as imputed depreciation, imputed and appropriate interests, and costs according to different policies of asset maintenance (Friedrich, 1988). Financial costs are also part of the costs.

The financial costs can occur in several ways:

1. They may be linked to special factor inputs in current production and treated as part of the factor prices, e.g. by special inventories, price conditions, waiting time costs, etc.
2. Financial costs occur for fixed factors such as amortization of buildings or equipment.
3. Some costs are due to the special financial conditions for ways of finance. They constitute additional fixed costs or they are variable costs changing with production and factor inputs.
4. Covering those costs and raising the necessary financial means is unavoidable.

The FOCJ II model shows a specified member number of municipalities which have to pay a levy to cover the costs of current operations. For the sake of simplicity, the authors imagine that the members of the FOCJ receive services only from the FOCJ. The FOCJ II possesses a monopoly.[17]

The following assumptions are introduced into the monopoly model:

There is a fixed factor *A* and two variable factors *L* and *C* and a factor finance *F*. The production function is

$$X = f\,(A,L,C,F) \tag{10}$$

The fixed costs are determined by factor *A* and its factor prise *a*. According to the assumptions above the fixed costs can also comprise a financial part $\hat{a}$, which stems from the special financial costs of financing buildings, equipment, rights, patents, etc. through special amortization conditions, leasing, etc. or as a negative part of finance costs if the FOCJ receives special grants or subsidies to finance such finance costs. The relevant costs are $\mathrm{KfA} = (\mathrm{a} + \hat{a})^{*}\mathrm{A}$.

Special variable costs are from finance *F* if production time is influenced through production, procurement and delivery for which prises *f* has to be paid. Special reserves are necessary to overcome difficulties $\digamma$. $\mathrm{KvF} = (\mathrm{f} + F) * \mathrm{F}$.

There are variable labour costs related to *L*, which are due to the amount of labour, the labour costs (prise *l*) and that part of finance costs $\hat{I}$ related especially to labour, such as fulfilling requirements with respect to labour guarantees, social expenses, etc. $KvL= (1 + \hat{l})*L$.

The variable capital costs for material and pre-services are linked to the material prices *i*, etc. and special financing to store and transport materials related to finance costs *í*. The respective costs are $KvC = (i + í)*C$.

Therefore, the cost function reads as follows:

$$KfA + KvF + KvL + KvC = (a + \hat{a})*A + (f + F)*F + (1 + \hat{l})*L + (i + í)*C \quad (11)$$

As costs should be covered through a levy, the contribution payment must equal these costs. The volume of services demanded from the FOCJ II depends on the contribution level.

$$P = P(X) \quad (12)$$

The amount to finance through the contribution equals:

$$P(X)*X = (a + \hat{a})*A + (f + F)*F + (1 + \hat{l})*L + (i + í)*C \quad (13)$$

Managements' utility function expresses the aims of FOCJ depending on output and the volume of labour inputs.[18]

$$U = U(X,L) \quad (14)$$

Management maximizes utility leading to the following Lagrange formulation:

$$\Lambda = U(X(A, L, C, F), L) - \lambda(P(X)*X - (a + \hat{a})*A - (f + F)*F - (1 + \hat{l})*L - (i + í)*C) \quad (15)$$

The first order conditions show:

$$\frac{\partial \Lambda}{\partial F} = \frac{\partial U}{\partial X} * \frac{\partial X}{\partial F} - \lambda \left(\left(\frac{\partial P}{\partial X} * X + P\right) * \frac{\partial X}{\partial F} - (f + F)\right) = 0 \quad (16)$$

$$\frac{\partial \Lambda}{\partial L} = \frac{\partial U}{\partial X} * \frac{\partial X}{\partial L} + \frac{\partial U}{\partial L} - \lambda \left(\left(\frac{\partial P}{\partial X} * X + P\right) * \frac{\partial X}{\partial L} - (1 + \hat{l})\right) = 0 \quad (17)$$

$$\frac{\partial \Lambda}{\partial C} = \frac{\partial U}{\partial X} * \frac{\partial X}{\partial C} - \lambda \left(\left(\frac{\partial P}{\partial X} * X + P\right) * \frac{\partial X}{\partial C} - (i + í)\right) = 0 \quad (18)$$

$$\frac{\partial \Lambda}{\partial \lambda} = P(X)*X - (a + \overset{\frown}{a})*A - (f + F)*F - (l + \overset{\frown}{l})*L - (i + \acute{\imath})*C = 0 \tag{19}$$

Two basic financial optimality conditions result. One concerns the input of production factors and implies the following financial guidelines:

The relation between marginal utilities from changing finance and material marginally equals the relation of the marginal profits connected to finance and materials (20). The relation of marginal productivities equals the marginal profits that stem from the production factor variations (21).

$$\left(\frac{\partial U}{\partial X}*\frac{\partial X}{\partial F}\right) / \left(\frac{\partial U}{\partial X}*\frac{\partial X}{\partial C}\right) = \left(\left(\frac{\partial P}{\partial X}*X + P\right)*\frac{\partial X}{\partial F} - (f + F)\right) / \left(\left(\frac{\partial P}{\partial X}*X + P\right)*\frac{\partial X}{\partial C} - (i + \acute{\imath})\right) \tag{20}$$

$$\frac{\partial X}{\partial F} / \frac{\partial X}{\partial C} = \left(\left(\frac{\partial P}{\partial X}*X + P\right)*\frac{\partial X}{\partial F} - (f + F)\right) / \left(\left(\frac{\partial P}{\partial X}*X + P\right)*\frac{\partial X}{\partial C} - (i + \acute{\imath})\right) \tag{21}$$

For labour variation results:

The relation between marginal utilities from changing finance and labour marginally equals the relation of the marginal profits connected to finance and labour.

$$\left(\frac{\partial U}{\partial X}*\frac{\partial X}{\partial F}\right) / \left(\frac{\partial U}{\partial X}*\frac{\partial X}{\partial L} + \frac{\partial U}{\partial L}\right) = \left(\left(\frac{\partial P}{\partial X}*X + P\right)*\frac{\partial X}{\partial F} - (f + F)\right) / \left(\left(\frac{\partial P}{\partial X} + P\right)*\frac{\partial X}{\partial L} - (l + \overset{\frown}{l})\right) \tag{22}$$

The other financial optimality conditions refer to the size of the contribution:

$$P(x) = \frac{(a + \overset{\frown}{a})*A + (f + F)*F + (l + \overset{\frown}{l})*L + (i + \acute{\imath})*C}{X} \tag{23}$$

The contribution has to equal the average costs.

The finance conditions are reflected in the factor prices through $\hat{a}$, $_F$, $\hat{I}$, $\acute{\imath}$. F raises as well, because the fixed costs and their financial part have to be covered by the contribution of the members.

If the size of finance-related factor prices changes with factor inputs because of $\hat{a}$ (A), $_F$ (F), $\hat{I}$ (L) and, $\acute{\imath}(C)$ the optimal conditions vary.

$$\frac{\partial \Lambda}{\partial F} = \frac{\partial U}{\partial X}*\frac{\partial X}{\partial F} - \lambda\left(\left(\frac{\partial P}{\partial X}*X + P\right)*\frac{\partial X}{\partial F} - \left(f + \frac{\partial F}{\partial F}*F + F\right)\right) = 0 \tag{24}$$

$$\frac{\partial \Lambda}{\partial L} = \frac{\partial U}{\partial X}*\frac{\partial X}{\partial L} + \frac{\partial U}{\partial L} - \lambda\left(\left(\frac{\partial P}{\partial X}*X + P\right)*\frac{\partial X}{\partial L} - \left(l + \frac{\partial \overset{\frown}{l}}{\partial L}*L + \overset{\frown}{l}\right)\right) = 0 \tag{25}$$

$$\frac{\partial \Lambda}{\partial C} - \frac{\partial U}{\partial X} * \frac{\partial X}{\partial C} - \lambda \left(\left(\frac{\partial P}{\partial X} * X + P \right) * \frac{\partial X}{\partial C} - \left(i + \frac{\partial í}{\partial C} * C + í \right) \right) = 0 \tag{26}$$

$$\frac{\partial \Lambda}{\partial \lambda} = P(X)*X - ((a + \widehat{a}(a))*A - (f + F(F))*F - (l + \widehat{l}(L))*L - (i + í(C))*C = 0 \tag{27}$$

A change in the financial parts of costs (factor prices) does not change the basic conditions as mentioned but it changes the relation of marginal utilities and marginal profits and therefore the inputs, outputs, and the levy.

The case shown refers to an FOCJ II, which is graphically demonstrated in Figure 2. For the sake of simplicity, only the case of two factors, labour and material, is depicted. In the second quadrant a demand curve (12) for the services provided by the FOCJ can be seen. Moreover, there is the turnover curve, which results (13) and reflects contribution revenues, which are used to cover all costs. The fixed costs get deducted, which causes an inward reduction of the revenue curve, resulting in a net revenue curve demonstrating which financial means are achieved to cover the variable costs. In the fourth quadrant the respective budget is marked which is available to cover variable costs. The budget lines show how the budget (net revenue) may be allocated to labour and material. The output-labour curves in the first quadrant show which combination of labour and output can be achieved when the respective budget is totally used for alternative combinations of labour and material. To achieve such a net budget of the FOCJ the FOCJ members have to order services. The municipalities demand a specified volume of services, which can be produced in two ways under covering costs. There is material- or labour-intensive production possible when the whole budget (net-revenue) is used to cover costs. At the respective output-labour curve two points become relevant. For each service volume such two points are situated at the thick line. If they are connected, the thick line in the first quadrant appears as a possible solution curve. FOCJ management wants to maximize utility according to relation (14) and (15). Respective indifference curves with respect to output and labour are introduced in the first quadrant. The solution *S*

Figure 2. Determination of fees and contributions for FOCJ type II
Source: Authors' illustration

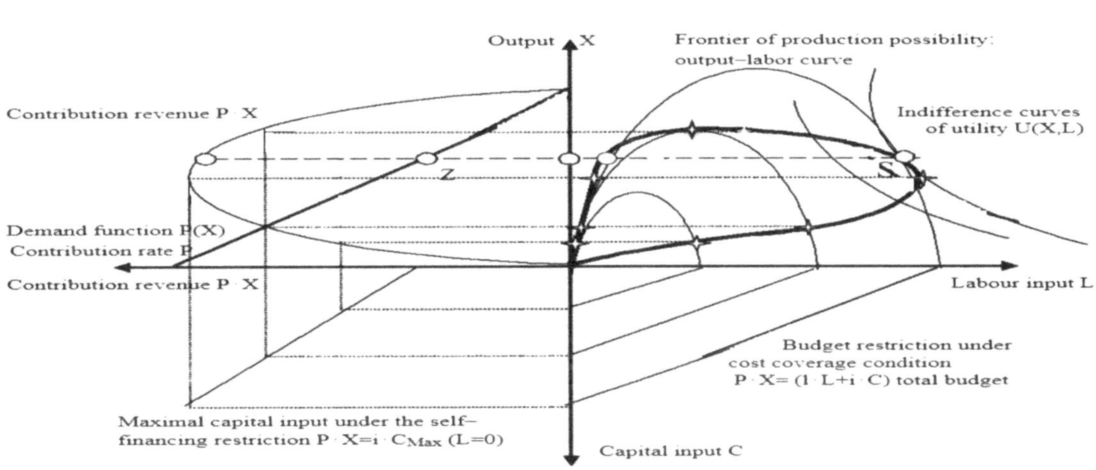

with fee Z is found where an indifference curve of the highest level touches the possible solution curve shown in the first quadrant again.

Effects of finance are demonstrated in different ways. If the finance costs related to the fixed costs increase $\hat{a} * A$ then the deduction of fixed costs from turnover increases. The budgets for variable costs decrease in parallel. The respective output labour curves move to the right and shrink. The possible solution line moves inward and the optimal solution moves in the direction of a higher contribution and a lower service volume. If the financing of variable costs become more expensive, that means $F^{*}F$, $\hat{I}^{*}L$, and $\acute{\imath}^{*}C$ grow and the budget curves shrink and move to the left at the labour or other variable factor axes. The output-labour curves shrink and the possible solution curve move inwards. The contribution increases and the service level shrinks.

This is the case of a self-financing FOCJ II. Under special restrictions, other sources of financing might be used too. If an FOCJ receives a grant or donation to the members by a higher jurisdiction, the scope of financing increases. The equation (13) becomes:

$$\mathrm{P(X)^{*}X} + \mathrm{G} = (\mathrm{a} + \overset{\frown}{a})^{*}\mathrm{A} + (\mathrm{f} + F)^{*}\mathrm{F} + (\mathrm{l} + \overset{\frown}{l})^{*}\mathrm{L} + (\mathrm{i} + \acute{\imath})^{*}\mathrm{C} \tag{28}$$

The first optimality conditions are not changed if G is a constant sum independent of x. However, the second condition varies to:

$$\mathrm{P(x)} = \frac{(\mathrm{a} + \overset{\frown}{a})^{*}\mathrm{A} + (\mathrm{f} + F)^{*}\mathrm{F} + (\mathrm{l} + \overset{\frown}{l})^{*}\mathrm{L} + (\mathrm{i} + \acute{\imath})^{*}\mathrm{C} - \mathrm{G}}{\mathrm{X}} \tag{29}$$

The contribution gets decreased and corrected by G/X. In Figure 2 the turnover curve gets a push to the left. The net revenues become higher. The budget lines move downward in parallel. The output-labour curves move to the right and turn out to be bigger.

Consequently, the solution possibility curve moves up and to the right. The solution will be with a lower contribution and a larger service volume. If the grant depends on the size of the volume of services such as $G(x)$, the first optimality conditions change as well. The relation of marginal utilities must equal the relation of marginal profits plus the marginal grants. In Figure 2 the turnover curve shifts with higher service volume even more, and through enforced movements of budgets and output-labour curves, the possibility curve as well and the contribution becomes smaller and the service volume bigger. Such grants and donations might also be paid directly to the FOCJ II from non-member jurisdictions, from private donors, etc. or indirectly from member municipalities.

Similar effects are caused by the sale of unused buildings and equipment B or by unneeded financial assets. Bank credit to the FOCJ II is already considered when the FOCJ II or the members have to pay back the credits and cover amortization. This is considered by the $(f + F) * F$ term and the other financial terms.

The contribution of the members to the financing of the FOCJ II's current operation might be zero if the self-financing takes place through sales to non-FOCJ II members. The abovementioned optimality rules for management apply. If the members pay a fixed amount, we end up with the grant solution in analogy. Sometimes there are also opportunities to split the market into a market for members and non-members. Then there might be two demand curves, two turnover curves, one joint turnover curve and one net revenue curve, and for resulting budgets respective output-labour curves and one possible solution curve. The optimum is where it touches the indifference curve. The service volume is determined and the price for non-members is fixed at their demand curve and contribution for members at members' demand curve.

The model points again to two basic microeconomic finance FOCJ principles:

1. The relation between marginal utility of finance and of another factor has to equal the relation of marginal FOCJ profits related to finance and the other factor.
2. The contribution or fee has to equal average costs.

The Model for Member Allocation between FOCJs According to FOCJ Financing

There is also a model (Friedrich, 2006) that shows the distribution of members between two or more already established FOCJs. This approach includes different forms of financing for the FOCJ. The authors firstly concentrate on two Russian FOCJs II, which have chosen the same financial policy. A net-benefit to a municipal member results from the service and the contribution paid. The member net-benefit might increase and decrease with the number of FOCJ members. For one FOCJ this development is reflected in curve TL in Figure 3 in the left-hand section. The middle section shows the situation for the competing FOCJ. The curve KP varies for the second FOCJ with the development of net-benefit if the number of members increases. Left of assignment G it makes no sense for possible members of the first FOCJ to stay with members of the competing FOCJ. The same is true for possible members of the first FOCJ right of point G. Therefore, the size of first FOCJ turns out to be N1 and that of the second FOCJ becomes N2 if both FOCJ apply the same financing policy.

The two FOCJs can also apply different policies. For example, the first FOCJ practices financing through contributions, and the second one through a mix of fees and public debt. Now, the curve TL reflects the contribution financing by members. The curve KP refers to the mix of fees and debt financing. The different ways of FOCJ financing determines the allocation point G as well.

Figure 3. Distribution of FOCJ members if FOCJ apply same financial policy
Source: Friedrich 2006, pp. 150

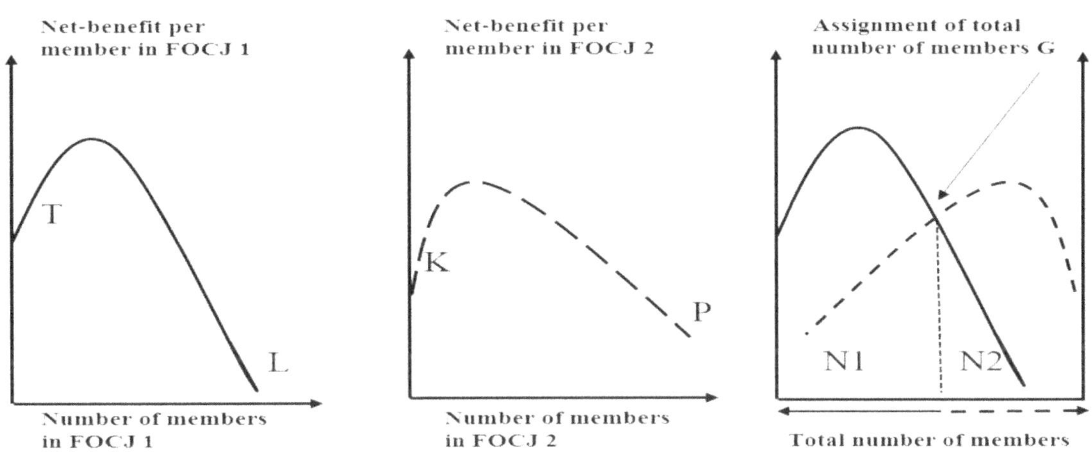

To each financial policy, one can draw a net-benefit curve for members. The financial policies can be different or the same in the two FOCJs. The members still choose that FOCJ which allows the highest net-benefit. As in Figure 4 in the left hand part of the picture the two curves depicted don't cross. The potential FOCJ members decide following the higher curve. The same is true with the second FOCJ. If the upper curves are the same as in Figure 4, the allocation point G results again. If only less advantageous financial opportunities are available, point M is the solution in the right section graph.

If many financial policies are available and the net-benefit curves cross with respect to one FOCJ, then the curves between the crossing points have to be compared. The curves of highest net benefit between the crossing points show the most favourable financial policy. A curve of optimal financing results for the respective FOCJ. For the other FOCJ a similar construction of the optimal financing curve takes place. The optimal finance curves are transferred into the far right hand picture. The crossing point of the two optimal curves reflects the resulting allocation of members to the two FOCJs and the related optimal financial policy (see Figure 5). A change in condition of financing will vary the net-benefit curves. In all three cases new allocation points appear.

A given number of possible FOCJ members should be allocated between two FOCJs according to the following microeconomic finance FOCJ principles. If the FOCJs are ordered by the net benefits per member, which stem from optimal finance of FOCJs (member sizes), the split between the two FOCJ is determined by the member size of equal net benefit per member.

SOLUTIONS AND RECOMMENDATIONS

Principles of Finance for FOCJs

The public finance-oriented recommendations and the management rules of finance resulting from the microeconomic theory are too general to guide the financial management of FOCJ. They do not explicitly mention the role of financial instruments in the phases of establishment and operation. In FOCJ II

Figure 4. Distribution of FOCJ members in case of different financial policies
Source: Compiled by the authors

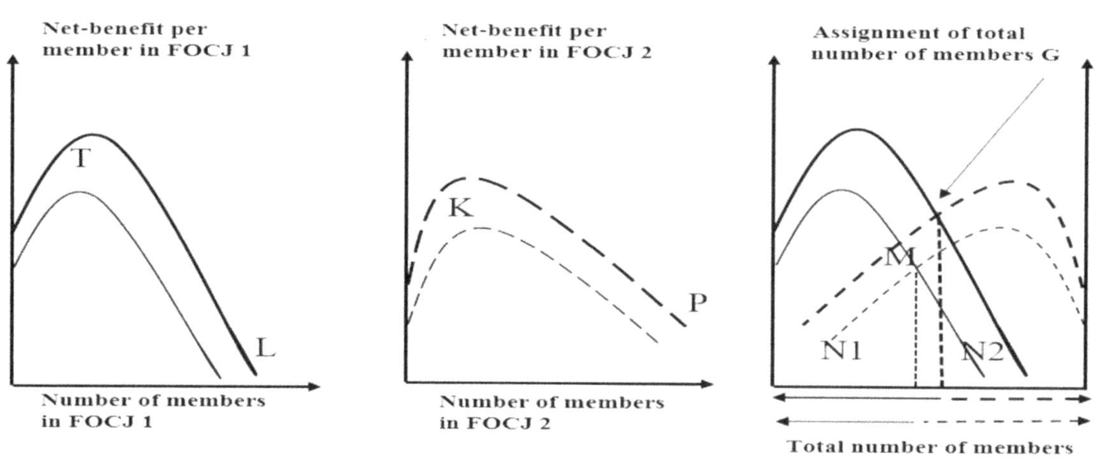

Figure 5. Distribution of FOCJ members in case of different financial policies, crossing net-benefit curves and adjustment curves ABCD and FHIJ
Source: Compiled by the authors

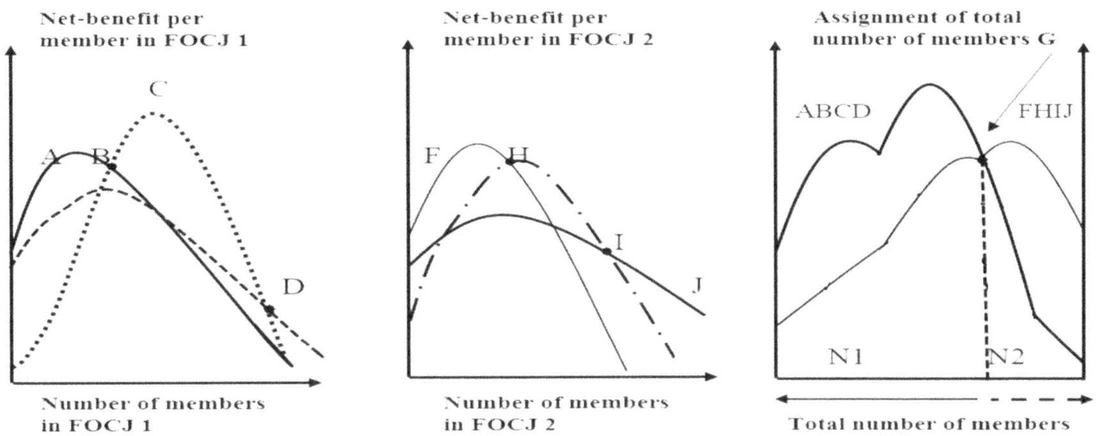

self-financing (Perridon & Steiner, 2009) occurs during the phase of establishment and the municipalities have to provide equity capital. The capital can refer to different assets, for example, cash, materials, equipment (buildings and machines), real estate, patents, usage rights, exploitation rights, claims, etc. What kind of equity capital is provided depends on the need to reach a specified equity capital amount and on the functions the assets have to fulfil within the FOCJ, e.g., locations, buildings and resources needed for production. The mode of equity capital participation also depends on what kind of capital and related assets the other members can contribute. In addition, the functions of capital for guaranteeing liquidity and financial stability in the future have to be considered, including the so-called "golden rule of finance", which says that long-term assets should be financed by equity capital (Bodenhoff, 2002). What a municipality offers also depends on the effects of the transaction, the production activity, and its financial and budget situation, etc. The effects turn out differently because the equity capital stems from different sources, such as taxation, grants, profits of public enterprises, revenues from sales, property and real estate business, citizens' and firms' donations, or grants the municipality receives. A network of finance connecting the financial sphere of the FOCJ with that of the member municipalities influences the establishment finance. Other forms of capital are grants from jurisdictions or donations contributed by other economic units in the form of cash or other kinds of contributions. A Russian FOCJ of private law may receive special subsidies paid by non-member jurisdictions to the FOCJ. FOCJ establishment financing from a third party who is not an FOCJ II member is welcomed.

In addition to equity financing, *debt financing* is chosen for FOCJ establishment. The debt capital stems from third parties, other jurisdictions, or other firms like banks, insurance companies, etc. through long-term or short-term credit. Private capital owners, citizens and others economic subjects from abroad can serve as FOCJ creditors.

Self-financing instruments from internal sources are member contributions, sales, liquidating reserves, profits, member grants and non-cost municipal pre-services. The *outside financing* of current operations happens in the form of grants, property usage, and transfers. Long-term credit consists of different forms like real estate credit, loans, municipal credit, commercial papers, etc. and short-term credit is cash lending, credit from suppliers, or pre-payments from municipalities. With respect to all these financial instruments, well-known doctrines of financing exist in business administration (Perridon & Steiner, 2009).

Therefore, the authors concentrate on a few means of financing. If the utility function of the managers in the model above is substituted by individualistic welfare functions, the *marginal cost pricing rule* (Nelson, 1964; Brown & Sibley, 1986) results when the FOCJ II faces no cost-covering restrictions. The Russian FOCJ II is under private law and cannot endure long-term losses because its equity capital is limited. Subsidization to compensate for the losses becomes necessary if marginal costs are lower than average costs. The introduction of a cost-covering constraint leads to péage systems (Boiteux, 1951;), commercial rules, and other cost-coverage approaches results (Nelson, 1964; Friedrich, 1969; Rees, 2006). For fulfilling political goals, other solutions are gained (Bös, 1986). Two tariff systems comprising a working price and a base price connect the financing with the volume of services received and with amount of periodical fixed costs need to be covered (Rees, 2006). For special investments, managers cover costs by levy according to the efficiency principle. An FOCJ II competing for clients considers the market forms on the sale side and sales market solutions as well. Even the market forms on the procurement markets determine the results and finance behaviour.

On occasion municipalities use FOCJ as a mean of *financial outsourcing* through achieving additional turnover (revenues), profits, special grants, credit, and participation in the financial potentials of other municipalities. FOCJ III or IV types in the form of a public-private partnership (PPP) raise funds from the private partners as well. A special kind of financial outsourcing comprises the establishments of plants and instalments by the PPP and the service production by the municipality in the framework of a contract between a PPP and a municipality.

FOCJ literature mostly neglects *public debt*. But promissory notes, loans, and commercial papers (Budget Code of the Russian Federation, Article 100, 2016) serve as financial instruments for FOCJ II as well, in particular, if municipalities achieve special favourable bank conditions. The FOCJ II groups several municipalities offering guarantees to banks. Derivatives such as swaps and options should be not allowed for FOCJ to discourage speculative behaviour by FOCJ management. Public finance specialists argue that initial investments financed by taxes lead to a double burden if the services are paid using fees, which also comprise depreciation considering the replacement value (Zimmermann, 1999). Through financing, mix-oriented calculation of fees, such as double burden, should be avoided in FOCJs. Public debt allows financing according to the pay-as-you-use principle (Musgrave & Musgrave, 1989). Through public debt, the financial consequences of expenditure waves could be allocated over time. Public debt of FOCJ assists in avoiding higher costs from fiscal stress and facilitating synchronisation. Introduction of sector-specific debt limits contribute to avoiding risky FOCJ financial management. The cash flow could represent an appropriate debt limit indicator. An aggregated indicator also comprising the debt limits of the member municipalities points to long-term FOCJ service capability. An FOCJ's founders should evolve a priority hierarchy among financial instruments, such as grants, sales revenues, and public debts, integrated into a financial management concept, which relates to the internal management of FOCJ and to the financial relation to the FOCJ members.

With respect to Russian FOCJs, not many rules for enforcing the financially sound behaviour of management exist, in particular, for achieving public goals. Some countries applied so-called *cooperate governance codes* for public enterprises of private law forms (Corporate Governance in Estonia, 2011; Leixnering & Bramböck, 2013; Papenfuß, 2013; Corporate Governance of State-Owned Enterprises, 2014). Such codes need adaptation to the Russian law conditions, to the tasks, and to the sector conditions of FOCJ II, FOCJ III or FOCJ IV. Governance codes could be inserted into the FOCJ statute or in the by-laws of the company. The usefulness of cooperate governance codes is still questioned, in particular, if the management fears no sanctions for violating the code (Leixnering & Bramböck, 2013; Papenfuß,

2013). To elaborate such a code, the experts may use management stipulations fixed in German laws for FOCJ II (Baden-Württemberg, 2016). Financial obligations of the municipalities with respect to financing and FOCJ II management requirements of task fulfilment are formulated there. Management activities underlie the same requirements as those for municipal administration. Estonia introduced a specified *management by objective* through the State Asset Act and foundation Act of public enterprise (OECD, 2011) for central state public enterprises. The board of directors handles everyday management (the executive function). Business or investment planning occurs in cooperation with the supervisory board staffed by the central government. Such a concept leads to financial management mainly determined by the Ministry of Finance. Apart from the question whether such a solution is legally possible in Russia with respect to limited liability and stock companies, its practice shows many weaknesses. An explicit municipality FOCJ finance policy must be available. The municipalities need to establish FOCJ goals and responsibilities and the functions of the supervisory board. The roles of the municipal city councils are not clear. Capital increases for the FOCJ II need to be based on profound business plans, normally not available or understandable to municipal council parliamentarians. Instead of a very sophisticated financial concept, it might be wiser to orientate FOCJ financial management to the financial regulations for municipalities in Russia (Budget Code of the Russian Federation, 2016).

FOCJs under private law have to be checked by the municipal auditing offices. The FOCJs have to follow some regulations and rules concerning the municipal budget which do not conflict with the special rules of a public joint stock company, for example. FOCJs should also employ the **so-***called principles of budgeting* for preparation and execution of municipal financial plans. Not all the principles of FOCJ financial management need to be followed. But their application would support adequate financial decision-making. Municipal planning deals with short-term planning and sometimes middle-term and long-term planning (Kraan, Bergvall, Hawkesworth, Kostyleva & Witt, 2008; Истомина, 2015). Of special importance are the principle of efficiency (Rehm & Matern-Rehm, 2003; Budget Code of the Russian Federation 2016, Article 34) and the principle of targeting and targeted nature of budget (Budget Code of the Russian Federation, 2016, Article 38). The principle of completeness claims to include all revenues and expenditures in the budget and to demonstrate how a budget deficit is financed (Budget Code of the Russian Federation, 2016, Article 32). Following the principle of a balanced budget requires that all revenues equal all expenditures, including those revenues necessary to finance a budget deficit, the principle of general budget cost covering must be followed (Budget Code of the Russian Federation, 2016, Article 35).

Budget principles applied in Germany point to further requirements. They are related to the principle of planning in advance and the principle of speciality in terms of quantitative correctness and in terms of time appropriateness and non-affectation. The latter means that all revenues are available to finance all expenditures (Heinig, 1951; Andel, 1998; Nowotny, 1999; Zimmermann & Henke, 2001). Further principles concern the principle of continuous performance of duties, the principle of permanent solvency (Vogt, 1984) and the principle of ensuring stability and growth (in Germany § 19 StabWG).

Following these principles effects the choice of the financial instruments and in particular the debt policy (Zimmermann, 1999; IMF, 2001). The FOCJ needs a positive cash flow for security, amortisations, or credits. The method of financing is monitored by the organs and the members of the FOCJ, by the local supervisory authorities, and by the audit offices of other jurisdictions. They check compliance with the budget principles, regulatory requirements, and special control principles (Eichhorn & Friedrich, 1976). To detect the implications of these principles of financial planning, liquidity planning models or debt management models serve only partly. Financial decision-makers can apply linear programming and debt management models (Eichhorn, 1974; Rehm, 2001). There are some quantitative attempts, especially for private firms, to verify such principles by formulating terminal value models (Albach, 1962; Hax, 1964, Blumentrath, 1969; Augusto & Reis, 2013). However, those models don't deal with public enterprises like FOCJs to much extent. Operations research models serve to determine optimal budgeting in the sense of the efficiency principle. Investment planning models, capital market-oriented financial models, and medium-term financial planning models help to consider some of the principles in project and medium-term financial planning. Although these models allow combining some FOCJ management principles and budget principles (Mertens, 2006), they are not able to integrate financial decisions and all activities necessary to reach FOCJ goals. Researchers still need to develop financial concepts, doctrines, and principles to avoid financial mismanagement.

Advantages and Disadvantages of FOCJ in Finance

When discussing basic financing and organization of FOCJs and the respective public finance literature on fiscal federalism, the authors looked at FOCJ from a national economic point of view (see section on public finance for FOCJ). Further discussions dealt with FOCJ finance related to municipalities and FOCJ and their benefits (see section on principles of finance for FOCJ). The results are the basis for presenting the pros and cons of FOCJ. The authors note general advantages and disadvantages and those with particular respect to finance (Table 2).

The extent to which general advantages can be realized by municipalities depends on the sectors of public production and the socio-economic, political and fiscal circumstances. Therefore, the importance of general advantages differs from case to case. When focusing on type II FOCJ, the advantages (3), (5), (7), (8), (10) and (11) become relevant in particular in Russia. The general disadvantages also vary according to the sectors, etc. In Russia, disadvantages (2), (3), (4), (8), (12) and (13) would hamper the management of type II FOCJ. By adopting legislation and formulating appropriate statutes when establishing FOCJ II as a public joint stock company, limited liability company, association, non-commercial partnership, and autonomous non-commercial organisations, general disadvantages may be reduced or avoided.

How far fiscal advantages can be realized depends also on the circumstances mentioned above. One can expect that most of the fiscal benefits are related to (1), (2), (3), (4), (5), (9), and (10) when employing Russian FOCJ II. All financial disadvantages (1)-(7) might occur with FOCJ II in Russia as well. Through suitable legislative changes in intergovernmental fiscal relations, statute and financial cost-reducing management, the fiscal disadvantages may be eschewed.

FUTURE RESEARCH DIRECTIONS

Further studies should focus on special behavioural oligopoly solutions when establishing an FOCJ. In models of contributions for operating FOCJs, further market forms should be introduced with respect to competitive relations to other FOCJs prevailing with clients in delivery and on procurement markets. In addition, public choice approaches and negotiation theory will serve to explore the relation between FOCJ management and members. Financial principles for achieving non-financial public goals in planning, financial realization, and auditing of the FOCJ are required. In particular, the integration of FOCJs in financial markets needs more attention.

Table 2. Advantages and disadvantages of FOCJ in Russia

General Advantages	General Disadvantages
(1) FOCJs combine the efforts of municipalities using the synergy effects of knowledge, management skills, capacities, etc. (2) Clustering is supported in energy provision, transportation, education, etc. (3) The services can be more specific and differentiated. Even innovations can be expected. (4) Compensation possibilities between municipalities increase with type II and with firms, households, and other jurisdictions with type III and IV. (5) FOCJs may change market forms in regional competition. (6) An FOCJ under public law can use different and extended instruments of action. (7) Co-operation in different fields is feasible without having too many interventions from other governments, e.g. political ones. (8) FOCJs could employ a more highly skilled management team and improve the quality of provided. (9) The option to leave an FOCJ promotes efficient production. (10) Combined procurement market power increases the tendency to higher efficiency. (11) The autonomy of municipalities against central governments is better protected, since they are able to organize additional political support through the FOCJ. (12) Some private legal forms are available for the formation of FOCJs of type II to type IV. (13) FOCJs help to develop and finance inter-municipal projects.	(1) There is no legal form under public law for the Type-I FOCJ in Russia. (2) FOCJ company formation requires a long period of negotiation for cooperation and competition. (3) Transaction costs for creating and operating an FOCJ might compensate for efficiency gains, e.g. because of members entering and leaving, complicated rules, cost of information, publicity, and monitoring. (4) The structure of FOCJs shows rigidities, as rules of equal treatment and member rights have to be considered. (5) The establishment of several competing FOCJs might lead to multiple infrastructures, increasing capacity costs, e.g. waterlines, etc. (6) FOCJs imply partly a loss in flexibility of political and economic parameters. (7) Finding rules determining decisions within FOCJs is complicated. (8) FOCJs may overlap with competing regions, which might lead to conflicts with municipalities. (9) There is a lack of management theory relating to FOCJs, particularly in goal determination, management phases, levels, tools, and management styles. Microeconomic analysis of plans, competition solutions, etc. needs development. (10) A general scheme for FOCJ evaluation does not exist yet. (11) A loss of citizens' influence on service management and democratic power can occur. (12) The member municipalities lose instruments of their own economic and social policies, which are now under the competence of the respective FOCJ. (13) FOCJs face difficulties integrating into the Russian legal, political, and fiscal intergovernmental relation system. (14) The political conflicts between FOCJs with members and in member municipalities may increase.
Special financial and fiscal advantages:	**Special financial and fiscal disadvantages:**
(1) Through FOCJs, municipalities may acquire the necessary fiscal means more easily than individually. (2) Appropriate means of finance and kind assets, e.g. real estate, are detected more easily by several cooperating municipalities. (3) Cost savings, because of economy of scale, better production techniques, etc. lower financial needs. (4) FOCJs allow for improved medium- and long-term financial planning. (5) More favourable financial conditions for FOCJs stem from increased market power on factor markets. (6) Credit is easier to guarantee and to achieve more cheaply. (7) In member municipalities, cost savings occur. (8) The FOCJ realizes advantages when acquiring credits from abroad. (9) An FOCJ II might increase its financial opportunities by joining a PPP or a type III or IV FOCJ. (10) Municipal public debts possibilities can be increased or shifted to the FOCJ. (11) The fiscal federalism principles as well as the FOCJ financing principles are applicable.	(1) High transaction costs lead to higher financial requirements because fiscal advantages turn out to be smaller. (2) The system of FOCJ internal financial equalization creates additional costs of financing. (3) The transaction costs of entering and exiting FOCJ members as well as market competition increase financial needs. (4) The individual municipalities may lose financial flexibility because cross-subsidization possibilities within the municipality budget become fewer. (5) If FOCJs comprise considerable networks, high fixed costs occur. Costs cannot easily be adopted and reduced. The financial costs turn out to be higher. (6) The member municipalities may lose special grants, which they would get if they offer services individually and themselves. (6) There are double costs of auditing the financial management of FOCJ and of the member municipalities. (7) Municipalities where the FOCJ is not located and where the workers do not live will experience negative economic and fiscal effects.

The application of FOCJs in different public service sectors in Russia is not yet explored; neither is the financial implication. The role of FOCJs in Russian intergovernmental relations needs to be debated.

CONCLUSION

The authors conclude that FOCJ as a form of municipal cooperation is only partly applicable in Russia. To facilitate the discussion, four types of FOCJ are distinguished. They are classified according to which kind of members the FOCJ possesses. An important result is that type I with citizens as members and forming a municipality is not feasible in Russia. Type II with municipalities as members seems appropriate for Russia if company forms under private law are used. Type III with participation of municipalities and firms and type IV with public jurisdictions, firms, and citizens might exist in the form of private law company forms as well.

The Russian legislator should formulate a special public law for FOCJ II, which considers principles for financing FOCJ II and management powers of subordination and superordination against members, e.g. enforcing contributions. Also general management financing rules for FOCJ II establishment, operations, and member competition are missing in Russia. The authors develop them based on microeconomic establishment, operation and competition models. Moreover, public finance principles, institutional requirements for public finance of public enterprises and public offices, and recommendations to finance them are applied to FOCJ II. In this way, the authors point to future principles of FOCJ II financing for Russia, thus enriching theoretical and management literature on FOCJ. The authors outlined the advantages and disadvantages of FOCJs as well as special financial advantages and disadvantages of FOCJ financial management.

REFERENCES

БутоваТ. В.Пухова,М. М.Щукин,И. А. (2013). Проблемы и перспективы становления института межмуниципального сотрудничества в России. *Управленческие науки, 3*(8), 4-15.

Гельман, В. (2006). Возвращение Левиафана? Политика рецентрализации в современной России. *Политические исследования, 2*, 90-109.

Ирискина, О. В. (2010). Современные формы управления межмуниципальным сотрудничеством в Российской Федерации. *Известия Саратовского университета. Новая серия. Серия Социология. Политология, 2*(10), 60–65.

Истомина, Н. А. (2015). Среднесрочное бюджетное планирование в регионе: этапы эволюции и перспективы развития. *Вестник ВГУ. Серия: экономика и управление, 2*, 77-82.

ЛадыгинВ. В. (2011). Возможности создания муниципального бизнес-инкубатора с использованием механизмов межмуниципальной кооперации. *Вопросы государственного и муниципального управления, 2*, 25-37.

Ляпин, И. Ф. (2015). Конституционные особенности децентрализации государственной власти в российской федерации. *Вестник Нижегородского университета им. Н.И. Лобачевского, 1*, 152–154.

Петров, Н. (2000). Федерализм по-российски. *Pro et Contra*, *5*(1), 7–32.

Проект Федерального закона N° 252441-6 «О публично-правовых компаниях в Российской Федерации и о внесении изменений в отдельные законодательные акты Российской Федерации». (n.d.). Retrieved June, 2016, from http://www.consultant.ru/law/hotdocs/46731.html

Винницкий, А. В. (2010). Участие публичных образований в имущественных отношениях: проблемы сбалансированности административно-правового и гражданско-правового регулирования. *Административное и муниципальное право, 11*, 82-91.

ВинницкийА. В. (2013). *Публичная собственность*. Москва: Статут.

Albach, H. (1962). *Investition und Finanzierung*. Wiesbaden: Gabler.

Andel, N. (1998). *Finanzwissenschaft*. Tübingen: Mohr.

Baden-Württemberg (2016). *Gesetz über kommunale Zusammenarbeit (GKZ) in der Fassung vom 16.* September 1974 GBl. 1974, 408, ber. 1975 S. 460, ber. 1976 S. 408 Gliederungs-Nr: 2805-1.

Bartholomae, F. W., & Popescu, A. M. (2008). Technologischer Fortschritt als Entscheidungskriterium für Kooperation: Europäische Regionale Zusammenarbeit mittels FOCJ. In P. Friedrich, & F. Bartholomae (Eds.), FOCJ als Mittel regionaler Kooperation (pp. 62-72). Universität der Bundeswehr München, Fakultät für Wirtschafts- und Organisationswissenschaften Lehrstuhl für Finanzwissenschaft, Discussion paper No. 50. Neubiberg.

Blankart, Ch., & Borck, R. (2004). Local Public Finance. In J. G. Backhaus, & R. E. Wagner (Eds.), Handbook of Public Finance (pp. 441-476). Boston: Kluwer Academic Publisher.

Blankart, Ch. B. (2002). Steuern als Preise: Eine finanzwissenschaftliche Untersuchung mit einer Anwendung auf die EU-Zinsbesteuerung. *Schweizerische Zeitschrift fur Volkswirtschaft und Statistik*, *138*(1), 19–38.

Blöchliger, H., & Charbit, C. (2008). Fiscal Equalisation. *OECD Economic Studies,* 44. Retrieved June, 2016, from https://www.oecd.org/eco/42506135.pdf

Blumentrath, U. (1969). *Investitions- und Finanzplanung mit Hilfe von Endwertmodellen*. Wiesbaden: Gabler. doi:10.1007/978-3-663-02599-3

Bodenhoff, H. (2002). *Einführung in die Finanzierungs- und Investitionsrechnung*. Verlag München.

Boiteux, M. (1951). Le revenue distribuable et les pertes économique. *Econometrica*, *19*(2), 112–133. doi:10.2307/1905729

Bös, D. (1986). Public Enterprise Economics: Theory and Application. Amsterdam: Oxford.

Brown, S. J., & Sibley, D. S. (1986). *The Theory of Public Utility Pricing*. Cambridge, UK: Cambridge University Press. doi:10.1017/CBO9780511666773

Brümmerhoff, D., & Büttner, T. (2015). *Finanzwissenchaft*. München: De Gruyter, Oldenbourg.

Budget Code of the Russian Federation. (2016a). Retrieved May, 2016, from http://www.consultant.ru/document/cons_doc_LAW_19702/

Civil Code of the Russian Federation. (2016b). Retrieved June, 2016, from http://www.consultant.ru/document/cons_doc_LAW_5142/

Cornes, R., & Hartley, R. (2001). *Joint Production and Share Functions*. Working Paper, Department of Economics. Keele University.

Cyr, H. (2014). *Autonomy, subsidiarity, solidarity: foundations of cooperative federalism*. Constitutional Forum 20. doi:10.2139/ssrn.2512429

Dafflon, B. (2006). The assignment of function to decentralized government: from theory to practice. In E. Ahmad & G. Brosio (Eds.), *Handbook of Fiscal Federalism* (pp. 271–301). Cheltenham, UK: Edward Elgar. doi:10.4337/9781847201515.00020

Detig, S. (2004). *Die kommunale Anstalt des öffentlichen Rechts als Wirtschaftsförderungsinstitution*. Baden-Baden: Nomos.

Detig, St., Feng, X., & Friedrich, P. (2002). FOCJ als Grundlage für Förderungs-institutionen. Aufbau-Ost und Bevölkerung-Ost. In P. Friedrich (Ed.), Öffentliche Unternehmen im Standortwettbewerb für den Aufbau Ost (pp. 82-115). Diskussionspapier Nr. 41, Lehrstuhl für Finanzwissenschaft, Universität der Bundeswehr München, Neubiberg.

Eckardt, M., & Gritsch, M. (2016). Governance of cross-border regional cooperation by the EGTC – An explorative analysis with a focus on Hungarian EGTCs. *Public Governance Journal for Public and Non-Profit Services*, 39(1-2), 200-215.

Eichhorn, P. (1974). *Liquiditätsplanung und Gelddisposition in öffentlichen Haushalten*. Baden-Baden: Nomos.

Eichhorn, P., & Friedrich, P. (1976). *Verwaltungsökonomie*. Baden-Baden: Nomos.

Filion, L. (2011). Defining the entrepreneur. In L.-P. Dana (Ed.), *World Encyclopedia of Entrepreneurship* (pp. 41–52). Cheltenham, UK: Edward Elgar. doi:10.4337/9781849808453.00011

Fink, A. (2012). The Hanseatic League and the Concept of Functional Overlapping Competing Jurisdictions. *Kyklos*, *65*(2), 194–217. doi:10.1111/j.1467-6435.2012.00534.x

Fladung, L., & Friedrich, P. (2008). Finanzierung eines Wirtschaftsförderungs-FOCJ. In P. Friedrich, & F. Bartholomae (Eds.), FOCJ als Mittel regionaler Kooperation (pp. 21-46). Universität der Bundeswehr München, Lehrstuhl für Finanzwissenschaft, Discussion paper No. 50, Neubiberg.

Frenkel, M. (1986). *Federal theory. The Australian National University*. Canberra: Centre for Research on Federal Financial Relations.

Frey, B. (1999). *The New Democratic Federalism for Europe*. Cheltenham, UK: Edward Elgar.

Frey, B. (2000). *A Utopia? Government without Territorial Monopoly*. Institute for Empirical Research in Economics of the University of Zurich, Working paper No. 47. University of Zurich.

Frey, B. (2005). Functional, Overlapping and Competing Jurisdictions: Redrawing of geographic boarders of administration. *European Journal of Law Reform*, *5*(3/4), 543–555.

Frey, B., & Eichenberger, R. (1995). Competition among Jurisdictions: The Idea of FOCJ. In L. Gerken (Ed.), *Competition among Institutions* (pp. 209–229). London: Macmillan. doi:10.1007/9781349242627_8

Frey, B., & Eichenberger, R. (1996). FOCJ: Competitive Governments for Europe. *International Review of Law and Economics*, *16*(3), 315–327. doi:10.1016/0144-8188(96)00019-1

Frey, B., & Eichenberger, R. (2002). Democratic Governance for a Globalized World. *Kyklos*, *55*(2), 265–287. doi:10.1111/1467-6435.00186

Frey, B., & Eichenberger, R. (2006). Functional, Overlapping and Competing Jurisdictions (FOCJ): A Complement and Alternative to Today's Federalism. In E. Ahmad & G. Brosio (Eds.), *Handbook of Fiscal Federalism* (pp. 154–181). Cheltenham, UK: Edward Elgar.

Frey, B. S. (1997). *Ein neuer Föderalismus für Europa: Die Idee der FOCJ.* Tübingen: Mohr.

Friedrich, P. (2006). *FOCJ: Ein neuer Körperschaftstyp im Gesundheitswesen? Liberalisierung im Gesundheitswesen. In Schriftenreihe der Gesellschaft für öffentliche Wirtschaft* (Vol. 53, pp. 133–154). Baden-Baden: Nomos.

Friedrich, P. (2011). *Legal European Company Forms to Realize FOCJ – Functional Overlapping Competitive Jurisdictions*. Retrieved May, 2016, from https://www.researchgate.net/publication/254457385_Legal_European_Company_Forms_to_Realize_Cross_Border_FOCJ_asA_Functional_Overlapping_Competing_Jurisdictions

Friedrich, P., & Eckardt, M. (2014). *European grouping of territorial cooperation (EGTC) as a legal form to realize cross border FOCJ. In Public Sector in Transition* (pp. 106–141). University of Tartu.

Friedrich, P., Kaltschütz, A., & Nam, C. W. (2004). *Significance and determination of fees for municipal finance*. CESinfo Working Paper Nr. 1357. München: Info-institute.

Friedrich, P., Nam, C. W., & Reiljan, J. (2009). *Local fiscal equalization in Estonia: is a reform necessary?* CES Ifo Working Paper No. 2800. Munich: Ifo-institute.

Friedrich, P., & Popescu, A. M. (2006). FOCJ as institution for population policy. In T. Markowski & M. Turala (Eds.), *New members – new challenges for the European regional development policy* (pp. 228–242). Warsaw: Polish Academy of Sciences.

Friedrich, P., & Reiljan, J. (2011). *FOCJ as model for financing the Estonian general Education*. Retrieved June, 2016, from http://papers.ssrn.com/sol3/papers.cfm?abstract_id=1998055

Friedrich, P., & Ukrainski, K. (2013). Quangos as a tool of public management to promote public sector entrepreneurship. *Zeitschrift für öffentliche und gemeinwirtschaftliche Unternehmen*, *43*, 37–51.

Friedrich, P., Ukrainski, K., & Timpmann, K. (2014). *Teaching Public Management: Between Academic Disciplines*. Chair of Public Economics and Policy, University of Tartu. Retrieved from https://ec.europa.eu/epale/sites/epale/files/tu_avaliku_sektori_pm_book_friedrich_ukrainski_timpmann_isbn.pdf

Gabbe, J. (2008). *Legal status of cross-border co-operation structures – past, present and prospects. Gronau: Associations of European border regions*. AEBR.

Hax, H. (1964). Investitions- und Finanzplanung mit Hilfe der linearen Programmierung. Zeitschrift für betriebswirtschaftliche. *Forschung*, *16*, 430–446.

Heinig, K. (1951). *Das Budget*. Tübingen: Mohr.

Hirschman, A. O. (1970). *Exit, voice, and loyalty: responses to decline in firms, organisations, and states*. Cambridge, MA: Harvard University Press.

Hyman, D. N. (1993). *Public Finance: A Contemporary Application of Theory to Policy*. Fort Worth, TX: The Dryden Press.

International Monetary Fund. (2010). *Toolkit manual. Inter-municipal cooperation*. Retrieved June, 2016, from http://www.municipal-ooperation.org/images/4/4c/IMC_Toolkit_Manual.pdf

International Monetary Fund (IMF). (2001). *Guidelines for public debt management*. Retrieved May 2016, from http://www.imf.org/external/np/mae/pdebt/2000/eng/

Josselin, J.-M., & Marciano, A. (2004). Federalism and Conflicts over Principalship. Some Insights into the American Constitutional History. *Constitutional Political Economy*, *15*(13), 281–304. doi:10.1023/B:COPE.0000040433.37364.3e

Klein, P. G., Mahoney, J. T., Mcgahan, A. M., & Pitelis, C. N. (2010). Toward a theory of public entrepreneurship. *European Management Review*, *7*(1), 1–15. doi:10.1057/emr.2010.1

Köllner, W. (2001). *Der interne Finanzausgleich in den niedersächsischen Samtgemeinden* (Dissertation). München: Universität der Bundeswehr München.

Kraan, D., Bergvall, D., Hawkesworth, I., Kostyleva, V., & Witt, M. (2008). Budgeting in Russia. *OECD Journal on Budgeting, 8*(2). Retrieved June, 2016, from https://www.oecd.org/russia/42007227.pdf

Leitenstorfer, A., & Török, A. (2008). *Geeignete Aufgaben und Rechtsformen für FOCJ*. FOCJ als Mittel regionaler Kooperation. Lehrstuhl für Finanzwissenschaft, Universität der Bundeswehr München Discussion Paper, 50, 5-22.

Leixnering, S., & Bramböck, S. (2013). Public-Corporate-Governance-Kodizes: Die Kupenickiade der Beteiligungsverwaltung. *Zeitschrift für öffentliche und gemeinwirtschaftliche Unternehmen, 36*(2), 170-190.

Mertens, D. (2006). *Portfolio-Optimierung nach Markowitz*. Frankfurt am Main: Frankfurt School Verlag.

Metis. (2014). *Monitoring report 2013, Towards the New Cohesion policy*. European Union. Retrieved June, 2016 from a.http://cor.europa.eu/en/documentation/studies/Documents/EGTC-monitoring-report-2013.pdf

Musgrave, R. A., & Musgrave, P. (1989). *Public Finance in Theory and Practice*. McGraw-Hill.

Napp, H. G. (1994). *Kommunale Finanzautonomie und ihre Bedeutung für eine effiziente lokale Finanzwissenschaft*. Frankfurt am Main: Lang.

Nelson, J. R. (1964). *Marginal Cost Pricing in Practice*. Englewood Cliffs, NJ: Prentice-Hall.

Neumark, F. (1970). *Grundsätze gerechter und ökonomisch rationaler Steuerpolitik*. Tübingen: Mohr.

Nowotny, E. (1999). *Der öffentliche Sektor*. Berlin: Spinger. doi:10.1007/978-3-662-06538-9

OECD. (2011). *Corporate Governance Corporate Governance in Estonia*. Retrieved June, 2016, from https://books.google.ru/books?id=199U7QplPzIC&pg=PA31&lpg=PA31&dq=corporate+governance+code+in+estonia&source=bl&ots=XFz_du0lsl&sig=85GbWAaa8z64kn3h2fKuK_Z0Wmo&hl=en&sa=X&ved=0ahUKEwiezbGbv8rNAhXFDiwKHTJ3A7gQ6AEIRjAG#v=onepage&q=corporate%20governance%20code%20in%20estonia&f=false

OECD. (2016). *Fiscal federalism 2016: making decentralization work*. Paris: OECD Publishing. Retrieved May, 2016, from https://books.google.ru/books?id=FiJ3DAAAQBAJ&pg=PA115&lpg=PA115&dq=the+solidarity+principle+fiscal+federalism&source=bl&ots=890qZl2pPM&sig=5Xd9cekGJ32kQePiL9hTG

Olson, M. (1969). The principle of fiscal equivalence. The division of responsibilities among different levels of government. *The American Economic Review*, *59*(2), 479–487.

Papenfuß, U. (2013). Ein Public Corporate Governance Kodex ist aus theoretischer wie praktischer Perspektive nützlich– Eine Replik zu einer Köpenickade. *Zeitschrift für öffentliche und gemeinwirtschaftliche Unternehmen/Journal of Public and Nonprofit Services, 4*, 302-322.

Perridon, L., & Steiner, M. (2009). *Finanzwirtschaft der Unternehmung*. München: Vahlen.

Postlep, R. (1993). *Gesamtwirtschaftliche Analyse kommunaler Finanzpolitik*. Baden-Baden: Nomos.

Rees, R. (Ed.). (2006). *The economics of public utilities*. Cheltenham, UK: Edward Elgar.

Rehm, H. (2001). *Management der öffentlichen Schuld – Befund, Probleme, Perspektiven. Wissenschaft für die Praxis Abt. 3 Analysen* (Vol. 14, pp. 24–60). Stuttgart: Deutscher Sparkassen Verlag.

Rehm, H., & Matern-Rehm, S. (2003). *Kommunale Finanzwirtschaft*. Frankfurt am Main: Lang.

Reingewertz, Y. (2014). *Fiscal Decentralization - a Survey of the Empirical Literature*. University of Haifa. Retrieved June, 2016, from https://mpra.ub.uni-muenchen.de/59889/1/MPRA_paper_59889.pdf

Reis, M., & Augusto, G. (2013). The Terminal Value (TV) Performing in Firm Valuation: The Gap of Literature and Research Agenda. *Journal of Modern Accounting and Auditing*, *9*(12), 1622–1636.

Russian Federation Act N° 3085-1 "On Consumer Cooperation in the Russian Federation". (n.d.). Retrieved May, 2016, from http://www.consultant.ru/document/cons_doc_LAW_608/

Russian Federation Federal Act N° 131 "On General Principles of the Organization of Local Self-Government in the Russian Federation". (n.d.). Retrieved November, 2016, from http:// http://www.consultant.ru/document/cons_doc_LAW_44571/

Russian Federation Federal Act N° 14 "On Limited Liability Companies". (n.d.). Retrieved May, 2016, from http://www.consultant.ru/document/cons_doc_LAW_17819/

Russian Federation Federal Act N° 161 "On the State and Municipal Unitary Enterprises". (n.d.). Retrieved May, 2016, from http://www.consultant.ru/document/cons_doc_LAW_39768/

Russian Federation Federal Act N° 208 "On Joint Stock Companies". (n.d.). Retrieved May, 2016, from http://www.consultant.ru/document/cons_doc_LAW_8743/

Russian Federation Federal Act N° 41 "On Production Cooperatives". (n.d.). Retrieved May, 2016, from http://www.consultant.ru/document/cons_doc_LAW_10286/

Sinn, H. (1997). Das Selektionsprinzip und der Systemwettbewerb. In A. Oberhauser (Ed.), *Fiskalföderalismus in Europa* (pp. 9–60). Berlin: Duncker & Humblot.

Smith, D. (2011). *Functional, Overlapping, Competing Jurisdictions among Fractionalized Agents: Medieval Spain*. Retrieved October, 2015, from http://papers.ssrn.com/sol3/papers.cfm?abstract_id=1937955

Spindler, J. (1998). *FOCJ – ein Konzept zur Neuordnung der Zusammenarbeit öffentlicher Gebietskörperschaften*. Bern: Haupt.

Spindler, J. (2008a). Die Zweckgemeinde – auch ein waldpolitisches Referenzmodell. *Schweizerische Zeitschrift für das Forstwesen*, *12*(12), 427–434. doi:10.3188/szf.2008.0427

Spindler, J. (2008b). Die Grösse allein ist nicht entscheidend. *Kommunalmagazin*, *3*, 4–6.

Ter-Minassian, T. (1997). Intergovernmental Fiscal Relations in a Macroeconomic Perspective: An Overview. In T. Ter-Minassian (Ed.), *Fiscal Federalism in Theory and Practice* (pp. 3–24). Washington, DC: International Monetary Fund.

Tiebout, Ch. (1956). A Pure Theory of Local Expenditures. *Journal of Political Economy*, *64*(5), 416–424. doi:10.1086/257839

Vogt, G. (1984). Staatliches Haushaltsrecht. In A. v. Mutius, K. H. Friauf & P. Westermann (Eds.), Handbuch für die öffentliche Verwaltung (pp. 519-584). Neuwied, Darmstadt: Luchterhand.

Wagner, A. (1890). *Finanzwissenschaft*. Leipzig: Winter'sche Verlagshandlung.

Wilhelm, B. (2013). *Inter-Municipal Cooperation (IMC). New Perspectives for IMC in the field of Energy Efficiency (EE)*. Retrieved June, 2016, from http://www.sustainable-energybih.org/wp-content/uploads/2013/07/BWilhelm_Presentation-on-IMC.pdf

World Bank. (2014). *Toolkit Corporate Governance of State-Owned Enterprises*. Retrieved June, 2016, from https://openknowledge.worldbank.org/bitstream/handle/10986/20390/9781464802225.pdf?sequence=1

Zimmermann, H. (1999). *Kommunalfinanzen*. Baden-Baden: Nomos.

Zimmermann, H., & Henke, K.-D. (2001). *Finanzwissenschaft*. München: Oldenbourg.

KEY TERMS AND DEFINITIONS

Competition Among FOCJ: Concerns the struggle of various FOCJ for clients and members.

Financial Management: Deals with financial decision making within the three phases: establishment, operation, and competition.

Functional Overlapping and Competing Jurisdiction (FOCJ): A public economic unit performing one or several functions and competing (overlapping) for members or clients. FOCJ has a democratic structure and can levy contributions on their members.

Establishment Phase of FOCJ: Concerns the founding period of the FOCJ and its initial finance with basic capital.

Municipal Cooperation: Takes place "when two or more municipalities agree to work together on any of the tasks assigned to them in order to gain mutual benefits." (IMC Toolkit Manual, 2010, p. 7).

Microeconomic Finance FOCJ Principles: Stem from finance rules gained from microeconomic models of establishment, operation and competition.

Operation Phase of FOCJ: Concerns the operation period of the FOCJ and its current finance.

Public Finance Principles for FOCJ: Refer to basic propositions of financing jurisdictions (governments) discussed in public finance.

Public Business Administration FOCJ Finance Principles: Concentrate on applying practical financial instruments.

Types of FOCJ: According to the kind of members, one can distinguish between different types of FOCJ. Members may be citizens (type I), governments (type II), governments and firms (type III), governments, private households, firms, etc. (type IV).

ENDNOTES

1 The authors acknowledge the support of the Estonian Science Foundation's grant 8580 and doctorial grant No. 544916-EM-12013-1-FI-ERA MUNDUS-EMA21.

2 Public entrepreneurship refers to a public actor who recognizes entrepreneurial opportunities, undertakes investment decisions with a view to innovating, and takes action by using resources to implement a differentiated vision that adds public value (Filion, 2011; Friedrich & Ukrainski, 2013). Public entrepreneurship deals with the coordination of private and public actors to achieve public goals, the formulation of tasks and goals, the design and development of public organisations, the ways of management of resources, and the control of external effects (Klein et al., 2010; Friedrich & Ukrainski, 2013). One can distinguish social entrepreneurship, political entrepreneurship, knowledge entrepreneurship, technology entrepreneurship, and public sector oriented entrepreneurship (Friedrich & Ukrainski, 2013).

3 Sometimes the plural of FOCJ is expressed as "FOCJ" and the singular as "FOCUS" (Frey, 1999). For sake of simplicity, the authors of this article use for the plural "FOCJs" and for the singular "FOCJ".

4 A Zweckverband (special purpose associations of public law) has to operate covering costs.

5 Sometimes municipal cooperation refers only to neighbouring communities or it's defined with respect to some special municipal functions only (Wilhelm, 2013).

6 Individual entrepreneurs and commercial companies can be members of the general partnership (Civil Code of the Russian Federation (CC of RF), 2016, Article 66). All the members of the general partnership take part in its management. Members have to participate in its activity and can freely quit partnership (CC of RF, 2016, Article 69). The majority of decisions can be made voting, fixed in its Foundation Agreement (CC of RF, 2016, Article 70, 71).

7 In the limited (commandite) partnership alongside the participants, using their entire personal property as a guarantee, are participants-investors (commanditaires) (CC of RF, 2016, Article 82). They guarantee only with their capital share. The number of commanditaires is restricted to twenty (CC of RF, 2016, Article 82). In limited partnerships participants-investors can be natural persons, legal persons, and governmental units (CC of RF, 2016, Article 66). The majority vote prevails according to the Foundation Agreement and members can quit (CC of RF, 2016, Article 70, 71, 85).

8 A limited liability company (LLC) may perform all free or licensed activities (Russian Federation Federal Act N° 14 "On limited liability companies", 2015, Article 2). Participants may sell and buy shares (CC of RF, 2016, Article 94). The main organ of a limited liability company is a general meeting of shareholders (CC of RF, 2016, Article 67.1). The company's statute regulates entering and exiting. A Russian limited liability company may have or may not have a board of directors (Russian Federation Federal Act N° 14 "On Limited Liability Companies", 2015, Article 32). The number of participants must be less than 51. The statute of the LLC may include other rights for its members, such as the right to fix fees and contributions (Russian Federation Federal Act N° 14 "On Limited Liability Companies", 2015, Article 12).

9 A non-public joint stock company can carry out licensed and other functions (Russian Federation Federal Act N° 208 "On Joint Stock Companies", 2016, Article 2). The foundation document of a joint stock company is a statute (Russian Federation Federal Act N° 208 "On Joint Stock Companies", 2016, Article 11). The supreme governing organ is the general meeting of shareholders (Russian Federation Federal Act N° 208 "On Joint Stock Companies", 2016, Article 47). The board of directors executes the general management of a company (Russian Federation Federal Act N° 208 "On Joint Stock Companies", 2016, Article 64). Shareholders decide about changes in equity capital approved by internal documents and the board of directors. Shares of non-public joint stock company cannot be publicly purchased (Russian Federation Federal Act N° 208 "On Joint Stock Companies", 2016, Article 7).

10 With public joint stock companies, membership competition is easier because their shares can be publicly sold. Most of the management function is concentrated in the hands of the board of directors (Russian Federation Federal Act N° 208 "On Joint Stock Companies", 2016, Article 64). Apart from competition for members a competition for clients can exist. Natural persons, legal persons, and governmental units (for example, municipalities) can be founders of a joint stock company (CC of RF, 2016, Article 66).

11 A production cooperative can be established for providing public services. Membership is mainly based on personal participation (CC of RF, 2016, Article 106.1). Decisions are made by the general meeting of the members (Russian Federation Federal Act N° 41"On production cooperative", 2011, Article 15). The foundation document serves as a statute (Russian Federation Federal Act N° 41"On production cooperative", 2011, Article 5). The executive bodies are the chairman and the board of directors (Russian Federation Federal Act N° 41"On production cooperative", 2011, Article 17). Members can leave the production cooperative and transfer their share to another member (Russian Federation Federal Act N° 41"On production cooperative", 2011, Article 8).

12 State and municipal enterprises operate in the form of unitary enterprises, using state and municipal property. A combination of public property belonging to different jurisdictions is not allowed (Russian Federation Federal Act N° 161 "On state and municipal enterprise", 2016, Article 2).

13 The legal status of a consumer cooperative is very similar to that of a production cooperative. Members can be natural persons or natural persons and legal persons. It is aimed at satisfying the needs of its members. It may perform different functions. The statutes of the consumer cooperative should regulate the volume, structure, and size of shares, responsibilities, and obligations of members and organs as well as democratic decision-making competences (CC of RF, 2016, Article 123.2). Members can enter and leave (Federal Act Nº 3085-1 "On consumer cooperation in the Russian Federation", 2013, Article 11).

14 The statute of associations (units) includes information about the purpose of its activities, the conditions of entrance and opt-out, competence and decision-making procedures, and the rights of participants (members), etc. (CC of RF, 2016, Article 123.9). Members of an association (unions) may use its services and have to pay contributions (CC of RF, 2016, Article 123.11). The general meeting of members makes decisions about membership fees (CC of RF, 2016, Article 123.10). Associations (unions) have legal persons and/or natural persons as members. An association (union) is established to achieve social and other non-commercial aims (CC of RF, 2016, Article 123.8). Non-commercial partnerships and self-regulated organisations belong to associations (unions) (CC of RF, 2016, Article 50).

15 Public organizations are voluntary associations of natural persons aimed at satisfying spiritual or other non-material needs (CC of RF, 2016, Article 123.4). The statute of the public organization provides information about its object and purpose, members' entrance and opt-out options, the competence of its organs, decision-making procedures, the rights of the participants, etc. (CC of RF, 2016, Article 123.5). Members of a public organization use its services and pay membership fees determined by a general meeting of the members. Members also have the right to quit the public organization at any time (CC of RF, 2016, Article 123.6).

16 Establishments provide management, cultural, or other non-commercial functions (CC of RF, 2016, Article 123.21). Establishments can be state, municipal, or private. State and municipal establishments are divided into state-owned, budgetary, and autonomous categories (CC of RF, 2016, Article 123.22, 123.23). There are no members, but founders representing a natural person, or a legal person, or the Russian Federation, a subject of the Russian Federation, or a municipality. There cannot be several co-founders for an establishment (CC of RF, 2016, Article 123.21).

17 If the FOCJ is not only competing for members, but competes on sales markets, delivering services to non-members, other market forms prevail, e.g. an oligopoly or a duopoly. A duopoly model should be formulated to evolve optimal financial rules (Eckardt & Friedrich, 2014). According to the 6 manager group (see footnote 15) present in the FOCJs results a matrix comprising 36 solutions in the case of an FOCJ duopoly acting autonomously. The combination of group II managers in both FOCJs allows for an efficient (cost minimizing) financial solution. That is also true for profit-maximizing management of group V in both FOCJs. All the other solutions are inefficient, and the finance solution as well.

18 The utility functions of managers differ. A group I values output and labour positively. A group II maximizes output. A group III wants to maximize labour only. A group IV values labour positively, but output negatively. A group V aims to maximize profit, whereas group VI managers value profit positively as well as labour. According to the prevailing group of managers in an FOCJ, different financial management solutions appear (Fladung & Friedrich, 2008).

Chapter 5
The Business Model of a Public Social Partnership:
Contextual Determinants

Aldona Frączkiewicz-Wronka
University of Economics in Katowice, Poland

Monika Tkacz
Mondragon Innovation and Knowledge (MIK), Spain

Saioa Arando
Mondragon Innovation and Knowledge (MIK), Mondragon Unibertsitatea, Spain

ABSTRACT

The authors discussed the dynamic evolution of the public management manifesting innovative and collaborative methods of social service delivery transcending the public authorities' individual boundaries. The chapter purposely presented the inter-organizational cooperation and public social partnership as an effective business model for public/social value creation unleashing the potential of public organizations and empowering social actors in democratic and collaborative policy/decision making. The chapter identified two public social partnerships in Poland and Spain. The research was conducted in the regime of constructivism. As suggested by R. Rorty (1982), the researchers posed a question as to why the research was being conducted and then adapted the research strategy to it (a case study). In order to analyse obtained data, the so-called thick description was used revealing the contextual determinants that influence the architecture of the researched models.

INTRODUCTION[1]

A fundamental problem faced by the developed countries is to find such practices for public funds management that will allow for the effective achievement of social goals defined in the process of political choices made by citizens. As a consequence, the governments of many countries undertake initiatives aimed at the reform of the public sector (Pollitt & Bouckaert, 2000). One of the key issues for the public

DOI: 10.4018/978-1-5225-2215-7.ch005

sector is the effective provision of social services to citizens. Innovative solutions in this area often involve the cooperation between organisations that adopts the institutional format of a public social partnership.

The effective operations of a public social partnership rely heavily on the selection of organisations that will act as partners. The resources, values and goals of these organisations should be matched in such a way that the outcomes they produce can contribute to the creation of social value expected by stakeholders. The business logic of the public services where production and consumption occur simultaneously and involve co-production between professionals and users, leads the public management models into a new direction of inter-organizational network and sustainability growth mechanisms. Public services, as the result of complex service delivery systems being relational and process-based, require functional architecture and sustainable business models (Osborne, Radnor, Vidal, & Kinder, 2014, pp. 167-169). The reform strategies and business models for public services delivery need cultural change within public sector (Osborne, Radnor, Vidal, &Kinder, 2014, p. 169) and thus, this chapter aims to identify and compare the building blocks of the business models for public social partnerships established to provide social services. Therefore it examines some examples from Poland and Spain.

The chapter structure starts with a brief summary about the current need for innovation in the social service delivery systems, which gradually leads to the public social partnerships overview as a new significant model of inter-organizational collaboration for more effective social service identification, design and the social needs fulfilment. While the new social delivery systems are becoming highly complex, the authors discuss the business model definition and architecture that is relevant for public social partnerships. The authors refer to the Osterwalder's and Pigneur's (2002, p. 5-23) definition that describes the business model as a firm's architecture and its network of partners for creating, marketing and delivering value and relationship capital to one or several segments of customers. For better understanding of the building blocks of the networking architecture, the authors present two examples from Poland and Spain on the basis of the proposed Business Model Canvas that enables to analyse the sustainable business models of the sophisticated public social partnerships.

BACKGROUND

The harmonious development of a state is only possible if social needs that are satisfied with public funds are identified correctly. The institutional state ensures the fulfilment of social needs recognised by citizens as important and its obligation is to build the effective system for the provision of social services that meet these needs. It is in the process of determining social policy goals that the decisions on what services are necessary and how they should be delivered are made, while practical measures that allow for the implementation of these decisions are part of public management.

Social policy is a concept commonly used in public space to describe the socio-economic outcomes of a state's activity as a public authority. The scope and depth of the measures adopted in social policy practice reflect the economic, political and legal doctrines pursued by a given state. Additionally, social policy practice is also the product of historical and ideological conditions. Contemporary researchers increasingly often endorse the idea that the way in which social policy goals are pursued is strongly affected by a society's organisational culture and its manifestations in collective and individual activity.

The analysis of the Polish and Spanish social policy – or rather the way in which it is implemented – supports the assumption that it should gradually evolve towards the solutions that embrace institutional change defined as social innovation, the essence of which is cooperation on all organisational levels

– between organisations representing all sectors of the economy – and empowerment. All European trends in social policy management recognise the dominant role of initiatives aiming at the inclusion of a variety of actors in the process of pursuing social policy priorities in creating the rules for socio-economic growth (Grewiński, 2014). This means the necessity to build models embracing the cooperation between different organisations.

In terms of the scientific discipline of public management, the proposed thesis translates into the necessity to create organisational conditions conducive to the effective delivery of social services.

So far, the observation of practice supports the statement that organisations working together, choosing common goals and jointly deciding on how to use their resources often develop an ability to create new structures. The new structures, taking advantage of adopting the idea of collaboration, are capable of generating more positive outcomes than each of the organisations operating on its own.

In the public sector, the entities that create network relationships are a relatively young organisational phenomenon, the origin of which should be linked with the growing popularity of the idea of collaboration between organisations from different sectors. The emergence of network organisations offers a response to the challenges of practice, where it has been noticed that the solving of complex problems requires varied skills and resources, often held by many different actors of the economic life – not only businesses, but also public institutions and non-governmental sector organisations (Czakon, 2005, pp. 429-437).

New forms of cooperation between different actors, in particular partnerships and alliances, require such management practices that will facilitate the processes involving clashing interest and coordinate joint activities – hence the need for research into and reflection on them.

Social services – as services of general interest SGI, of which they are one component – underpin human dignity and guarantee the universal right to social justice and to full respect of fundamental rights, as set out in the *Charter of Fundamental Rights of the European Union (2000)* and in international commitments such as the revised European Social Charter and the Universal Declaration of Human Rights. The purpose of social services is to ensure social, territorial and economic cohesion by implementing the principles of social solidarity. The significance of social services for building the foundations of a modern state's growth was first strongly emphasised in the Lisbon Strategy, only to return in the *European Parliament resolution of 14 March 2007 on social services of general interest in the European Union*,[2] and Europe 2020 strategy

In terms of theory, the last conclusion the researchers observes the shift of the research focus from an individual organisation to the emerging multi-actor cooperation formations treated as social innovations (Wronka-Pośpiech, Frączkiewicz-Wronka, & Lara Montero, 2015, pp. 27-43).).

Social innovations play a crucial role especially in the sectors where the existing models of innovation are unsuccessful, obsolete or unable to take advantage of the opportunities arising in the environment (Mulgan, Tucker, Ali, & Sanders, 2007). These sectors are primarily education (Loogma, Tafel-Viia & Ümarik, 2012, pp. 1-19), health care (Beinare, McCarthy, 2012, pp. 889-893), environmental protection (Maruyama, Nishikido, & Iida, 2007, pp. 2761-2769), public finance (Philips, Laforest, & Graham, 2010, pp. 189-199), public and social services (Teets, 2012, pp. 15-32) and urban planning (Moulaert, Martinelli, Swyngedouw, & González, 2005, pp. 1969–1990). The literature review (Nicholls & Murdock, 2012; Caulier-Grice, Davies, Patrick, & Norman, 2013) reveals that social innovations emerge in the context of the following five areas:

- Social transformation (the role of civic society in the process of social changes and the role of social economy and social entrepreneurs in achieving economic growth and social integration;

the role of business in social change – corporate social responsibility and the role of enterprises in implementing the next wave of innovation and productivity through focusing on 'social' areas such as education or health care);

- Organisational management model (developing business strategies, embracing changes in human, institutional and social capital, which enhance organisational efficiency and improve competitiveness; organisational restructuring, the modernisation of industrial relations, and improved human resource management; non-profit management);
- Social entrepreneurship (developing new and innovative ways to deal with social challenges by engaging 'socially sensitive' entrepreneurs. Social enterprises are, first of all, enterprises with a social purpose and their surpluses are reinvested in the organisation);
- The development of new products, services and programmes (fulfilling social needs, innovations in the public sector and public services delivery by social enterprises and organisations of civic society; the delivery of public services and other redistribution means in the manner that can ensure budgetary savings in the welfare state);
- Governance as a management model manifesting the strengthened position and increased capacity of social institutions (enhanced relationships between different social actors, the improved skills, competences and social capital of social actors engaged in the development and implementation of social and economic programmes and strategies) (Wronka-Pośpiech, 2015, p. 128).

In further considerations, social innovation will be understood as a management model, clearly manifested in the concept of governance. In this particular case, the model of governance that is slowly becoming a new paradigm for organisations delivering social services and, more widely, for social policy the tasks of which remain within a state's remit, but are nowadays increasingly frequently performed by actors from different sectors.

There are numerous attempts to introduce changes in the public sector policy framework and architecture, both in the European Union (e.g. Joinup, the common portal for e-Government solutions) and globally (e.g. the OECD's Observatory of Public Sector Innovation) (European Commission, 2013a, p. 5). The necessity for innovation in the public sector has been underlined by the European Union in the recent research proving that there is a call for the 'New Public Movement' (European Commission, 2013b, p. 1). The latest comprehensive report of the Expert Group on Public Sector Innovation (European Commission, 2013b) describes the motives and significance of change and the crucial steps in the powering of the public sector and the creation of its new architecture. Relating to the report's findings, 'there is an urgent need to power innovation within the public sector itself in order to unlock radical productivity improvements and efficiency gains, to foster the creation of more public value and a better response to societal challenges' (European Commission, 2013b, p. 1). The European Union proposes four principles of the public sector innovation movement: (European Commission, 2013a, p. 6)

- Co-design and co-creation of innovative solutions (with other Member States, other
- Parts of government, businesses, the third sector and citizens);
- Adopting new and collaborative service delivery models (across public, private and non- governmental actors, both within and across national borders);
- Embracing creative disruption from technology (the pervasive use of social media,
- Mobility, big data, cloud computing packaged in new digital government offerings);
- Adopting an attitude of experimentation and entrepreneurship (government itself

- Needs to become bolder and more entrepreneurial).

Moreover, the change in the paradigm of public/social services delivery and changes in the environment cause that the process of providing complex and innovative services requires today inter-organisational collaboration irrespective of the sector in which an organisation operates (Agarwal & Selen, 2009, pp. 431-475; Lewandowski, 2013a).

The key drive of the public sector innovative actions is the proper value delivery determined by the frequently changing society requirements. From the general viewpoint, each organisation that strives for success in the modern competitive global market needs to create new value embedded in its products or services. Accordingly, public organisations aiming to operate effectively in the so-called new economy should seek a model that will allow them to deliver high quality services to citizens. The understanding of the notion of "value" may vary. Koźmiński and Jemielniak (2008) state that "value is something precious, something that people desire, strive for and act in order to get" (p. 419). The simplest way – typical of market organisations – is to define value in terms of above-average profit. Obłój (2002) proposes that a firm creates new value when it generates profits that are above the industry average and when they exceed alternative, safe ways of investing. In the situation when an explicit measure of success such as profit is not available, it becomes much more complicated to clearly define what constitutes profit for the customers of public organisations and what they perceive as desired value. This difficulty also translates into problems involved, on the one hand, in the identification of the factors contributing to the success of public organisations, on the other hand, of the ways of achieving this success.

The analysis of the reference literature confirms that value creation in every organisation stems from the application of such management methods that ensure concentrated effort, bring the focus on key value creating factors and their adequate integration, and facilitate the process of making strategic and operational decisions. As a result, an organisation's success relies on whether it can act in an integrated manner allowing for the flexible adjustments to the requirements of the market and environment, in other words, whether it is able to design and implement a business model for its operations. To perform this task, it has to identify value creating factors, define the areas where value is created, and discover the method for its creation. Every organisation creates value for its key stakeholders, while each participant of the economic and social process creates specific added value, contributing to the value chain. The key stakeholders, that is everybody who has a vested interest in the value created by the organisation, can be defined as: partners who benefit from their contacts with the organisation in terms of an increase in the value that they achieve; the organisation as such or rather different groups of its internal stakeholders; intermediaries who deliver value between the organisation providing the product and its end users. In this context, Moore's concept of public/social value is of particular relevance (Moore, 1995; Moore, & Khagram, 2004). Although the notion itself comprises the word "public", it is worthwhile to mention that in his texts the author uses it interchangeably with "social". The concept focuses more on understanding what creates this value and how it can be applied to the public management perspective rather than defining the final outcome. In this article, the Authors concentrate on the approach proposed by Moore in 1995, which proposed what the managers of institutions that are not profit-oriented should do in order to deliver value to society. Based on a series of interviews that he conducted with public administration workers, Moore concluded that the greatest difficulty lied in defining the goals, determining the methods to achieve them and measuring the outcomes in such a way that local communities perceived them as valuable. The conceptualisation of the goals important for all the stakeholders appears to be a critical factor for initiatives undertaken in the public sector and determines the overall framework for a relevant business model. The observation of the practice of public organisations indicates that the collaboration of actors in the area of public/social service provision is becoming a value creating factor increasingly often (Frączkiewicz-Wronka, Szołtysek & Kotas, 2012, pp. 231-255). In public management practice, the collaboration between organisations from different sectors – aimed at the effective pursuit of shared goals – is becoming necessary (Frączkiewicz-Wronka, 2011, pp. 72-81). The well-designed configuration of partners within the collaboration framework creates an opportunity to develop and implement such a business model for a public social partnership that will respond to the needs emerging in a local community and that the outcomes achieved as a result of its operations will contribute to the survival and growth of the organisations making up the partnership though the creation of the desired public/social value.

THE SIGNIFICANCE OF INTER-ORGANIZATIONAL COLLABORATION FOR THE EFFECTIVE FULFILMENT OF THE NEEDS IDENTIFIED AS SOCIALLY IMPORTANT AND THE EFFICIENT SOCIAL SERVICES DELIVERY

The observed practice of the public sector confirms that public institutions tend to deliver social services within relationships with other autonomous private, quasi-public and voluntary organisations more and more frequently. This trend is one of the symptoms of social, political and economic changes taking place in the last two decades in the concepts concerning the management of the public sphere. Another shift concerns the role of public authorities, which move from their position of a service provider to the role of an institutional facilitator and controller of the process of service delivery as well as a collaboration moderator.

Modern public organisations undertake actions in the public interest and pursue their goals mainly by establishing relationships of collaboration with other organisations or groups of actors in the socio-economic life (Kożuch, 2011, p. 14), because this increases the value of an organisation, as numerous studies confirm (Conteh, 2013, pp. 501-521; Lewandowski, 2013b, pp.172-182). This means that the outcomes of their operations are becoming increasingly dependent on external partners (Frączkiewicz-Wronka, 2012, p. 275; Kożuch & Sienkiewicz-Małyjurek, 2013, pp. 82-95), with whom the inter-organisational bonds are created (Kastberg, 2016, pp. 33-56). As a consequence, the public sector is, to a growing extent, organised and steered by the relationships with external entities (Peters, 1998) and inter-organisational bonds (Hood, 1991).

The evolution – as a consequence of seeking a premium on the value created to meet stakeholder expectations – of the principles underpinning the public sector towards embracing first market mechanisms (NPM) and then network mechanisms (Governance) is a driving force behind the change in the actors operating in the domain of the public sector and it contributes to the development of public management theory. In their search for more effective methods of social services delivery, organisations seek solutions transcending their individual boundaries. An innovation and relational premium can be earned when organisational projects based on the collaboration between actors from different sectors are implemented in practice.

As a result of the emergence of new ways for delivering services, the level of interdependence between actors establishing institutional forms of collaboration is increasing (Klijn & Teisman, 2008). One of the reasons is the willingness, or sometimes even the necessity, to acquire resources that remain at the disposal of other organisations (Frączkiewicz-Wronka, 2014), which creates the need to build new organisational projects and, in consequence, the new architecture of collaboration – unique for a given configuration – is developed (Fjeldstad, Snow, Miles & Letti, 2012, pp. 734-750). Bringing together organisations varying in terms of their legal form and ways of operating causes a number of problems in the process of negotiating shared goals and values that need to be defined in order to ensure the proper definition of collaboration principles and the development of a business model.

The fact of collaboration means that the organisations involved in it can be perceived as structures building a unique form of the operation, or rather co-operation, of private and/or social and/or public actors or as formations that are new structures created by the mentioned actors in the pursuit of a shared goal (Kożuch, 2011, pp. 109-134), which is social/public value creation. As Bryson argues (2004, pp. 28), a public organisation that wants to achieve success needs to present to its customers such a set of goals that respond to the questions about what the organisation is, what it does, why it does it, and what value its operations offer to end users. This realisation means moving towards the creation of such a business model that will contribute to the organisation's success in a local community and win the approval for its actions granted by local authorities. The accomplishment of the goals adopted requires the combination of the organisation's decisions and actions such as: participation, designing strategic actions, building a winning coalition, or implementing a strategy. Bardach (1998) defines it as "a method of interaction in a given situation that ensures the achievement of the goals and helps to seize some hidden opportunities to create value" (p. 36).

The process of building collaboration/networks increases the need to identify the conditions and components essential to designing such a business model for collaboration that will give the partners the foundations for long-term growth. The formal expression of such collaboration may be a public social partnership, which may be treated as an innovation.

The Overview of Public Social Partnerships

The new vision of the public sector and the recognition that the government does not hold monopoly in the delivery of social services require a more entrepreneurial and participatory approach of public institutions. Public collaboration with a wider range of partners 'can help unleash the potential of an innovative public sector with the capacity for generating new ideas and implementing them to create value for society, covering new or improved processes (internal focus) and services (external focus)' (European Commission, 2013b, p. 1). Therefore, the OPSI[3] highlights that public results are strongly determined by the contribution from multiple actors (OECD, 2016a), of which public social partnerships are a very significant mode of overcoming the public sector barrier to be transformed into an effective 'growth engine' for the local and European economy (European Commission, 2013b, p. 1).

Partnerships are widely accepted as an invaluable instrument for stimulating more sustainable socio-economic development and they are crucial to achieve the EU 2020 targets in the area of employment, social cohesion, economic development, environmental sustainability and the quality of life within the EU (European Commission, 2013c, p. 2). Nevertheless, the term is not commonly used and understood across Europe. In the literature discussing relations between the state and social partners, a number of terms are used to describe similar arrangements. The social partnership is the most widespread and commonly used term in Austria, Denmark, Germany, Ireland, France, and the Netherlands, whereas the more specific 'concertation' is prominent in Italy and Spain. In Great Britain, the expression that is more common is 'corporatism', which is the form of a social organisation where key political and social decisions are made by corporate groups in conjunction with the government (Boyd, 2002, p. 3).

"In Austria and many other Western European countries, public policy making continues to be the outcome of explicit negotiated formal or informal agreements between the government and social partners" (Boyd, 2002, p. 4). The social partnership can take different institutional forms across countries but its purpose is the same: "the government is prepared to offer the social partners a role in public policy making" (Boyd, 2002, p. 4).

Through the involvement, adequate consultation and participation of stakeholders, partnerships develop cross-cutting perspectives and integrated approaches to multidimensional problems that are 'too complex and interdependent for any one institution to effectively respond to alone' (European Commission, 2013c, p.1). Thus, public social partnerships are perceived as strategic partnering arrangements enabling more efficient delivery of public services and 'with more person-centred outcomes for users of services, by putting co-production at the heart of service design' (The Scottish Government, 2013, p. 4). With reference to European Economic and Social Committee, (European Commission, 2013c, p. 2) partnerships have become a key instrument of collective commitment and an important policy tool that contributes to better efficiency of public expenditures and public policies. According to ILO,[4] the partnership approach evokes social dialogue that leads to democratisation of social and economic policy-making and reduction of social conflicts (ITC, 2012, p. 13). It reflects the democratic principle, included in Article 11 of the Treaty on the European Union, that representative associations should be able to express their views, to be consulted by, and hold dialogue with the public authorities', and that workers and employers should be 'involved in decision-making on issues that affect them closely.'(European Union, Social Europe Guide, 2012, p. 12). The European Social Fund (ESF) Regulation refers to partnerships as a governance mechanism (ESF Reg., Art. 5) as well as territorial alliances (ESF Reg., Art. 3) 'through stimulating the involvement of diverse actors and approaches; clearly defining target groups, objectives and priorities; and building participatory democracy through collaborative decision-making' (European Commission, 2013c, p. 2). The Copenhagen Centre for Partnership Studies (Andersen & Mailand, 2002) defines the social partnership as a tri or multi-partite arrangement involving employers, trade unions, public authorities and/or others (e.g. the voluntary sector), which is concerned with areas of economic and social policy and might be based on a binding agreement or declaration of intent. Moreover, it does not necessarily require an institutional setting and can be described as a mode of policy making characterised by actively seeking compromise (Boyd, 2002, p. 3).

Nevertheless, public social partnerships are not always easy to set up as they are not suitable for every organisation or service, and they demand accurate level of trust or commitment and proper knowledge or tools to efficiently manage inter-organizational network. (The Scottish Government, 2013, p. 4; Daglio, Gerson, & Kitchen, 2015, p. 32). The European Commission (2013c) underlines the significance of partnerships in the labour market, where public employment services (PES) need to adapt to changing environments and work in close cooperation with other stakeholders to 'maximise job matching, secure pathways for vulnerable groups and improve labour market intelligence in the context of the new skills agenda' (European Commission, 2013c, p. 2). The partnership approach in the labour market area, inter alia, concerns partnerships at the local level between local employment offices and other stakeholders to ensure a co-ordinated strategic service offer and easy referral of individuals to support services and partnerships with employers and providers of education and training services in order to improve the way in which the training offer is tailored to demand (European Commission, 2013c, p. 2). Partnerships enhance competences and resources, which creates the mechanism that helps to solve problems in the labour market more readily than by institutions working in isolation. (European Commission, 2013c, p. 3). Partnerships are also crucial in the design and administration of public pension schemes. They play a formal (public pension schemes through tripartite bodies) or informal (permanent, tripartite, or bipartite consultative bodies) advisory role in many European countries. In Belgium, social partners jointly manage the National Pension Office, while in France, they actually supervise the day-to-day management of the national pension fund in the private sector (Duchemin, & Weber, 2013). The more limited role of social partner organisations in tripartite decision-making may be seen in Cyprus, the Czech Republic, Malta and the UK, which can be attributed to different traditions of industrial relations. However, in some countries, such as France, Italy, Latvia, Spain, Poland, Slovenia, or Slovakia, the government is legally obliged to consult social partners via established bodies (Duchemin, & Weber, 2013). Furthermore, Innovation Labs as entities facilitating the development of public social innovation (PSI) models have become an important trend in the recent years. Labs are cross-government bodies or independent organisations whose mission is to generate ideas for the renewal of the government's operations by bringing together small interdisciplinary teams, bridging organisational and sector boundaries, with experience and expertise that reflect the complex nature of policy challenges. Denmark is a leading example in Europe through the establishment of MindLab in 2002 (Leon, Simmonds & Roman, 2012, p. 15; Daglio, Gerson, & Kitchen, 2015, p. 33).

Public Social Partnerships in Poland

In the Polish society, the culture of acting through partnership is growing very slowly and reluctantly, only after three decades following the political transformation. The associations involving inter-organisational collaboration tend to be negative and often relate to such expressions as "connections" or cronyism. This concerns both the social and economic sphere of life and is obviously rooted in the low level of social capital. In the Polish practice of developing inter-sectoral collaboration, Rymsza (2007) indicates three major periods of adopting relevant legislation. The first one is the phase of building space for civic initiatives while at the same time creating a set of "privileges" for formalized forms of civic activity (pp. 23-35). The phase comprised the years 1989-1993 and it was concluded with the adoption of the Act of Law on Associations (1989)[5], creating the legal framework for the realization of the freedom of association, and the granting of tax incentives to non-governmental organisations, which boosted the growth of the third sector organisations. The second period – the years 1993-2001 – was the phase of the declarative support given to the third sector by the state, while in fact its role was marginalized when it came to the social policy reforms.

At that time, the third sector organisations that did not have enough clout to be invited as partners in the social reforms stalled, though it needs to be stressed that the principles constituting cooperation – of subsidiarity and social dialogue – were included in the Constitution of the Republic of Poland. The adoption of this legal norm resulted in the first attempts to launch cooperation on the local level. The third period, comprising the years 2001-to date, has been the phase of defining the principles of inter-sectoral collaboration. The momentous event of the period was the adoption of Act of Law on Public Benefit and Volunteer Work (2003)[6] – the first legal act that regulated the issues relating to non-governmental organisations in a comprehensive manner and imposed the obligation of cooperation between non-governmental organisations and public administration. At the same time, efforts were undertaken to lay the foundations for institutional collaboration and develop standards for its management rules and practices. Collaboration between organisations, as a rule, is initiated to solve social problems through the delivery of social services. The legal act mentioned above addresses the issue of collaboration in a comprehensive way, providing its definition, listing its forms, and describing the rules for its implementation. The main principles determining the model of the cooperation between public administration and non-governmental organisations are the principles of subsidiarity, sovereignty, partnership, effectiveness, and transparency.

The contents of the principles determines the main forms of collaboration between non-governmental organisations and public administration, stipulated by the act of law, and it is these two types of actors that enter public social partnerships. The chart below shows the actors that can form public social partnerships in Poland.

Ontologically, public social partnership is the essence of the constitutional principle of subsidiarity, the special importance of which is conveyed by Luhmann's (1994, p.88) statement: as little state as possible, as much state as necessary. Public social partnership is an expression of public participation and new form of cooperation between local authorities and NGOs.

Public Social Partnerships in Spain

Spain does not have a long tradition of social partnership. For the greater part of the previous century, Spain's historical development was characterised by deep, violent social and political conflicts and confrontation, rather than co-operation and compromise, in industrial relations and politics, which caused the absence of the partnership approach. The constructive dialogue between the government and social bodies has been sporadic since the demise of Franco's dictatorship in 1974. Nevertheless, there was a period when agreements between the government, trade unions and employers were widely used in Spain leading to changes in policy. From 1977 to 1986, a number of income policy "grand pacts" were signed, which was the result of employers' willingness to respond to union strength at a time of immense political uncertainty (Boyd, 2002, p. 37).

Figure 1. Organisational actors with a legal capacity to cooperate in the framework of a public social partnership
Source: own elaboration

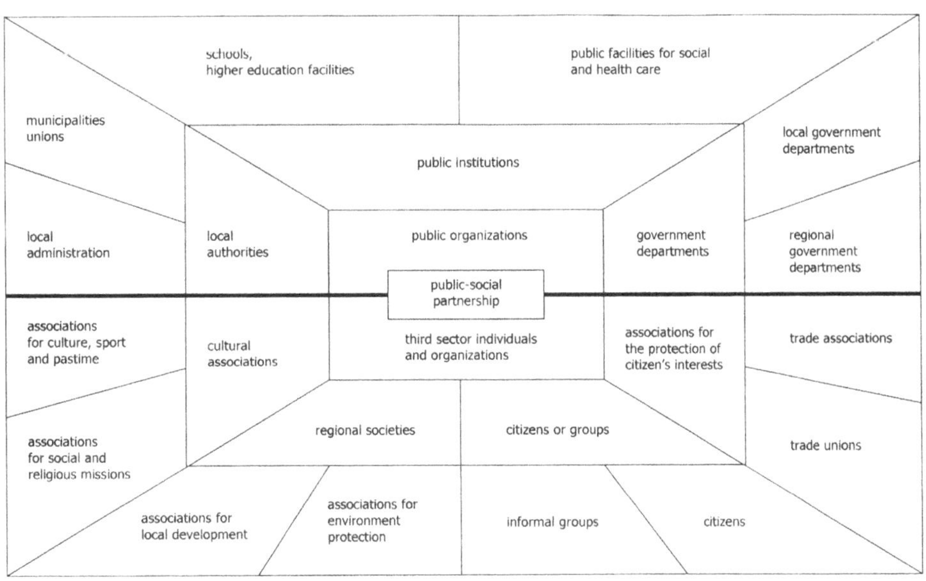

The Spanish Economic and Social Council established in 1991 played a significant role in allowing the economic and social agents to participate in economic and social policy decision-making. The ESC was established to provide a formal channel for the participation of the economic and social partners in the process of introducing laws concerning socio-economic and labour matters, provide a permanent institutional forum for dialogue and debate between social partners and encourage social partners to review and consider economic and social issues (Boyd, 2002, p. 38). The number of partnerships on inclusion as well as on broader employment related issues increased in Spain during the 1980s and especially the 1990s. However, the multiplicity of levels, together with unclear divisions of responsibility led to very little co-ordination. (Boyd, 2002, p. 40). On 20 June 2002, Spanish trade unions called a general strike that was mainly triggered by opposition to the newly announced (May 25, 2002) government reform of the unemployment compensation scheme. One of the main oppositions related to the authoritarian decision making process without consultation of trade unions and social partners (Chabanet & Faniel, 2012, p. 167).

Political conflicts, confrontations, the lack of social dialogue or the participatory approach have caused distrust in the government in Spain. According to the 2013 Eurobarometer, the overall tendency to distrust the government reached 91% of respondents in Spain, which is 26 percent points above OECD-EU countries. The same high level of distrust is also evidenced at the regional/local level, where 78% of respondents do not trust the local authorities. As a result, Spain ranks third among OECD countries with the highest level of distrust of the government, which indicates a significant shortage of involvement of social partners in the policy-making process and the lack of transparency of the government (OECD, 2014, p. 229).

Responding to social challenges, Spain's government became a member of the Open Government Partnership (OGP),[7] a platform aiming to increase standards of Open Government sharing experiences at an international level. The government of Spain considers transparency as a current core principle of public action where only with accessible Public Institutions the citizens can participate in public decisions that concern them with better understanding of the government and higher trust in public decisions (OGP Spain, 2016). The main goal is to transform the bureaucratised, formalist, process-centred model of administration to a new kind of public management with mechanisms that facilitate citizens' direct supervision of public action increasing transparency, accountability and democracy (OGP, Action Plan, 2012, p. 1-2). Within the policies being carried out by the government of Spain, two action plans were published for the periods of 2012-2013 and 2014-2016 (OGP Spain, 2016). The recent commitments relate, inter alia, to public integrity increase through the Preliminary Draft of the Transparency, Access to Public Information and Good Governance Act, which has a three-fold purpose: 'expand and strengthen transparency in public activity, which is set forth as an obligation of the public administration and public agencies at every level; recognise and guarantee access to information, which is regulated as a far-reaching right; and establish the good governance obligations to be met by public officials' (OGP, Action Plan, 2012 p. 4). Additionally, the Spanish government has been recently working on channels and tools to engage the participation of social bodies in policy making and a number of proposals concerning IT projects that have been developed to enable citizen interaction through social networks and web-based technology (OGP, Action Plan, 2012, p. 7).

In recent years, the actions towards public sector innovation have become a strong tendency in Spain and, according to the Observatory of Public Innovation (OPSI), the Spanish government has been promoting innovation activities in the General Framework for the Improvement of Quality of Public Services in Central Administration. A number of initiatives have been implemented, such as the Spanish Inter-administrative Network, created in 2005 as a forum composed of central, regional and local administrations, as well as the National Agency for the Quality Evaluation and Accreditation. It indicates inter-organizational cooperation to share innovative experiences and knowledge aiming at the development of innovative initiatives, public sector management and 'the common framework of continuous improvement and innovation on the road towards excellence' (OECD, 2016a).

A BUSINESS MODEL: THE GENERAL CHARACTERISTICS OF THE CONCEPT

In the reference literature, a business model has not been defined in clear terms yet. The notion of a business model was first used in 1957 to consolidate in the late 1990's. Originally, the notion referred only to the practice of business organisations, but at the beginning of the 21st century it became a management instrument in organisations operating in the sectors that are not profit-oriented. Both in the public and social sectors, this is a new management concept that struggles to win acceptance, because for many years nobody has recognised the necessity of outcome oriented management in such organisations. In the authors' opinion, if an organisation has a proven method of creating, delivering and applying value, it also has a business model. The conviction that a business model is a concept created solely with reference to business organisations is wrong. Every organisation that wants to achieve results, create value and survive in its industry has to clearly formulate and develop a business model as it will not operate effectively without a stable model. It is of no significance whether it operates in the profit-oriented or not-for-profit sector, since all organisations have a business model, although some managers may still be unaware of this. The business model can be treated as a general concept that provides the framework for how the business is run and determines such characteristics of this business as innovativeness or competitiveness. The identification of the components of the business model contributes to the successful definition of value offered to customers, resources, operations, and relationships with partners, which all serve the creation of this value and ensure the success of the organisation. The skills involved in the suitable definition of the business model in not-for-profit organisations are crucial for their success, as they help to concentrate on the elements that contribute to value creation, while a well configured system of connections and causal links between particular components of the business model ensures the accomplishment of the goals designed by the organisation.

While analysing the meaning of the words making up the concept of a business model, the Authors conclude that a model is a simplified image of the reality, while business is an activity involved in value creation (Palo & Tahtinen, 2011, p. 378). The approaches to the business model rely on three basic ways of defining this category.

The first group of definitions relates to the concept of the value chain, the second one to the resource-based view, whereas the third one focuses on the interpretation of the business model as business architecture with the value creation process as its key component. Rudny (2013), quoting Demil and Lecocq, states that, "in general, the category of the business model refers to the description of relationships between elements the co-existence and collaboration of which ensure the creation and delivery of value for the customer and, as a consequence, for the enterprise" (pp. 12-13). Amit and Zott (2007, p. 181) argue that the business model can be treated as a kind of matrix that illustrates how the organisation runs its operations, how it delivers value to stakeholders, and how it links factor and product markets. The design of a business model is primarily affected by the sector since it is the way in which the organisation runs its activity in the environment that it identifies and diagnoses itself. This way embraces a unique combination of resources, actions and products and services, offered by the organisation, that ensures the creation of value for the customer.

The successful design of the business model requires the knowledge of the sectoral context in which the organisation operates, as the model has to be rooted in a specific environment that determines the key assumptions adopted for the design and implementation of the activities planned by the organisation. This comprehensive understanding of the context is related to the definition proposed by Johns (2006, pp. 384-388) and places emphasis on such situational opportunities and constraints that affect the occurrence and meaning of the behaviour inside the organisation and of the organisation itself as well as shape the relationships between variables. The context in a narrower sense, on the other hand, can be defined as the aspect sensitizing an organisation to the way how particular environmental factors can act as temporal or spatial determinants of the business model (Cappelli, Sherer 1991, pp. 55-110). The complexity of the context affects the nature of actions undertaken both in business and not-for-profit sectors. Rudny (2013) and Nogalski (2009) alike argue that it is the character of the environment that gives meaning to business models and determines their usefulness.

The Business Model According to A. Osterwalder and Y. Pigneur

Originally, Doubosson-Torbay, Osterwalder and Pigneur (2002, p. 5-23) defined the business model as a firm's architecture and its network of partners for creating, marketing and delivering value and relationship capital to one or several segments of customers in order to generate profitable and sustainable revenue streams. Later on, they offered a more rigorous version of the definition, describing the business model as the conceptual and architectural implementation of the strategy and as the foundation for business processes.According to the scholars, the business model illustrates the fundamental premises for the organisation to create value and earn profits from this value. In general terms, the business model determines the ways of generating revenue in the organisation. A good business model aims to achieve or maintain competitive advantage. The structure of the business model Canvas proposed by the scholars quoted above is built from the sum of resources and actions that a given organisation manages and implements in order to deliver specific value to the specific customer.

The central part of the model is a value proposition. The value proposition for specific groups of customers is the core around which the revenue and cost parts of the business are built. They are arranged against the background of nine key elements, describing the model of every organisation's operations, irrespective of its market position and growth phase, such as: key partners, key activities, key resources, value proposition, channels, cost structure, revenue streams, customer segments, and customer relationships. While analysing the activities of a number of organisations, those authors argued that every business consists of the same areas presented above and they are most suitably represented as a graphic template which clearly shows all the relationships. The business model, on the other hand, is the sum of all the identified and mutually depended elements. When a business model is constructed, the adequate order of accounting for every element has to be maintained – the first one is always the value proposition, marked with digit 1, which is followed by the identification of the customer (who is the target for the value proposition developed by the organisation). The final element in the construction of the model is the cost structure, marked with digit 9. The first two elements are the key coordinating the composition of the resources and actions that are necessary for the creation and delivery of value for the customer. However, more extended business model frameworks have been also elaborated, like for circular economy for instance (Lewandowski, 2016).

The corporate link of business model has been recently reinvented into social area and public entrepreneurship and its various partnerships in social services delivery. So far, the Lean Model Canvas[8] has been very broadly used among social entities, however, the dynamic development in social economy sector and the turbulent environment has influence the emergence of various types of social bodies, survival strategies, networks or partnerships whose complexity has started to require different and more sophisticated models (Osborne, Radnor, Vidal, Kinder, 2014). The public social organizations have moved beyond traditional and narrow perspective as they are no longer just single organizations but 'systems' that need interactive sustainable models, complex architecture and knowledge that to be successfully transformed into effective public services delivery must by processed through relational mechanisms rather than discrete and transactional ones (Osborne, Radnor, Vidal, Kinder, 2014, pp. 167-170). As the single organizational profitability is no longer a measure of public entrepreneurship sustainability, the Business Model Canvas helps clearly articulate the architecture of the mutually inclusive partners or activities and other symbiotic elements in the inter-organizational network within public services delivery. The social economy development and the cultural change in social organisations has influenced numerous business model frameworks within social sector worldwide. The level of functionality of the social partnerships models and their architecture vary among countries, which encourages to observations and further deeper insight into the models active in Spain and in Poland.

The Identification of the Business Model for a Public Social Partnership Operating in Poland

The partnership that is the subject of this case study was launched in Rybnik, a city located in Upper Silesia (Sielsia Voivodeship, Poland) with the population of 141,400. Despite its dynamic socio-economic growth, it still has the districts where most residents are at the risk of social exclusion, hence the need to undertake initiatives aimed at combating poverty and social exclusion. The observed practice reveals that the highest chances for success lie with activities that are initiated in cooperation with various organisations (public and social) and the local community.

The leader of the partnership is the Social Welfare Centre (OPS), whose employees launched the project "Seize the chance", co-funded by the European Union within the EU Social Fund. The prime goal of the partnership was to raise the level of activity among the residents of the target districts and local institutions so that they could address the problems related to the risk of social exclusion themselves. The aspiration was to develop the model of community work based on cooperation with residents, volunteers and institutions operating in a given district. The conviction was that the initiative should contribute to a change occurring in the local community. In order to formalize the cooperation between the organisations operating for the good of the residents in the district, a public social partnership was established. The leader of the partnership was a public organisation – the Social Welfare Centre – that initiated collaboration with 6 public and 11 social organisations. The partnership was based on the following principles: voluntary participation, the equality of partners, shared goals, the sharing of risks and benefits, the involvement of residents, and supporting local initiatives. The analysis of the operations run by the partnership allowed for the creation of the description of the business model, based on the assumptions developed by Ostewalder and Pigneur (2010) and presented below.

Table 1. The public social partnership business model in Rybnik, Poland

<table>
<tr>
<td rowspan="4">8. Partners
Who are our key partners and suppliers who enable us to create value for our customers?

• Municipal authorities.
• Local community.
• Other actors operating in the public space on a municipal level, interested in combating and preventing social exclusion.</td>
<td>6. Key activities
What activities do we have to undertake to make sure our business model works effectively?

• Active integration through work-related, educational, health0related and social activation.
• Community focused activities – integration oriented initiatives, such as group meetings, picnics, arts, educational, integrational, and cultural workshops, tourist trips.
• Community social work – steps, taken by a social worker (or another person), aimed at families and groups excluded from society to help them return to society; creating conditions that will allow them to perform adequate social roles.
• Purpose-specific and temporary benefits, educational grants.
• The management, organisation and promotion of the partnership.</td>
<td rowspan="2">1. Value proposition
What specific value proposition do we address to our customer segment?

• Delivery of high quality social services, according to the standards stemming from the principles of social work.
• Activating the local community to cooperation and working together towards identifying the needs and developing ways to fulfil them.
• Mobilizing grass-root resources and creating the participatory model for solving social problems.
• Involvement of people suffering from or at the risk of social exclusion in the life of the local community.
• Building social capital as a result of launching Inter-organisational collaboration within the partnership.
• Stronger involvement of authorities in the process of the diagnosis of social needs and the innovative delivery of social services.</td>
<td>3. Customer relationships
How can we acquire, increase and retain the value of our customer in the long term?

• Professionalism.
• Customer satisfaction.
• Status.
• Trust.
• Reliable information.
• Inviting customers and their families to the process of need identification.
• Inviting customers and their families to the process of service delivery.</td>
<td rowspan="2">2. Customer segment
Who is our customer and for whom do we create value?

• People suffering from and/or at the risk of social exclusion.
• City residents living in the area where the partnership operates.</td>
</tr>
<tr>
<td>7. Key resources
What resources do we need to make sure our business model works effectively?

• Skills and qualifications of employees in partner organisations.
• Expertise and information held by partner organisations.</td>
<td>4.Channels
What channels can we use to deliver our product or services to our customers?

• The website of the partnership.
• Social media.
• Local media.
• Field social workers.
• Local social activists and animators.</td>
</tr>
<tr>
<td colspan="2">9. Cost structure
What are the fixed and variable costs of our business model?

• Wages.
• Costs of training and professional development of people employed in the organisations constituting the partnership.
• Operational costs related to the operations of the organisations constituting the partnership.
• Cost of the delivery of social services, e.g. training workshops, integration trips.</td>
<td colspan="2">5. Revenue streams
In return for what, how much and in what way do we receive financial means from our customers?

• Initiatives assigned to the partnership are financed within the system project which gave the basis for the establishment of the partnership.
• Potential additional funds provided by the municipality.
• Potential additional funds provided by sponsors.</td>
</tr>
</table>

Source: Osterwalder, A., & Pigneur, Y. (2010). Business Model Generation. New Jersey: John Wiley & Sons, Inc

The Identification of the Business Model for a Public Social Partnership Operating in the Basque Country, Spain

The analysed partnership was established in Gipuzkoa province in the Basque Country in northern Spain. The Basque Country is divided into three provinces: Araba, Bizkaia and Gipuzkoa. The number of population in Gipuzkoa reaches 709,991 in total, which exceeds Araba (321,777), but it is lower than in Bizkaia (1,141,442). Gipuzkoa is the second biggest province in the Basque Country (EUSTAT, 2015).

Currently, there are many initiatives aiming at social innovation in the Basque Country. The Basque Government, for example, has included the promotion of social innovation in several of its plans – the Science, Technology and Innovation Plan and the Public Innovation Plan. The three Provincial Governments of Bizkaia, Gipuzkoa, and Araba have developed different policy instruments for the generation and selection of social innovation initiatives. Importantly, Basque authorities underline that social innovation is a strategy 'tackling the issues that really matter and doing so in a different way' (Agencia Vasca de la Innovacion, 2016, p. 1). Thus, they are exploring a participatory system for democracy and developing models for citizen participation and involvement in discussions and public decision making to combine different resources and processes to create public value. The focus is on shared social learning with the objective of generating new perspectives, new partnerships and new solutions that result in experimental projects. 'Basque institutions are helping to create a "market" which supports social innovation by incorporating it into their management process and into government structures and instruments (legislation, institutional development, etc.)' (Agencia Vasca de la Innovacion, 2016, pp. 1-2).

In general, social innovation initiatives have strong roots in the Basque history, which provide the solid base for the future participatory approach and partnerships. The model developed through an organisation called GUREAK was initiated four decades ago by CSO Atzegi Association sponsored by the local government. The association is aimed at rights and social integration of people with disabilities in Gipuzkoa (Mair & Noboa, 2003, p. 3). The main mode of the mission accomplishment is through the employment dimension. In the first phase, 1970-1975, Atzegi acted alone, but the mission of the social integration of disabled people through work required extensive resources. Acting in isolation, the CSO Atzegi Association did not have a capacity to manage the process effectively, therefore, in 1975, they chose to create a corporate structure called Gureak, with the non-profit corporation as their legal form, which developed into one of the foremost organisations in the field of social and work integration. It provided 2,008 jobs in 1997, of which only 174 where for people without disabilities (Mair & Noboa, 2003, p. 3; Vidal, 2014, p. 6). Relating to the latest data from 2015 (Gureak, 2016), it currently provides more than 5000 jobs, of which 84% are for people with disabilities including intellectual (36.7%), physical (23.9%), sensory (6.8%) and mental (15.5%). Today, it forms the Gureak group with investors and partners from public, social and private sectors, creating multi-sectorial cooperation to solve employment related issue in an effective manner.

Gureak has established strong cooperation with 2 local public authorities – the Gipuzkoa Government and the San Sebastian City Hall. The partnership is crucial in terms of financial support and the legal agreement to jointly manage public facilities and share activities in the same area. Additionally, Gureak has 3 shareholders in the third sector, namely the Kutxa Foundation, the Atzegi Association, and ILUNION. Atzegi, the pioneer of the Gureak group, is the leading shareholder with a 32.19% stake (Gureak, 2016).

The Gureak group has had a cooperation agreement with the province of the Gipuzkoa Government renewed annually since 1986. The government delegates the right to Gureak to manage a public service towards inclusion in the employment dimension (Vidal, 2014, p. 7). Furthermore, the local government funding accounts for 4.2% of the Gureak group's income, which is mainly devoted to the itinerary service focusing on the activities related to disable people in the area of personalized job match process, further professional training, adaptation, and the monitoring of the work integration success. Additionally, the Basque employment service cooperates with Gureak with respect to their mutual agreement providing the grant that accounts for 8.4% of the Gureak group's income (Gureak, 2016).

To facilitate the mission of work integration, the Gureak group has established strong cooperation with large and medium-sized companies operating in the Basque Country and it performs business activities in the following areas: industry, marketing, and services. The business corporations cooperating with

Gureak comply with the legislation, which requires them to directly or indirectly create employment for people with recognized disadvantages (Vidal, 2014, p. 7).

The Gureak networking structure is unique in its form as it has been initiated by the social body rather than the public one and it has gradually involved public authorities to obtain the legal right to operate in the area of social services delivery and support local authorities in tackling the issue of exclusion. It acquired shareholders from the third sector and business partners from the private sector to increase resources and capabilities to entrepreneurially reach the political goal of work integration and fulfil legal obligations stemming from the public social agreement.

SOLUTIONS AND RECOMMENDATIONS

The capacity to fulfil the undertaken objectives of the analysed partnership is related to the ability to design such a configuration of partners, in which skills, values, and the sense of community are used in the right way, and within which organizations gather and share resources, activities and protocols, procedures and infrastructure that enable effective cooperation between multiple actors.

The two researched models have unveiled that the value proposition and the shortage of resources or knowledge crucial to design or deliver proper service or value are the key and primary determinants that precede the partnership model and its architecture building. The agreed value pursues the selection of partners, whose participation affects the process of service delivery to target customers. Additionally, the need for high quality service requires professionals, so the analysed networking structures involve entities specialized in the area of social inclusion in order to avoid potential limitations of proper service delivery. In both cases, the engagement of public authorities in the networking structure results in the transparency of public institution positively influencing the process of democratic decision-making and collective social services delivery.

Value Proposition

Both partnerships generate value in the employment dimension aiming at social integration. The partnership in Rybnik in Poland aims at the social integration of people excluded from the labour market through interactions with the local community evoking participatory approach and co-production of social services for the effective inclusion of the target group into the market. The partnership in Gipuzkoa, the Basque Country, Spain, is aimed at the social integration of disable people through work. It pursues its objectives by engaging social bodies to share responsibilities and knowledge and working with business partners to facilitate the achievement of its mission, which, in turn, leads to the creation of political and social value through collaboration with partners in the private sector. While in Rybnik initiatives are mainly focused on activation and mobilisation of the local community and the target group in seeking solutions, the Basque partnership focuses on delivering ready-made solutions and the effective implementation of the strategy for stable work opportunities for people with disabilities. The strategy requires collaboration with partners from public and social partners as a foundation and private partners in the area of the idea enforcement through advanced business activities. Both partnerships create value that would not be effectively delivered by organisations acting in isolation.

Table 2. GUREAK – the business model of the public social partnership in Gipuzkoa, the Basque Country, Spain

8. Key Partners Who are our key partners and suppliers who enable us to create value for our customers?	**6. Key Actions** What activities do we have to undertake to make sure our business model works effectively?	**1. Value Proposition** What specific value proposition do we address to our customer segment?	**3. Customer Relationships** How can we acquire, increase and retain the value of our customer in the long term?	**2. Customer Segments** Who is our customer and for whom do we create value?
• Public institution (Gipuzkoa Government, San Sebastian Town Hall) as investors and partners that collaborate through grants • Third sector: (Atzegi Association, Foundation Kutxa, ILUNION) as investors and as collaborating organisations • Private sector: as organisations that demand the Gureak services, and as organisations that comply with the mission of social integration of disabled people through work	• Active collaboration with public institutions to share activities in the area of work integration of disable people • Active collaboration with social bodies in terms of knowledge sharing towards accurate design of social services in the area of work integration of people with disabilities • Active collaboration with business partners to achieve the goal of social integration of disabled people through work • Active collaboration with business partners that are obliged to comply with the mission through employment • Service of professionals to fit the demand and supply of jobs and to adapt jobs to each person's needs with reference to the disability • Professional service of monitoring the process of job adaptation and further integration of disabled people • Service of professional trainings to prepare disabled people for work • The generation of five main advantages of the work integration process of disable people: self-esteem enhancement, relationship building, social inclusion, professional development and organisational participation	• The generation and the management of stable employment opportunities for people with disabilities • Bottom-up approach towards work integration of people with disabilities • Sharing activities in the area of work integration of disabled people with the Gipuzkoa Government • Delivery of social services in the employment dimension to disabled people through advanced business/entrepreneurial activities	• Professionalism • Trust • Personalized and individualized relationship tailored to the specific needs of each person/worker with respect to different disabilities	• People with disabilities: intellectual (36.7 %), physical (23.9%), sensory (6.8 %), mental (15.5 %)
	7. Key Resources What resources do we need to make sure our business model works effectively?		**4. Channel** What channels can we use to deliver our product or services to our customers?	
	• Entrepreneurial knowledge to conduct business activities • Professional knowledge of working with disabled people (CSO Atzegi Association as an initiator that originally aimed at work		• Families of the disable people • Different associations • Public Social Services • Department of Mental Health • Professionals that cooperate individually with disabled people	

continued on next page

Table 2. Continued

	integration of disabled people was/is the key source of knowledge while closely cooperating with the group helping in the accurate design and social service delivery) • Legal agreement with the Gipuzkoa Government to obtain the right to deliver social services • Legal agreement with shareholders to comply with the Gureak mission and to regulate the relationship • LISMI Law (Law 13/1982) Law on Social Integration – it states that all Spanish companies (..) whose workforce exceeds 50 workers, must have the 'booking fee for persons with disabilities'. Namely, the booking of 2% of the positions for people with a disability equal to or greater than 33% (DEMA, 2016)		• Professionals that provide training to disabled people	

9. Costs Structure What are the fixed and variable costs of our business model?	**5. Revenue streams** In return for what, how much and in what way do we receive financial means from our customers?
• 42.8% purchases and external expenses • 12.6% wages for people who are not disabled (professionals and managers managing the Gureak group (operational costs of managing the network inside the group) • 30.3 % wages for disabled people • 2.9% social security contributions • 11.5% other expenses	• 84.3% sales • 4.2 % local government grant • 8.4% grant of the Basque employment service • 3.2% other revenues

Source: Osterwalder, A., & Pigneur, Y. (2010). Business Model Generation. New Jersey: John Wiley & Sons, Inc

Customer Segments

The two partnerships models act in the employment dimension supporting groups excluded from the market or at the high risk of exclusion. The partnership in Rybnik, Poland, is aimed at people excluded from the labour market in the city districts that are affected by poverty. Additionally, it focuses on the local community, which is constantly being mobilized and stimulated to create its own initiatives seeking solutions to its problems. The partnership model in Gipuzkoa, the Basque Country, is aimed at people with intellectual, mental, sensory and physical disabilities in the Gipuzkoa Province – the created networking structure employs 5000 people and 84% of people with disabilities are directly or indirectly employed by the partners within the chain.

Customer Relationships

Professionalism and trust are the key customer relationship traits shared by the two partnerships in Poland and Spain. It might be assumed that trust is the effect of the bottom-up approach involving families, local communities and the affected group into the co-production of the delivered services. The networking structure in Poland strongly highlights the participatory approach of the local community and families during the process of creating the best strategy for the affected group to support them in the adaptation to the current labour market needs. The model in Gipuzkoa is significantly focused directly on the affected group monitoring their skills and abilities or potential limitations to ensure effective participation of disable people in the labour market. Professionalism stems from knowledge sharing among partners in the chain. In Gipuzkoa the network involves, inter alia, the CSO Atzegi Association or the Department of Mental Health providing crucial information about disabled people, while in Rybnik it embraces, inter alia, the local community and local authorities providing comprehensive data about the affected region and group. Interestingly, the partnership in the Gipuzkoa Province focuses on the personalised relationship with respect to the specific type of disability, which requires accurate analysis, research and individualized approach/relationship increasing the effectiveness of the job match process and the productive participation of every single disabled person in an organisation

Channel

The channels of value delivery differ with reference to the specific customer segments. The partnership model in Poland focuses on active presence in social media, offering and promoting their services through social networks and social portals to the local community. In contrast, the partnership model in Gipuzkoa targeting disabled people does not consider social media as an effective mode of communication and value delivery with respect to varied limitations of specific disabilities (more than 30% of people with mental disability; see Table 2). Social media are used as a channel to reach or update families of disabled people about current actions, as the relatives are involved in the process of the individuals' professional development and self-esteem/relationships building. However, both partnerships underline the importance of direct contact, where social or welfare agents contact personally with the local community (Rybnik) or professionals that work with disabled people on an individual basis, matching their skill to the specific type of work and preparing them for active participation in the labour market (Gipuzkoa). Additionally, the partnership model in the Basque country uses the channels of institutions such as the Department of Mental Health, the Public Social Services, or other local associations that have direct contact with disabled people and their families, which facilitates effective communication and value delivery.

Revenue Streams

Both partnerships have varied streams of financial support. The partnership in Rybnik as an initiation based on European programme is primarily supported by the European Union grant. The Gureak group as an organisation active on the market generates a high percentage of its income from regular business activities. However, both partnerships are financially supported by local public authorities: the Rybnik organisation receives the grant from the municipality, while in Gipuzkoa the partnership obtains the government grant from the Gipuzkoa Government or the Basque employment service grant. Importantly, the key public and social partners of the Basque partnership are also investors holding stakes of varying size. In Rybnik, the additional stream of revenue relates to potential sponsors, which does not guarantee regular income in comparison to the Gureak group with official and stable investors from public and social bodies (e.g. Atzegi, the Kutxa Foundation or the local government[9]).

Key Activities

The key activities in both partnerships focus on work integration services of the chosen customer segments (see Table 2). The main activities of the Basque partnership are highly specialized as they focus on one concrete customer segment including people with four types of disabilities. The key activities of the partnership in Gipuzkoa can be divided into legal activities formalizing the partnership and assigning responsibilities to public authorities, social bodies and private actors, and issue-oriented activities aiming at the employment challenges in the chosen customer segments including professional training and development, the job match and adaptation process, or further monitoring. The partnership in Poland offers a wide of range of activities in the area of active integration through numerous events with the local community and the target group in a form of specific group events, formal or informal meetings or workshops, and, additionally, temporary welfare support with respect to the social welfare service regulations.

Key Resources

Both partnerships use knowledge as the key resource for building the 'architecture' necessary for value proposition delivery. Knowledge is the determinant of the partnership's success, the basis of and the driving force behind the partnership model creation as well as the important resource needed to pursue the agreed mission. It is especially evident in the partnership active in Poland, where qualified workers in the chain are crucial to effectively interact with the local community to build networks and develop the participatory approach. Additionally, the knowledge of the involved public institutions and social partners is necessary for the delivery of secondary data that can be further re-used and updated through direct interactions with the community resulting in higher reliability of the data affecting the process of social services delivery in the region. In Gipuzkoa, the primary source of knowledge for the analysed partnership is the Association CSO Atzegi (see section 6.3), which specializes in the social integration of disabled people. The professional approach, the database and the long existence of the association in the centre of the affected group are a significant base for the partnership and Gureak itself, as they allow the networking structure to effectively design the accurate social service towards the successful integration of disabled people in the labour market. Apart from the knowledge about the target group, Gureak's mission of social inclusion through work requires specific resources to fulfil the mission in the real market, demanding entrepreneurial knowledge to achieve and, even more importantly, maintain the value proposition through advanced business activities. Nevertheless, the whole process of the mission active enforcement would not be possible without significant legal resources in the form of the agreement between the Gureak group and the Gipuzkoa Government on sharing activities in social service delivery in the area of employment of disabled people and the legal agreement between Gureak and private enterprises that subscribe to its mission and comply with the Spanish LISMI law imposing the 2% employment of disabled people provided that the enterprise exceeds 50 workers – including alternative measures.[10]

Key Partners

The key partners are the strong determinant of the partnership's proper functioning and the creation of the networking structure. Different social services require specific 'architectures' of service delivery, so the selection of partners is preceded by the process of seeking additional resources by a single organisation aiming at overcoming limitations of value delivery. The two analysed cases present a pressing need for building partnerships allowing for capability enhancement. Both models involving public authorities, social bodies and private actors reveal the significance of network building in the successful achievement of the common mission through, inter alia, closer cooperation, risk and knowledge sharing among chosen partners. Interestingly, the partnership in the Basque Country was initiated by the social organisation (CSO Atzegi Association; see section 6.3) rather than the public one and it has gradually involved public institutions to increase resources and capabilities of the Gureak group to share activities and achieve the common goal. Comparably, in Poland the partnership was started by the public institution (OPS; see section 6.2) and it has been gradually involving social actors to increase the potential of public activities in social services delivery. In the two analysed areas, the created network adopts the participatory approach creating the bottom-up model of solving social issues, where the leading organisations (Gureak – Gipuzkoa, OPS – Rybnik) are at the centre of the target group and use social partners (associations, foundations, public authorities, volunteers) for, inter alia, knowledge enhancement and professional design and delivery of social services. Both network structures are characterized by mutual benefits where the public sector is 'unblocked' by sharing the responsibilities in the area of social integration with other partners and the social and business entities are empowered through financial and legal support by the public authorities.

Costs Structure

The costs structures of the two analysed partnerships cover the operational expenses for managing the network including administration and communication processes. Both models employ professionals, who need to be remunerated, that manage the whole architecture and interact with other partners in the chain. The Gureak group employs approx. 20% of people without disabilities including professionals and approx. 80% of people with disabilities (see section 6.3) and the partnership model covers the compensation for both of the group. Moreover, the leading organisation of the partnerships determines the costs structure. Therefore, the Basque partnership includes expenses for business activities as the Gureak group is a social economy enterprise with a strong presence in the market. The costs structure of the partnership in Poland mainly includes salaries and costs of the professional trainings of the employees involved in the network and costs of the social service design.

Value creation resulting from inter-organisational collaboration established in the public and social sectors is facilitated by democratization and decentralization, dynamic technological advancements, limited resources, and increasing organisational inter-dependencies, which generate the need to undertake initiatives stimulating cooperation between organisations. It is also relevant that information necessary to implement a given project is part of the resources held by many actors (a single organisation does not have knowledge on all the aspects of a specific problem) and organisations are becoming increasingly specialized (they are often specialists in their area, so projects require inter-sectoral cooperation). In this situation, non-governmental organisations are growing in importance, as their role in the democratic society is gradually becoming more significant. Responding to these challenges, the modern state builds formal foundations for non-governmental actors.

FUTURE RESEARCH DIRECTIONS

The dynamic evolution of the principles underpinning the public sector towards embracing first market mechanisms and the networking structure as driving force behind the change in the actors operating in the domain of the public sector, contributes to the new vision of the public management whose innovative approaches require future theoretical and empirical elaboration. The chapter presenting possible departure from the state monopoly in the public sector, and moving towards the search for effective solutions of delivering services to citizens, prepares and introduces future research relating to social innovation, its adaptation and outcome. The gradually developing phenomenon of public social partnerships provides significant base and data for further observations that should encourage researchers to deeper and wider insight to the appearing models architecture. The chapter limits the application of the concept of partnership only to the cases in which the objectives of collaboration go beyond the usual consultation process, without detailed descriptions of, inter alia, implementation process, administrative and management structures within particular partnership. Therefore, it is a perfect area of future analysis towards contextual external and internal determinants of the individual models efficiency and the potential risks in the turbulent environment. In addition, partnerships differ strongly from the perspective of inter alia, subject of cooperation, value proposition and cultural background, which limits the assessment process of universal success determinants of the partnerships and it rises a challenge for future researchers. The analysed partnerships in the chapter pose a question about the level of cultural interference into the public management development and partnerships in varied countries. Moreover, the two analysed models have unveiled differences in the partnership leadership, which pursues future research directions to analyze the influence of the partnership leader on the partnership effectiveness.

CONCLUSION

Generally speaking objectives formulated by a partnership are characterized by the fact that without the cooperation of stakeholders from different sectors these objectives could not be reached, and the source of the cooperation are the people and their behaviour. This means that although the public-social partnership is a relationship between institutions (more or less formalized, or curbed by rules), its actual results are most frequently determined by interpersonal relationships between individuals jointly performing a task for a local community, as it is the individuals who create institutions and determine their character. A partnership is thus a reflection of interpersonal relationships and, similarly to bonds created by people, it undergoes different phases: integration, relation, normalization and effective cooperation. When forming partnerships, these phases play an important role in defining it as a formalized cooperation between public and social actors, which involves integrating processes, including co-management of the process of formulating a strategy and creating organizational structures. The proposed approach limits the application of the concept of partnership only to the cases in which the objectives of collaboration go beyond the usual consultation process, that is to the cases without administrative structures and institutions aimed at implementing a particular purpose. In addition, partnerships differ strongly from other forms of commissioning, which represent a more traditional leadership-representative relationship, with a leader (e.g., department or public body) deciding upon the scope and methods of implementation of the plan on their own, and with the subject of cooperation (e.g., a private organization actively participating in local development) being responsible for the implementation of the decisions issued by the leader. Such a relation does not provide for co-operation in formulating a common policy, since the objectives are defined in a top-down agreement and created independently by a commissioner. Instead, partnerships assume joint objectives and their implementation, agreed and accepted by all parties involved in the decision making process, emerging as a consequence of a designed and deployed architecture of cooperation. The previous considerations clearly emphasize that cooperation between partner organizations emerges in response to identified social problems, and its organizational form might be described as an architectural response to these problems. The necessity to respond to disclosed needs by means of a certain structural form results in the fact that organizational work is undertaken, and the purpose of it is to design such an architecture of cooperation which will support the business model within which it will be possible to achieve the planned objectives.

Value creation resulting from inter-organisational collaboration established in the public and social sectors is facilitated by democratization and decentralization, dynamic technological advancements, limited resources, and increasing organisational inter-dependencies, which generate the need to undertake initiatives stimulating cooperation between organisations. It is also relevant that information necessary to implement a given project is part of the resources held by many actors (a single organisation does not have knowledge on all the aspects of a specific problem) and organisations are becoming increasingly specialized (they are often specialists in their area, so projects require inter-sectoral cooperation). The modern social public partnerships require strategic orientation and sophisticated sustainable business models that are value based and beyond single bottom line as they must face the challenge of sustainability of individual organization and simultaneously the sustainability of social service delivery systems and their governance mechanisms within inter-organizational networks. The new approach moves the public entrepreneurship from traditional models to specialized and effective systems that lead to sustainable innovation in social services delivery (Osborn, Radnor, Vidal, Kinder, 2014). The so far presented considerations head towards an observation that leaders direct and shape their organizations by adapting, accepting and refuting specific aspects of management theory and practice, as well as by creating new ways of managing, better suited to specific objectives. With regard to public-social partnerships, Authors claim that the leader of a partnership becomes an architect who, by rationally choosing elements, creates a properly collaborating whole and business model. The success of a partnership is determined by the knowledge of how to build and sustain a successful inter-organizational collaboration, reliable in the eyes of the partners, stakeholders and the local community. A partnership's leader ability to design such an architecture of cooperation, which will become accepted by potential organizational actors, that is the leaders of partner organizations and their members, affects the capacity to fulfill undertaken objectives and therefore contribute to the ultimate success of a collaboration. Such an approach enhances the status of the concept of *governance* together with its attributes, as the preferred model of management of organizations providing social services in the partnership formula. The ability to design such a configuration of partners, in which skills, values, and the sense of community are used in the right way, and within which organizations gather and share resources and protocols, procedures and infrastructure that enable cooperation between multiple actors, becomes crucial to achieving the intended results.

Summing up the threads running through this paper, Authors state that, in the public sector of many EU countries, the management of public programs aimed at providing social services by means of using partnerships has become more of a norm than an exception. This is a consequence of the appreciation of the view, emerging among both practitioners and researchers, that the integration of services through multi-stakeholder networks - by reducing fragmentation and enhancing coordination of services - leads to a more efficient system and thus to better results of operation: customers gain more individual benefit, and cooperating organizations have a better chance to develop.

REFERENCES

Act of Law on Associations. (1989). *Journal of Laws*.

Act of Law on Public Benefit and Volunteer Work. (2003). *Journal of Laws*.

Agarwal, R., & Selen, W. (2009). Dynamic capability building in service value networks for achieving service innovation. *Decision Sciences*, *40*(3), 431–475. doi:10.1111/j.1540-5915.2009.00236.x

Agencia Vasca de la Innovacion. (2016). *Social Innovation in Basque Country*. Retrieved July 1, 2016, from: https://webgate.ec.europa.eu/socialinnovationeurope/sites/default/files/sites/default/files/social_innovation_in_the_basque_country.pdf

Andersen, S. K., & Mailand, M. (2002). *Multipartite Social Partnerships. A new Role for Employers and Trade Unions*. Copenhagen: The Copenhagen Centre for Partnership Studies.

Bardach, E. (1998). *Getting Agencies to Work Together. The Practice and Theory of Managerial Craftsmanship*. Washington, DC: Brookings Institution Press.

Beinare, D., & McCarthy, M. (2012). Civil society organisations, social innovation and health research in Europe. *European Journal of Public Health*, *22*(6), 889–893. doi:10.1093/eurpub/ckr152 PMID:22117053

Bovaird, T., & Loffler, E. (Eds.). (2009). *Public management and governance*. London: Routledge.

Boyd, S. (2002). *Partnership Working: European Social Partnership Models*. Retrieved June 10, 2016, from: http://www.gov.scot/resource/doc/25954/0028681.pdf

Bryson, J. M. (2004). *Strategic planning for public and nonprofit organizations*. San Francisco: John Wiley & Sons.

Canvanizer. (2016). *Business Model Canvas vs Lean Model*. Retrieved October 10, 2016, from https://canvanizer.com/how-to-use/business-model-canvas-vs-lean-canvas

Cappelli, P., & Sherer, P. D. (1991). The missing role of context in OB. The need for meso-level approach. In L.L. Cummings & B.M. Staw (Eds.), Research in Organizational Behavior. Greenwich, CT: JAI Press.

Caulier-Grice, J., Davies, A., Patrick, R., & Norman, W. (2013). *Defining social innovation. A deliverable project "The theoretical, empirical and policy foundation for building social innovation in Europe". (TEPSIE), European Commission – 7th Framework Programme*. Brussels: European Commission, DG Research.

Chabanet, D., & Faniel, J. (Eds.). (2012). *The Mobilization of the Unemployed in Europe: from Acquiescence to Protest?* Palgrave Macmillan. doi:10.1057/9781137011862

Conteh, Ch. (2013). Strategic inter-organizational cooperation in complex environments. *Public Management Review*, *15*(4), 501–521. doi:10.1080/14719037.2012.674424

Czakon, W. (2005). Więzi rekurencyjne a więzi sieciowe przedsiębiorstwa. In J. Pyka (Ed.), Nowoczesność przemysłu i usług. Współczesne koncepcje i metody zarządzania przedsiębiorstwami. Katowice: TNOiK.

Daglio, M., Gerson, D., & Kitchen, H. (2015). *Building Organisational Capacity for Public Sector Innovation*. Paper presented at OECD Conference "Innovating the Public Sector: from Ideas to Impact", Paris, France.

DEMA. (2016). *Centro Especial de Empleo*. LISMI Law. Retrieved July 2, 2016, from: http://www.demacee.com/en/cumplo-la-ley-lismi/

Doubosson-Torbay, M., Osterwalder, A., & Pigneur, Y. (2002). eBusiness Model Design Classification and Measurement. *Thunderbird International Business Review*, *44*(1), 5–23. doi:10.1002/tie.1036

Duchemin, C., & Weber, T. (2013). Social Partners Involvement in pension reform in the EU. In *Eurofund; EurWork: European Observatory Of Working Life*. Retrieved June 20, 2016, from: http://www.eurofound.europa.eu/observatories/eurwork/comparative-information/social-partners-involvement-in-pension-reform-in-the-eu

European Commission. (2013a). *Powering European Public Sector Innovation, Towards New Architecture, Report of the Expert Group on Public Sector Innovation*. Luxembourg: Publications Office of the European Union. Retrieved June 13, 2016, from https://ec.europa.eu/research/innovation-union/pdf/psi_eg.pdf://

European Commission. (2013b). *Executive Summary. Powering Public Sector Innovation, Towards New Architecture*. Luxembourg: Publications Office of the European Union. Retrieved June 13, 2016, from http://psi_eg.pdf

European Commission. (2013c). *Successful partnerships in delivering Public Employment Services*. Retrieved June 13, 2016, from http://ec.europa.eu/social/BlobServlet?docId=11376&langId=en

European Union. (2012). *Social Dialog. Social Europe Guide, 2*. Luxemburg: Publication Office of the EU. Retrieved June 13, 2016, from: http://ec.europa.eu/social/main.jsp?catId=738&langId=en&pubId=6352&type=2&furtherPubs=no

EUSTAT. (2015). *Population of the Basque Country by province, according to place of birth*. Retrieved June 20, 2016, from: http://en.eustat.eus/elementos/ele0011400/ti_Population_of_the_Basque_Country_by_province_according_to_place_of_birth/tbl0011425_i.html#axzz4EOZjzwzM

Fjeldstad, Q. D., Snow, C. C., Miles, R. E., & Letti, C. (2012). The Architecture of Collaboration. *Strategic Management Journal*, *33*(6), 734–750. doi:10.1002/smj.1968

Frączkiewicz-Wronka, A. (2011). Efektywność sieci jako czynnik krytyczny rozwoju organizacji realizujących usługi społeczne. In J. Skalik (Ed.), Zmiana warunkiem sukcesu. Przełamywanie barier rozwoju i wzrostu przedsiębiorstw. Prace Naukowe Uniwersytetu Ekonomicznego we Wrocławiu.

Frączkiewicz-Wronka, A. (2012). Analiza interesariuszy jako metoda wspierająca formułowanie strategicznych zamierzeń w organizacji. *Metody badań problemów zarządzania strategicznego, Prace Naukowe Wałbrzyskiej Wyższej Szkoły Zarządzania i Przedsiębiorczości*, (17), 269-292.

Frączkiewicz-Wronka, A. (2014). *Zarządzanie usługami społecznymi Studium partnerstw publiczno-społecznych*. Warsaw: Difin.

Frączkiewicz-Wronka, A., Szołtysek, J., & Kotas, M. (2012). Key success factors of social services organizations in the public sector. *Management*, *16*(2), 231–255. doi:10.2478/v10286-012-0068-7

Grewiński, M. (2014). Social Services Mix State jako nowatorski i przyszłościowy model polityki społecznej w Polsce. In E. Trafiałek (Ed.), *Innowacje w polityce społecznej XXI wieku. Przedsiębiorczość, zmiana, rozwój, polityka socjalna* (pp. 90–111). Kielce: Wydawnictwo Uniwersytetu Jana Kochanowskiego w Kielcach.

Gureak. (2016). *Gureak website*. Retrieved July 10, 2016, from: http://www.gureak.com/es/quienessomos

Hood, C. (1991). A New Public Management for all seasons? *Public Administration*, *69*(1), 3–19. doi:10.1111/j.1467-9299.1991.tb00779.x

ILO. (2016). *About ILO*. Retrieved July 2, 2016, from: http://www.ilo.org/global/about-the-ilo/lang--en/index.htm

ITC. (2012). *Social Dialogue. A Manual for Trade Union Education*. International Trade Centre of International Labour Organization. International Training Centre of the ILO, Retrieved June 13, 2016, from http://www.lpsk.lt/lpsk-web/wp-content/uploads/2014/06/Social-dialogue-manual.pdf

Johns, G. (2006). The essential impact of context on organizational behavior. *Academy of Management Review*, (31): 384–388.

Kastberg, G. (2016). Trust and control in network relations: A study of a public sector setting. *Financial Accountability & Management*, *32*(1), 33–56. doi:10.1111/faam.12079

Klijn, E. H., & Teisman, G. R. (2008). Complexity theory and public management. *Public Management Review*, *10*(3), 287–297. doi:10.1080/14719030802002675

Koźmiński, A., & Jemielniak, D. (2008). *Zarządzanie od podstaw. In Podręcznik akademicki*. Warsaw: Wydawnictwa Akademickie i Profesjonalne.

Kożuch, B. (2011). *Skuteczne współdziałanie organizacji publicznych i pozarządowych. In Monografie i Studia Instytutu Spraw Publicznych*. Kraków: Uniwersytet Jagielloński.

Kożuch, B., & Sienkiewicz-Małyjurek, K. (2013). Mierzenie skuteczności współpracy międzyorganizacyjnej w zarządzaniu bezpieczeństwem publicznym. *Studia Ekonomiczne Uniwersytetu Ekonomicznego w Katowicach*, *168*, 82–95.

León, R. L., Simmonds, P., & Roman, L. (2012). *Trends and Challenges in Public Sector Innovation in Europe.* Thematic Report 2012 under Specific Contract for the Integration of INNO Policy Trend Chart with ERAWATCH (2011–2012). Inno Policy Trend Chart. European Commission. Retrieved June 13, 2016, from: http://www.technopolis-group.com/wp-content/uploads/2014/06/1447_INNO-ERATREND-2011-12.pdf

Lewandowski, M. (2013a). Introduction to Academic Entrepreneurship. In A. Szopa, W. Karwowski, & P. Ordóñez de Pablos (Eds.), *Academic Entrepreneurship and Technological Innovation: A Business Management Perspective* (pp. 1–28). Hershey, PA: Information Science Reference. doi:10.4018/978-1-4666-2116-9.ch001

Lewandowski, M. (2013b). *Zarządzanie strategiczne w instytucjach kultury.* Katowice: Wydawnictwo Con Arte.

Lewandowski, M. (2016). Designing the Business Models for Circular Economy – Towards the Conceptual Framework. *Sustainability, 8*(1), 2–28. doi:10.3390/su8010043

Loogma, K., Tafel-Viia, K., & Ümarik, M. (2012). Conceptualising educational changes: A social innovation approach. *Journal of Educational Change*, 1–19.

Luhmann, N. (1994). *Teoria polityczna państwa bezpieczeństwa socjalnego.* Warszawa: Wydawnictwo Naukowe PWN.

Mair, J., & Noboa, E. (2003). *Working Paper. The Emergence of Social Enterprises and Their Place in the New Organizational Landscape.* Navarra: University of Navarra, IESE Business School.

Maruyama, Y., Nishikido, M., & Iida, T. (2007). The rise of community wind power in Japan: Enhanced acceptance through social innovation. *Energy Policy, 35*(5), 2761–2769. doi:10.1016/j.enpol.2006.12.010

Moore, M. (1995). *Creating Public Value: Strategic Management in Government.* Cambridge: Harvard University Press.

Moore, M., & Khagram, S. (2004). *On Creating Public Value: What Business Might Learn from Government About Strategic Management.* Working paper of the Corporate Social Responsibility Initiative, Kennedy School of Government. Cambridge, MA: Harvard University Press.

Moulaert, F., Martinelli, F., Swyngedouw, E., & González, S. (2005). Towards alternative model(s) of local innovation. *Urban Studies (Edinburgh, Scotland), 42*(11), 1969–1990. doi:10.1080/00420980500279893

Mulgan, G., Tucker, S., Ali, R., & Sanders, B. (2007). *Social Innovation: what it is, why it matters, how it can be accelerated.* Retrieved July 10, 2016 from: http://youngfoundation.org/publications/social-innovation-what-it-is-why-it-matters-how-it-can-be-accelerated

Nicholls, A., & Murdock, A. (2012). *Social Innovation: Blurring Boundaries to Reconfigure Markets.* Palgrave Macmillan. doi:10.1057/9780230367098

Nogalski, B. (2009, March/April). Modele biznesu jako narzędzie reorientacji strategicznej przedsiębiorstw. *Master of Business Administration, 2*(97), 3-16.

Obłój, K. (2002). *Tworzywo skutecznych strategii.* Warszawa: PWE.

OECD. (2014). *Spain From Administrative Reform to Continuous Improvement. OECD Public Governance Reviews*. OECD Publishing. doi:10.1787/9789264210592-en

OECD. (2016a). *Country Profile: Spain*. Observatory of Public Social Innovation (OPSI). Retrieved June 15, 2016, from: https://www.oecd.org/governance/observatory-public-sector-innovation/country-profiles/spain/

OECD. (2016b) *About OECD*. Retrieved July 3, 2016, from: http://www.oecd.org/about/

OGP Action Plan. (2012). *Open Government Partnership. Action Plan: Spain*. Retrieved June 13, 2016 from: http://www.opengovpartnership.org/country/spain

OGP. (2016). *About OGP*. Retrieved July 2, 2016, from: http://www.opengovpartnership.org/

OGP Spain. (2016). *Open Government Partnership. Spain*. Retrieved June 13, 2016 from: http://www.opengovpartnership.org/country/spain

Osborne, S. P., Radnor, Z., Vidal, I., & Kinder, T. (2014). A Sustainable Business Model for Public Service Organizations. *Public Management Review*, *16*(2), 165–172. doi:10.1080/14719037.2013.872435

Osterwalder, A., & Pigneur, Y. (2010). *Business Model Generation*. John Wiley & Sons, Inc.

Palo, T., & Tahtinen, J. (2011). A network perspective on business models for emerging technology based services. *Journal of Business and Industrial Marketing*, *26*(5), 377–388. doi:10.1108/08858621111144433

Peters, B. G. (1998). Managing horizontal government: The politics of co-ordination. *Public Administration*, *76*(3), 295–316. doi:10.1111/1467-9299.00102

Philips, S. D., Laforest, R., & Graham, A. (2010). From shopping to social innovation: Getting public financing right in Canada. *Policy and Society*, *29*(3), 189–199. doi:10.1016/j.polsoc.2010.06.001

Pollitt, C., & Bouckaert, G. (2000). *Public management reform*. Oxford, UK: Oxford University Press.

Rorty, R. (1982). *Consequences of pragmatism*. University of Minnesota Press.

Rudny, W. (2013). Organizacja sieciowa jako model biznesu. In W. Rudny & B. Woźniak-Sobczak (Eds.), *Proces kreowania wartości w strukturach sieciowych. Katowice: Wydawnictwo Uniwersytetu Ekonomicznego*.

Rymsza, M. (2007). Polityka państwa wobec sektora obywatelskiego w Polsce w latach 1989-2007. In M. Rymsza, G. Markowski, & M. Dudkiewicz (Eds.), *Państwo a trzeci sektor. Prawo i instytucje w działaniu* (pp. 23–35). Warsaw: Instytut Spraw Publicznych.

Teets, J. C. (2012). Reforming Service Delivery in China: The Emergence of a Social Innovation Model. *Journal of Chinese Political Science*, *17*(1), 15–32. doi:10.1007/s11366-011-9176-9

The Scottish Government. (2011). *A Practical Guide to Forming and Operating Public Social Partnership*. Retrieved June 20, 2016 from: http://www.gov.scot/Resource/Doc/48453/0119024.pdf

Vidal, I. (2014). The Role of Third Sector in Local Welfare. *International Conference under the Patronage of Municipality of Bologna- Municipality of Ravenna*. Retrieved June 15, 2016, from. https://www.researchgate.net/publication/263442935_The_role_of_Third_Sector_in_Local_Welfare

Wronka-Pośpiech, M. (2015). Innowacje społeczne – pojęcie i znaczenie. Studia Ekonomiczne Uniwersytetu Ekonomicznego w Katowicach, 212, 124-136.

Wronka-Pośpiech, M., & Frączkiewicz-Wronka, A. (2016). Strategic orientation and organisational culture in Polish Public Organisations: Insights from the Miles and Snow typology. *Management*, *20*(1), 126–141. doi:10.1515/manment-2015-0029

Wronka-Pośpiech, M., Frączkiewicz-Wronka, A., & Lara Montero, A. (2015). Innowacje – przesłanki wyłaniania się nowego paradygmatu polityki społecznej. In M. Grewiński & A. Karwacki (Eds.), Innowacyjna polityka społeczna. Warszawa: WSP im. J. Korczaka.

Zott, C., & Amit, R. (2007). Business model design and the performance of entrepreneurial firms. *Organization Science*, *18*(2), 181–199. doi:10.1287/orsc.1060.0232

KEY TERMS AND DEFINITIONS

Business Model: The organizational architecture that allows to achieve new value, high quality services delivery, efficiency and flexible adjustments to the requirements of the market and environment.

Business Model for Public Social Partnership: The well-designed configuration of partners within the collaboration framework that responds to the needs emerging in a local community and that the outcomes achieved as a result of its operations contribute to the survival and growth of the organisations making up the partnership though the creation of the desired public/social value.

Collaboration: The network relationship of organisations choosing common goals and jointly deciding on how to use their resources developing an ability to create new structures.

Partnership: The networking structure that clashes interest and coordinate joint activities and resources to unleash the partners' potential to solve complex and multidimensional issues or challenges.

Public Management: The creation of organisational conditions conducive to the effective delivery of social services.

Public Participation: The involvement of those potentially affected by or interested in a decision-making process in public sphere.

Public Social Partnership: The multi-sectoral networking structure, taking advantage of adopting the idea of collaboration and capable of generating more positive outcomes and public/social value than each of the organisations operating on its own.

Social Innovation: The solutions that embrace institutional change and appear as a response to the unsuccessful and obsolete models that no longer apply to the changing environment. Social innovation can be understood as a management model, clearly manifested in the concept of governance.

Social Policy: The activity of a state and other actors pursuing pro-social goals, primarily subsidizing social groups that are marginalized or at the risk of marginalization and initiating actions aimed at improving the quality of life and stimulating growth.

Stakeholders: Everybody who has a vested interest in the value created by the organisation.

Value: The above- average profit with reference to market entities and the 'above-average' social services quality and delivery with reference to non- market entities resulting (in both cases) in satisfying the stakeholders' needs.

ENDNOTES

1 The article was financed within the framework of the research project "Public participation in the practice of public management," University of Economics in Katowice, 2015-2017 The article was financed within the framework of the research project "Public participation in the practice of public management," University of Economics in Katowice, 2015-2017

2 *European Parliament resolution of 14 March 2007 on social services of general interest in the European Union* (2006/2134(INI))

3 OPSI-Observatory of Public Social Innovation. The Observatory has been developed by OECD and it is led by a Task Force of OECD countries, chaired by Canada and France. It provides practical advice on public sector innovation on the basis of collected and analysed examples and experience (OECD, 2016a)

4 ILO- International Labour Organization, the only tripartite U.N. agency, since 1919 that brings together governments, employers and workers representatives of 187 member States (ILO, 2016).

5 Ustawa z 7kwietnia 1989 r. Prawo o stowarzyszeniach, Dz. U. 1989, nr 20, poz.104

6 Ustawa z 24 kwietnia 2003 r. o działalności pożytku publicznego i o woluntariacie, Dz. U. 2003, nr 96, poz. 873.

7 OGP was launched in 2011 to provide an international platform for domestic reformers committed to making their governments more open, accountable, and responsive to citizens. Since then, OGP has grown from 8 countries to the 69 participating countries. In all of these countries, government and civil society are working together to develop and implement ambitious open government reforms. It brings together governments and civil society organisations as true partners at both the national and international level. At the national level, governments work with civil society organisations to develop and implement their OGP national action plan. Countries are encouraged to institutionalize a mechanism for ongoing dialogue and collaboration between government and civil society (OGP, 2016).

8 Lean Model Canvas is a simple problem-solution oriented approach, it does not lay much emphasis on customer segments, key partners or assessment of the given solution to the identified problem (Canvanizer, 2016).

9 See more at Gureak website (Gureak, 2016).

10 Alternative measures: procure goods or services with a disabled worker self; make donations to non-profit entities that aim social integration of people with disabilities; procure goods and services with a special employment centre (DEMA, 2016).

Chapter 6
Applying Business Solutions to Social Problems:
Social Co-Operative and Its Business Model - Evidence from Poland

Martyna Wronka-Pośpiech
University of Economics in Katowice, Poland

ABSTRACT

Although research on social entrepreneurship has been expanding in recent years, the literature regarding business models in the context of social enterprises is still limited. This chapter explores the issues of business models in both profit and nonprofit contexts and discusses how value, being both the foundation and the overall outcome of the business model employed, is created. It begins with the background on the characteristics of social enterprises. This is followed by the conceptualization of business models and their components designed to account for their unique. Then a case to illustrate it is presented. The objective of the chapter is twofold. First, it is to analyse how nonprofit business models are different from its for profit counterparts (at the conceptual level) and to introduce relevant analytical framework. Secondly, the chapter aims at providing a case based on the particular model of a social enterprise in Poland (social co-operative) in order to investigate whatever the business models for profit firms could be applied in the social enterprises.

INTRODUCTION

Starting from the 1990s, a process of reorientation of social policy in the direction of activating support recipients might be observed in Europe. The new approach is referred to as active (or activating) social policy (APS) (Rymsza, 2007, p. 178). Such a shift is occurring both at the level of individual countries and in the structures of the European Union (Van Berkel & Hornemann Møller, pp. 15-44). The basic principles of APS comprise: (1) the preference towards the principle of public participation, i.e. the incorporation of beneficiaries of social programmes into active social life, even at the cost of limiting the level of social security; (2) the activation of the recipients of support through dissemination of vari-

DOI: 10.4018/978-1-5225-2215-7.ch006

ous forms of supported employment, especially social entrepreneurship; (3) concentrating the support on the unemployed who are able to work (or likely to regain such an ability); (4) the change in the type of delimited goods rather than the reduction of the scale of social transfers: the redistribution of work instead of the redistribution of benefits; (5) going beyond the traditional ideological divisions between supporters and opponents of the welfare state; (6) the cooperation with non-governmental organisations, in particular with nonprofit entities and organisations operating in the social economy sector (Szarfenberg, 2007, p. 412; Vis, van Kersbergen & Hylands, 2011, pp. 346-350; Rymsza, 2013, pp. 125-135). In addition, Europe's reorientation toward supporting programmes aimed at activating beneficiaries has led to the rediscovery of social entrepreneurship, and, together with it, of the achievements of the social economy movement of the late 19th and early 20th century and the interwar period. Currently, we are witnessing the second wave of interest in the social economy.

In the 1990s, both in the literature and public discourse, two concepts of the emergence of social economy entities appeared. The literature frequently refers to such a mechanism as social entrepreneurship. The entities of social economy, that is social enterprises, are defined as institutions and organisations that primarily wish to pursue a social mission and achieve social objectives (Shaw & Carter, 2007, p. 423). The concepts of social entrepreneurship in Europe were shaped under the influence of an intensive development of the third sector, its professionalization and the development of the so-called new social economy, focused on integration with labour market. In the U.S., the development of these ideas was subject to the influence of economic universities, such as Harvard Business School, and the theory of social change (Nicholls, 2006, p. 3). In the case of the American concept, entities of social economy are shaped by the market, as innovative solutions aiming at sustainable social change. The European concept focuses on a group, the collective process of creating social economy entities, in which an important role is played by democracy, and it is evident in the choice of leadership, management, and the participatory role of each member. An important issue that makes both approaches distinct is the role of a leader, or an initiator of a social enterprise. The American concept assumes utmost importance of an individual, a charismatic leader, the so-called social entrepreneur, who has solid, measurable characteristics and predispositions that increase the probability of success of such an undertaking in the future (Defourny & Nyssens, 2010, pp. 43-48).

The need for social economy and social entrepreneurship arises also from the fact that traditional political and economic solutions are no longer good enough to guarantee a minimal quality of life to a rising number of people. Therefore, social economy appears as an answer to the increasing disproportions between the rich and the poor and a response to the exclusion of large social groups from mainstream social life. Its goal is to create social enterprises, as well as new social entrepreneurship culture, oriented primarily at inclusion of marginalized groups through active participation in it. The specificity of this sector of the economy stems from the fact that it associates economic activities with social objectives (Perrini &Vurro, 2006; Rymsza, 2007; Thompson, Alvy & Lees, 2000). One of the objectives of social entrepreneurship is job creation, which, as well as being a key component of economic growth, is an essential tool for poverty alleviation. Social entrepreneurship can thus address two important aspects of the UN Millennium Development Goals: economic development and poverty alleviation. It also refers to the objectives of the Europe 2020 strategy, which sets out a vision of Europe's social market economy for the 21st century and puts forward three mutually reinforcing priorities:

- Smart growth: developing an economy based on knowledge and innovation;
- Sustainable growth: promoting a more resource efficient, greener and more competitive economy;

- Inclusive growth: fostering a high-employment economy delivering social and territorial cohesion (European Commission [EC], 2010).

The growing importance and recognition of social enterprise in recent years puts pressure on their success and efficiency in their governance. This in turn encourages taking a closer look at business model for social enterprises, as they not only may explain competitive advantage and performance, but also are the logic of the organisation, the way it operates and how it creates value for its stakeholders (Demil & Lecocq, 2010). Taking the above-mentioned facts into consideration, the purpose of this chapter to identify and explore the components of a business model for social enterprises. This will be done by: (1) analysing how nonprofit business models are different from its for profit counterparts at the conceptual level, (2) introducing relevant analytical framework, and (3) providing a case based on the particular model of a social enterprise in Poland (social co-operative) in order to investigate whatever the business models for profit firms could be applied in the social enterprises.

THEORETICAL PERSPECTIVE ON SOCIAL ENTERPRISE

According to Perrini and Vurro (2006, p. 74), the literature on social entrepreneurship might be divided into two main streams: one positions social entrepreneurship in the nonprofit theory, while the other characterizes it as a new interdisciplinary phenomenon. What connects these two points of view is that they both stress the aspect of tackling important social problems (Austin, Stevenson & Wei-Skillern, 2006, p. 13). And so, on the one edge, there are authors such as Boschee (1998), Boschee and McClurg (2003) or Dart (2004). They include social entrepreneurship within the scope of research on nonprofit organisations, arguing that this concept might be explained as "the latest innovation" in the functioning of social enterprises, and then justified by the interest of nonprofit entities in managerial competence and in the market-based approach, taken in order to improve their operational efficiency and productivity. In this perspective, a social enterprise and social entrepreneurship are treated as "a set of strategic answers to turbulence of environment and to situational challenges that nonprofit organisations face nowadays" (Dacin, Dacin & Tracey, 2011, p. 1206). Therefore, social entrepreneurship is seen as the third sector's response to an altered situation which might seem dangerous at the macro level: a predicted collapse of the welfare system, increasing financial pressures on organisations working for social goals, rising costs in many areas of the social sector, and the reduction of public and private grants and subsidies (Dees & Elias, 1998, pp. 165-178). At the other extreme, one might enumerate the advocates of extended theories of social entrepreneurship, such as Thompson, Alvy and Lees (2000), Hockerts (2005), and Mair and Martí (2009), who believe that the phenomenon may be regarded as a new, independent and largely interdisciplinary field of research. They highlight the quality of social entrepreneurial initiatives, treating it as the basis for the field. In this spirit, they define a social entrepreneur as an innovator, capable of making an active contribution to social change, through creativity and attitude towards innovation – that is, typical features of the classical entrepreneurial process (Perrini & Vurro, 2006, p. 76).

All researchers, however, seem to agree on the fact that the key to understanding this concept lies in the field of business entrepreneurship and in J. Schumpeter's theory, which forms the basis of social entrepreneurship (Mort, Weerawardena & Carnegie, 2003, p. 80). What social and business entrepreneurship have in common is a strong inclination for innovation and change, as well as the ability to perceive unmet needs and the resulting opportunities for entrepreneurial activities (Short, Moss & Lumpkin,

2009, pp. 172-176). In other words, projects in the field of social entrepreneurship are often innovative and – like purely business-like venture – they become agents of change, as well as the driving force of social and economic progress (Leadbeater, 1998, 58).

Considerable discrepancies exist as to the meaning of a social enterprise, which might be said to be the result of historical circumstances of the creation of this type of phenomena in different countries, as well as its economic nature. A social enterprise is a common example of a hybrid organisation (Battilana & Lee, 2014; Haigh & Hoffman, 2014; Michelini & Fiorentino, 2012). Battilana and Lee (2014, p. 400) define 'hybridity' as "the state of being composed through the mixture of disparate parts". According to Haigh and Hoffman (2014), social enterprises combine elements from for-profit and non profit fields as well as mixing market- and mission-oriented beliefs to address social issues. Based on the proposal included in the UK government document, social enterprise can be defined as an institution that runs business activity, sets primarily social goals, reinvests the earned surplus for those purposes in the business or in the community, rather than being driven by the need to maximize profits for shareholders and owners (DTI, 2002). According to Leś (2004, p. 7), it combines the characteristics of a non-governmental organisation and a market entity, creating new jobs and setting additional goals of sustaining these jobs and providing services to local communities. Pearce defines it as "all the economic entities that have a social objective, are not oriented towards the generation and distribution of capital, and have a democratic and measurable structure based on joint governance" (Pearce, 2003, p. 190). Haugh also sees a social enterprise as a collective term defining organisations acting for social objectives. They share the commitment to finding solutions, based on commercial experiences, which will allow them to pursue social goals and reinvest profits from the operations with the benefit for the community (Haugh, 2006, p. 5). Thompson and Doherty (2006, p. 403) define them as organisations that apply business solutions to address social problems, while Deloitte (2014, p. i) proposes a wide definition describing a social enterprise as "a corporate entity that exists primarily to promote public benefits using business strategies, and financial capital, and offering innovative ways of operating for social and/or environmental purposes". According to Yunus, Moingeon and Lehmann-Ortega (2010, p. 311), a social enterprise is a new organisational form, originating from somewhere between a for-profit and a nonprofit organisation, designed and operated just like a 'regular' business enterprise, with products, services, customers, markets, expenses and revenues (p. 311). The main difference, however, is the fact that social enterprises attempt to create social value instead of delivering a dividend to shareholders.

Given the complexity of the phenomenon and the numerous interpretative disputes throughout the world, the EMES network provides nine guiding criteria that an organisation should meet to be closer to 'an ideal' type of social enterprise (Defourny & Nyssens, 2013). These are: continuous provision of goods or services, undertaking economic risk, hiring paid employees, a clear goal oriented towards the benefits of a local community, bottom up initiative, high level of independence from other organisations, nonprofit distribution, participative nature of the enterprise and decision making not based on share in ownership (rule: one man-one vote). Generally, the characteristic organisational forms that social enterprises adopt depend on the existing legal frameworks, on the political economy of welfare provision, and on both cultural and historical traditions of nonprofit development in each country. As a consequence, the social enterprise sector nowadays may comprise both new typologies of organisations, as well as traditional third sector organisations. What is more, various different terms are used interchangeably such as social businesses (Molyneaux, 2004), social-purpose businesses (Cooney, 2011), community enterprises (Tracey, Phillips & Haugh, 2005), and social ventures (Sharir & Lerner, 2006). All that makes social enterprises difficult to define and categorize. However, regardless of the legal form, their principal purpose is not to generate commercial profit, but above all, to create workplaces for people at risk of social exclusion and professional marginalization and engage in delivery of social services and work integration services for disadvantaged groups (such as people with disabilities) and communities (Wronka-Pośpiech & Frączkiewicz-Wronka, 2014). As a key actor undertaking efforts to address many complex issues that the world faces today, it also focuses on providing a benefit to society as a whole rather than just the owners of the enterprise. Such an agenda makes the social enterprise an ideal vehicle for addressing the global concerns of the environment, healthcare, education, culture, economic growth, and poverty alleviation.

SOCIAL ENTERPRISES IN POLAND

In Poland, despite eight years of legislative effort, social enterprise is still an umbrella concept and neither official definition of social enterprise exists, nor a legal construct has been introduced into the system (Ciepielewska-Kowalik, Pieliński, Starnawska & Szymańska, 2014). An attempt to provide a definition of a social enterprise has been recently made in the draft National Programme of Social Economy Development (KPRES, 2014), which defines social enterprise as an entity demonstrating the following characteristics:

- Carrying out business activity with clear organisational boundaries and own financial reporting.
- The objective of business activity is either:
    - Social and employment integration of people at the risk of social exclusion (operationalized as employing a minimum of 50 per cent of staff from among those at risk of social exclusion or at least 30 per cent of disabled); or
    - Providing public benefit services while at the same time fulfilling pro-employment objectives.
- No distribution of profit or financial surplus among shareholders/partners; using profits or financial surplus to increase company capital, and in a certain share for:
    - Social and employment integration, in case of enterprises with pro-employment objectives; or
    - Providing public benefit services for the local community where an enterprise operates.
- Democratic governance or at least consultative role of employees and other stakeholders; and a salary cap for managers.

As highlighted in "A map of social enterprises and their eco-systems in Europe, Country Report: Poland" (EC, 2014, p. 2), The National Programme (KPRES, 2014) assumes the adoption of an Act on Social Enterprise which would introduce a social enterprise status enabling organisations to register as social enterprises on condition that they fulfil certain conditions. Nowadays in Poland a number of organisations with diverse institutional forms have many of the distinctive features of social enterprises and are often referred to as such in the public discourse. The broad spectrum of social enterprises consists of the following types of organisations (EC, 2014, p. 1):

- Non-profit organisations, including foundations, established in order to implement socially or economically beneficial activities and self-governing non-profit voluntary associations that perform revenue-generating activities.

- Co-operatives for the disabled and blind (being a specific subgroup of labour cooperatives) and social cooperatives established to promote the social and professional reintegration of their members into the labour market.
- Vocational enterprises for the handicapped (ZAZ), objective of which is carrying out rehabilitation activities (rehabilitation of disabled, especially with severe disabilities) along with economic activities.
- Vocational therapy workshops (WTZ), which can be established by associations, foundations and other organisations and its primary objective is therapy and rehabilitation of disabled.
- Social integration centres (CIS) and social integration clubs (KIS), which organise educational programmes for people at risk of social exclusion and support individuals and their families in re-establishing and maintaining their ability to participate in the life of the local communities.
- Limited liability companies or corporations that pursue a public benefit mission (so-called non-profit companies).

These legal and organisational forms, as the authors of "The first ICSEM Working Paper" remark, can be grouped in the three models of social enterprise (Ciepielewska-Kowalik, Pieliński, Starnawska & Szymańska, 2014, p. 8-9). Each of them derives from a different pillar of organisations that stay close to a set of EMES classical social and economic criteria for a social enterprise. These models of social enterprise are model 1 (traditional cooperatives and social cooperatives), model 2 (entrepreneurial non-profit organisations such as foundations, associations and other voluntary organisations), and model 3 (work and social integration social enterprises such as social enterprises for the disabled or social enterprises for wider group of persons threatened by social and economic exclusion). It should however be born in mind that, in spite of corresponding to the notion of social enterprise, most of existing in Poland legal and organisational forms do not fulfil all the EMES criteria. For instance many associations and foundations in Poland do not employ paid staff (relying on volunteers) and hardly do they conduct economic activity in a strict sense. Other entities such as ZAZ, CIS, KIS or WTZ do fulfil EMES criteria but they have no legal form and usually exist within third sector organisations, mainly associations and foundations (Ciepielewska-Kowalik, Pieliński, Starnawska & Szymańska, 2014, p. 8). They are also mostly relying on public funding instead of generating sufficient income being a result of collective efforts of their members and workers in the market. Therefore, in the light of EMES classical definition, only polish social co-operatives fulfil the criteria of a social enterprise. A social cooperative is an adaptation of the cooperative legal form and is characterized by the features of a social enterprise recognized at the European level:

- Combination of social and economic functions in their operation;
- Non-profit distribution constraint;
- The principle of democratic, shared management, based on a general rule of cooperative movement, stating that a cooperative is an association of people and not capital;
- High degree of autonomy. (EC, 2014)

In Poland, specific legal framework for social cooperatives exists that is Act on Social Co-operatives dated 27 April 2006. Polish social cooperatives are based on the models of social cooperatives in Italy, referred to as work integration social cooperatives and aiming at social and economic integration of individuals who are exposed to social and economic exclusion (such as long-term unemployed or the ones with low employability, the physically or mentally disabled, and other marginalized groups such as the homeless, alcohol and drug addicts, ex-convicts, refugees) (Starnawska, 2014). Many existing social cooperatives dedicate themselves to inner-city community renewal by providing sustainable employment – including living wages, job training, and benefits – to hard-to-employ individuals. Therefore, they work towards social and work reintegration of their members, via creation of new jobs to help the disadvantaged return to the labor market (Act, 2006). According to polish legislation, the minimum number of founding members is five in case these are natural people, and two in case these are legal persons (NGOs, public authorities, church units) (Starnawska, 2015, p. 83). Social cooperatives can benefit from state subsidies, particularly from the employment fund when created by or for socially excluded people. According to the most recent data, there are about 1312 social cooperatives in Poland. What can be observed is a significant growth in their number as in 2009 there were 187 social cooperatives, in 2010 – 276, in 2011 – 425, and in 2015 – 1266. This significant growth is the result of the support granted to setting up social enterprises with the investment of public money, mainly European funds. These funds, aimed to fund jobs for newly established social enterprises, have proven to be a great incentive for socially or economically excluded individuals.

SOCIAL ENTERPRISES AND VALUE CREATION

Nowadays the concept of value, which is of particular importance to business organisations, is also slowly gaining recognition among researchers of nonprofit organisations, because the ability to generate value, such as educational or cultural value, is the measure of an organisation's social utility, and, ultimately, it translates into support which the organisation obtains from its environment (Koźminski & Jemielniak, 2008, p. 430). Furthermore, this ability might be described as the energy driving a nonprofit organisation to take action. For this type of organisations, the original idea of value had to be modified, because of the need to characterize the activities of these entities by means of not only economic, but also social issues. The economic aspects belong to the classical canon of the basics of microeconomics and macroeconomics, and combining them with the social side of business is difficult, might be considered in various ways, and generates controversy; hence the need for research in that field. As a result, various attempts were made to answer the question of how to assess the results of the operation of nonprofit entities, not only financial, but also in the broader sense, including results hard to quantify, expressed by the value that they have for their clients, e.g. social cohesion or reducing marginalization, as well as how to configure the business model as a consequence of this assessment. It is clear that in the course of the development of society, the preferences regarding value alter; hence the need to keep re-determining what the genuine priority is at the time. With the knowledge of what is defined as value at a given time, organisations can plan their activities so as to obtain the support of stakeholders, and thus valuable resources. For organisations, this results in the need to correctly identify the needs of their stakeholders, and internal antecedents of the business model, in order to formulate it properly. Moore and Khagram (2004, p. 3) indicate that the identification of value is closely related to the task environment, because this is where the evaluation whether the value delivered to society is indeed significant actually happens.

Social enterprises, public organisations or non-governmental organisations often emphasize that they do not operate for profit, which means that the rules that govern typical businesses are not applicable. It is hard to fully agree with this statement, because if an organisation has a well-tried way to create, deliver and use value, it must also have a business model. Social enterprises do not generate financial returns for investors or owners, yet they do need to create value in order to finance their goals by means of donations, grants and revenue from services. In turn, public organisations receive financing from taxes, fees and income from services, but they are also responsible for creating value for citizens. Non-governmental organisations pursue their mission by focusing on generating social good rather than financial return on investment, but they also need a stable business model to achieve economies of scale. In the case of social enterprises it can be assumed that their primary goal is to create social value through the performance of tasks and the fulfilment of a social mission. Since they are different than commercial enterprises in, among others, their desire to support social change, the framework of organisational effectiveness will vary (Herman & Renz, 2008, 344). Performing the tasks of an organisation and completing its mission should result from the creation of fundamental decisions and actions that shape and indicate what an organisation does and why it does it. A social enterprise must be able to fulfil its mission and to have such legitimacy in the community that will enable it to reach a size sufficient for having a significant impact on society. Inevitably, the proper management of relationships with stakeholders plays a crucial role in the legitimacy and the settlement of obligations imposed (Ospina, Diaz & O'Sullivan, 2002, p. 9). Achieving financial stability, although important, is only a means to the ultimate objective, which is to achieve a positive impact on the current situation.

Therefore, in order to recognise how social enterprises build and use their resources to offer their customers and/or beneficiaries better value, it is necessary to focus on employed business models. Understanding how it determines the tactics available to the enterprise to cooperate with or compete against others derives from proper describing of the value offer to a group or groups of customers, including the definition of basic resources, processes (activities), as well as the external relations of the enterprise. Of course, dualism of the objectives of social enterprises must be reflected in the business model. Only by proper designing and employing business models can social enterprises achieve a competitive advantage position relative to their business, make the most of opportunities, be able to solve social problems, and contribute to the welfare of the people.

BUSINESS MODEL AND ITS CONCEPTUALIZATION FOR SOCIAL ENTERPRISE: DETERMINANTS AND COMPONENTS

The concept of 'a business model' – due to globalization, the use of information technology in activities, and the increasing complexity and turbulence of the world – permanently entered the practice and theory of management in the last decade of the 20th century. The concept itself is young and, because of its interdisciplinary nature, frequently challenged. Academics have started to study business models more scrupulously in the last 15 years. However, it is worth mentioning that in the fifties Drucker (1954) already emphasised the importance of a business model that could answer the following question: who is your customer? What does the customer value? How do you deliver value at an appropriate cost?

Several definitions of business models have been proposed in the literature. Firstly, taking into consideration its semantic meaning, model is the simplification of reality, and business is the activity related to value creation (Palo & Tähtinen, 2011, p. 378). According to Timmers (1998, p. 4), a business model stands for "an architecture of the product, service and information flows, including a description of the various business actors and their roles; a description of the potential benefits for the various business actors; a description of the sources of revenues". Chesbrough and Rosenbloom (2002, p. 529) speak of "the heuristic logic that connects technical potential with the realization of economic value". Amit and

Zott (2012, p. 42) define a business model as "a system of interconnected and interdependent activities that determine the way the company does business with its customers, partners, and vendors". Afuah and Tucci (2001, p. 3) propose the business model as unifying construct for explaining competitive advantage and firm performance and define it as "the method by which a firm builds and uses its resources to offer its customer better value and to make money by doing so". Casadesus-Masanell and Zhu (2013, p. 464) see business models as the "search for new logics of the firm, new ways to create and capture value for its stakeholders, and focusing, primarily, on finding new ways to generate revenues and to define value propositions for customers, suppliers, and partners". Markides (2006, p. 20) speaks of the discovery of fundamentally different business models in existing businesses: "to qualify as an innovation, the new business model must enlarge the existing economic pie, either by attracting new customers into the market or by encouraging existing customers to consume more." Teece (2010, p. 172) mentions "the essence of a business model is in defining the manner by which the enterprise delivers value to customers, entices customers to pay for value, and converts those payments to profit." Most scholars agree that the business model emphasizes a system-level, holistic approach to explaining how firms do business, that is how value is created and captured (Zott, Amit & Massa, 2011). It has also been stated that a well-tailored and successfully implemented business model can generate large amounts of money, while the wrong kind can lead to a business failure (Gibson & Jetter, 2014). In summary, the author follows Demil and Lecocq (2010) and define business model as the logic of the firm, the way it operates and how it creates value for its stakeholders.

Many researchers when defining a business model focus on specifying it main components. For example, Timmers (1998) points to:

- An architecture for the product, service, information flows, including a description of various business actors and their roles;
- A description of potential benefits for the various business actors;
- A description of the sources of revenues.

According to Afuah and Tucci (2001) components of the business model include customer value, customer segments, scope of products and services, pricing, revenue sources, connected activities, implementation, capabilities, and sustainability. Hedman and Kalling (2003) propose "causality related components" of a business model being: 1) customers, 2) competitors, 3) offering, 4) activities and organisation, 5) resources, 6) supply of factor and production inputs, and 7) the scope of management. The last component is supposed to cover the dynamics of the business model over time and the cognitive and cultural constraints that managers have to cope with. Newth (2013) suggests diversifying between qualitative and quantitative components of business model. In the first group she lists a value proposition, strategic resources, and dynamic processes, while in the second group – revenue sources, cost structure, and profitability. According to Johnson (2010), a business model consists of four elements: the customer value proposition, the profit formula, the key resources, and processes. One widely used framework for mapping business models is the Business Model Canvas (BMC) presented in 2004 (Osterwalder & Pigneur, 2004). Osterwalder and Pigneur (2010, p. 14) define a business model as "the rationale of how an organisation creates, delivers and captures value". In their conceptualization, a business model focuses on four major areas for consideration: infrastructure, offering, customers, and finances with nine sub-areas, which are in line with the process of making money by a company. These are:

1. Target customers, which include the segments of customers to whom a company wants to offer value;
2. Value proposition, which presents a general view of a company's products and services that create value for a specific customer segment.
3. Distribution channel, which explains the diverse means through which the company communicates with and reaches its customer segments to deliver a value proposition.
4. Customer relationship, which describes the type of links a company forms between itself and its different customer segments;
5. Key Resources, that allow an enterprise to create and offer a value proposition, reach markets, maintain relationships with customer segments, and earn revenues.
6. Key Activities, which explains what actions are required to create and offer a value proposition, reach markets, maintain customer relationships and earn revenues.
7. Key Partnerships, which outlines the network of cooperative agreements with other organisations that are essential to efficiently offer and distribute value;
8. Revenue streams, which portrays the way a company makes money through a variety of revenue flows.
9. Cost structure, which sums up the monetary consequences of the company's means employed in the business model (Osterwalder & Pigneur, 2010).

Each sub-area of the BMC template describes relevant activities. Infrastructure embraces key activities (i.e., value chain), key resources and the partner network. Similarly, the offering comprises the value proposition as proposed by the organisation. Customers include the description of the segments, channels, and type of customer relationship being used. Lastly, finances cover the cost structure and revenue streams. All of the above-mentioned elements compose a business model.

Although knowledge on business modelling in the profit sector is still at its early development stage, one can say it has recently gained ground. At the same time, the social enterprise business model was not given its well-justified attention. The business models of social enterprises are in the first place oriented to resolving social and environmental issues. In this vein, profit generation is not always the predominant concern of social entrepreneurs. Therefore, one of the fundamental questions every manager must ask: "How do we make money in this business?" (Magretta, 2002, p. 87) should be rethought and revisited. A variety of concepts relating to the notion of the business model indicate the need for understanding how value – which is both the foundation and the overall outcome of the business model employed – is created in organisations. In social enterprises, the equivalent of value may be social value, defined as "the degree to which a social enterprise is socially useful" (Głowacki & Jelonek, 2013, p. 11). Social usefulness, in turn, is the degree to which these entities accomplish the goals that are assigned to them. Wilson and Post (2011) acknowledge that understanding the business model of a social enterprise is what makes the achievement of the social mission possible. Although social enterprises create value and earn money at the same time, the goals might be prioritised differently compared to classic for-profit business models (Wronka-Pośpiech & Frączkiewicz-Wronka, 2014). Even though how social entrepreneurs manage to create value without always generating profit in their enterprises raises interest among researchers, the literature regarding business models in the context of social enterprises is still limited. Similarly, the experiences that can be set as good practices and used to create sustainable, positive action patterns for social enterprise are also limited. The matters of social enterprise business models, with their unique nature particularly highlighted, were presented by Czemiel-Grzybowska (2010, pp.

120-121), according to whom this nature results from: (1) the formulated social goals; (2) economic objectives expressed outright (understood as an attempt to obtain a certain level of revenue from sales of products and services, as well as bearing economic risk); (3) democratic management; and (4) social property. One of the few propositions is the one by Yunus, Moingeon and Lehmann-Ortega (2010), who identify four components of a social business model: (1) the value proposition (the stakeholders and the product/service), (2) the social profit equation (the social profit and the environmental profit), (3) the value constellation (the internal value chain and the external value chain), and (4) the economic profit equation (the sales revenues, the cost structure, and the capital employed). However, Wilson and Post (2011) criticize this approach and claim that this idea lacks the direct mechanism for organisations to be able to properly address social issues. Another approach is suggested by Lee (2015, p. 285), according to whom a social enterprise business model is "a decision-making framework for developing logically coherent activities to support social mission- driven value creation and formulating strategies to sustain competitive advantage". In his paper, he describes six logical business model components, which interact in order to generate social and economic value. These are: (1) Value Proposition, (2) Legal and Governance Structure, (3) Market Scope, (4) Networked Activities, (5) Resources, and (6) Sustainability (Lee, 2015, p. 286). The components are divided into two levels: the foundation level (value proposition, legal and governance structure, market scope, and networked activities) and the performance level (resources and sustainability). There are also several variants of the Business Model Canvas for social enterprise, one advantage of which is the fact that, due to its flexibility, it can be modified for specific applications or new types of business forms, such as nonprofit and social enterprises (Gibson & Jetter, 2014). Obviously, the BMC cannot completely capture the specificity of social enterprise business model. Both researchers and practitioners point out to the necessity of adding other components in order to understand how social enterprise creates, delivers and captures value. There are several variants of the Business Model Canvas for social enterprise, such as GoGreen Portland, which proposes adding Problem and Impact and Measurements. Problem component focuses on the specific problem the enterprise wants to solve, while Impact and Measurements addresses the social/environmental impact that social enterprise wants to achieve (GoGreen Portland). Smith (2012) suggests that in order to avoid confusion in having Donors and Beneficiaries in one canvas, a social enterprise should create two different variants of BMC. The first one bases on Osterwalder's Canvas but has specific focus on Donor's perspective, while in the second one Value Proposition is replaced by Impact Metrics. Another modification is the one by Social Innovation Lab (2013), in the proposition of which Value proposition is altered into Social and Customer Value Proposition, Impact Measures is added in Value Proposition component, Customer Relationship is replaced with Type of Intervention, and Surplus component is added to the BMC framework. The latter focuses on how surpluses have to be reinvested into the mission, addressing on of the specific features of social enterprise. Also, Michelini and Fiorentino (2012), in order to identify the main difference between the social and inclusive business models, propose to broaden BMC framework with the ecosystem (such as governance, the value chain, skills, and the network partners), the market (the customers and the distribution) and the economic features (revenue management). Another proposition of components by Brzóska and Jerzok (2014) who points out to (1) the generated social value, (2) human resources (employees), and – if applicable – (3) supporting the activities by social franchise (Brzóska & Jerzok, 2014, pp. 68-69). The latter might be treated as external resources available for use, but also as an offer to other businesses, constituting new social value. In Polish conditions, adopting the franchise model to social enterprises might be seen as an innovative approach. The proposition developed by these authors is presented in the figure below.

Figure 1. Social enterprise business model
Source: Brzóska & Jerzok, 2014

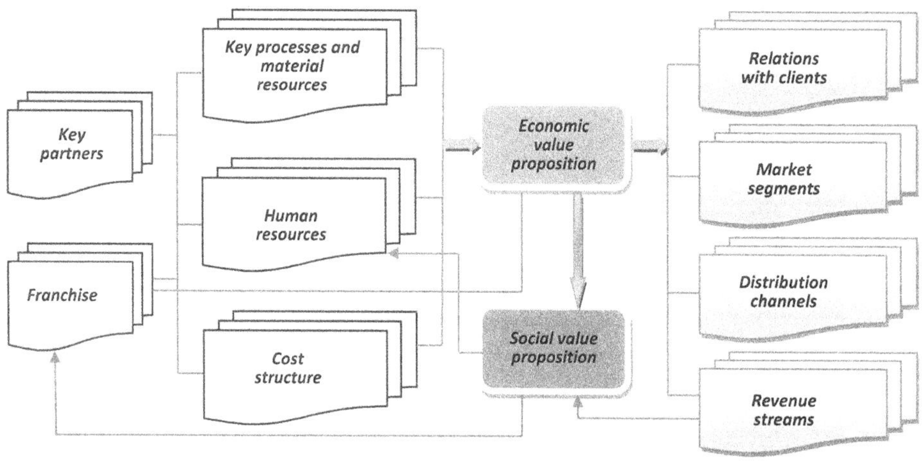

Even though Osterwalder and Pigneur (2010) claim that the Business Model Canvas includes all important elements of a business, one may argue that in the case of social enterprises one important aspect is overlooked, namely a social mission. In the model of Brzóska and Jerzok (2014), this issue is partly addressed by including a social value proposition as one of the components. This is a significant alteration, as many earlier studies have proved that a social mission is incorporated inside the entire value chain of a social enterprise, and that it creates the DNA of the enterprise (Austin, Stevenson & Wei-Skillern, 2006; Doherty, Haugh & Lyon, 2014; Wilson & Post, 2011). The following section illustrates the application of proposition by Brzóska and Jerzok (2014) using the example of a business model of the existing social enterprise.

THE IDENTIFICATION OF THE BUSINESS MODEL FOR A SOCIAL COOPERATIVE OPERATING IN POLAND

As noted previously, in the light of EMES classical definition, only social cooperatives fulfil the criteria of a social enterprise in Poland. Therefore, the author decided to provide a case to illustrate how the business models for profit enterprises could be applied for social enterprises. Due to the fact that specific features of social cooperatives (such as social and economic integration of its members) are taken into consideration it is reflected in the presented business model. As the topic of the business models for social enterprises generally lacks empirical research, the presented data could enrich existing research with a wider perspective.

The social enterprise that is the subject of this case operates as a social cooperative in one of the major Polish cities with the population over 300,000. The cooperative presently employs 7 people with a severe or moderate level of disability, those aged over 50 and/or possessing substantial life experience. They look after the elderly and disabled (but not bedridden due to carers' own limitations) and win contracts for distributing leaflets (by public tender), however the latter activity is treated as marginal. It is important to note, that the author has a good working relationship with the organisation having served there as an external strategy consultant for social cooperatives. Additionally, author's hands-on experience in empowering trainees working in the non profit sector through counselling, consulting, coaching and strategy development has equipped her with a perspective relevant to this study, allowing for successful data collection, on one hand, and facilitating communication with the cooperative, on the other hand. This approach is in line with Van de Ven's (2013) engaged scholarship, which emphasizes the co-production and co-ownership of the research process and challenges visions of academics as the sole knowledge producers and drivers of academic inquiry (Dempsey & Barge, 2013, p. 671). In addition to the data collected in the course of the study the author also used her own experiences and personal notes from the period of time when she had been working with the social cooperative. This is in line with Eisenhardt (1989) and Yin (2003), who value the possibility of the joint use of different data collection techniques. To sum up, in order to gather data for the case two semi-structured interviews with the management were conducted between March and April 2016 (one face-to-face, the other on the phone). In order to complement primary data, secondary data were used as well (the social cooperative website, newspaper articles, and annual reports).

The following section contains description and analysis of the social cooperative business model broken down into components. The author have used proposition of components presented by Brzóska and Jerzok (2014) basing on the Business Model Canvas template.

Customer Segment

Who is the social co-operative customer and for whom value is created?

The main ordering parties of care services are local government of the district (*powiat*) and the commune (*gmina*) as well as individual persons. In the case of the local government, the basic forms of concluding contracts are competitions and tenders, the completion of which provides social cooperative with regular orders (contracts are signed by communal social welfare centres for a year or two). Unfortunately, in this case, the price obtained for an hour of service is much lower than in the open market, as a result of the limitations of communal budgets, as they have to provide welfare services to a increasing number of clients. In the case of individuals, obtaining a direct request for a service from clients requires prior „reaching out" to them but at the same time it guarantees higher payment received for the same type of service. These are the most important clients, particularly because they are in most cases acquaintances (a neighbour's mother, a husband's aunt, etc.). The weight of caring for people in need of others' help permanently or temporarily is placed upon their family, and when the family cannot provide proper care, it is placed on entities prepared for this, including social cooperatives. An increase in demand for care services is also associated with the change of the family model, aging population, prolonged retirement age, and the migration of the younger generation.

Economic Value Proposition

What specific Value proposition does the social cooperative address to its customer segment?

Help and support in the form of care services is organised in the homes of ill, elderly or disabled people. In this way, the clients' basic physiological (biological) and psychosocial needs are fulfilled. By providing daily hygiene and care as well as preparing meals, social cooperative employees assist in meeting everyday needs. The offer of care and assistance services for people with disabilities alleviates the problems of old age, or visual, auditory and kinaesthetic deficits. A vast majority of other entities operating in the market of care services are individuals. Since social cooperatives write out bills for services they perform, expenditures on care for people with disabilities provided at their home may be partially refunded, as granted by the law.

Social Value Proposition

What specific public/social value does the social cooperative address to local community where it operates?

Initiatives taken by social enterprises are assessed positively as long as an organisation pursues values shared by a given community. This ensures the fulfilment of the needs that the community perceives as significantly improving the quality of life. Through the presence at and around clients' home, social cooperative employees respond to their need for security and stability, as well as the vital psychosocial need of social contact and feelings. They also provide contact with the outer world to people suffering from loneliness, who require assistance due to elderly age, illness or other reasons. Moreover, activities performed by social cooperative sometimes positively influence the changes in a family in the case of an illness of one of its members. However, the most important social value created involves providing employment to people with a severe or moderate level of disability, those having life experiences and/or 50+.

Key Resources

What resources the social cooperative needs to make sure their business model works effectively?

The provision of care services requires no additional equipment, other than client's personal objects. Anti-pressure sore mattresses, grab bars, booms, climbing frames, toiletries and household equipment composes a client's equipment. Social cooperative employees do not need special means of transport to provide this kind of services, and if some matters have to be settled at the office or a medical facility, public transportation is used. Office work, accounting, and managing require the equipment present in most households, i.e. a computer, a printer, a phone.

Human Resources (Employees)

What competences and personality traits should the social cooperative's employees possess to make sure their business model works effectively?

The fundamental and indispensable resource is people with great empathy. No one knows more about the needs of the elderly or the disabled than these people themselves, or people in a similar situation. For each client, one caretaker is required. Therefore, there need to be at least 5 caretakers (in Poland the legal status of a social cooperative requires that at least 5 people join it as members at its inception), however currently the cooperative gives employment to 7 people. The ladies who had had other occupations so far or had not had the appropriate qualifications to become a professional caretaker or a disabled person's assistant all volunteered at a local social care institution. Many free trainings are available on caring for persons permanently immobile or assistance to a disabled person. Social welfare centres, projects with ESF, Red Cross and other professional training institutions organise such courses. As it was already mentioned, the social cooperative employs people with a severe or moderate level of disability. Thus, this business model can be used to activate the disabled, and those aged over 50 and possessing substantial life experience.

Key Activities

What activities the social cooperative has to undertake to make sure their business model works effectively?

The basic premise of the cooperative's activity is „self-sufficiency". Accounting is performed within a part-time employment contract (1/2 FTE). The cooperative's manager is employed part-time (1/2 FTE). The establishment of the board consisting of 3 members might be seen as almost direct management, when considering 7 people on the staff. No activity is outsourced. One could say that key activities are connected with management, organisation and promotion of the cooperative. The key process in the operations of the cooperative are the monthly meetings of all members of the cooperative, which might also be seen as the board meetings extended to comprise other staff members. It is the time to speak their minds, discuss their concerns, analyse a monthly financial situation of the cooperative, exchange information on current and future potential clients, etc. It is important that these meetings take place outside the cooperative's premises. Independent public space (facilities made available particularly for non-governmental organisations) and the time of appointments (beyond service hours) make the meetings effective, integrating, and facilitative in solving emerging conflicts but also in fostering mutual trust.

Key Partners

Who are the social cooperative's key partners and suppliers who enable to create value for their customers?

Both partners and stakeholders of the cooperative are set within the local environment, as the co-operative originates from it. This local community provides the cooperative with potential customers. Municipal authorities and local administration are one of the most important stakeholders, as has already been proven in the past – it was the local administrator who has announced a competition for the provision of services related to the distribution of advertising materials for tourist information centers, airports, etc. The cooperative won the competition and we have been providing these services for 5 years. Therefore, good relationships with the local government might make the municipality the cooperative's key ordering party in the future.

Relations with Clients

How does the social cooperative has to build relations with clients to make sure their business model works effectively?

Relationships with major clients are based on trust and emotional bond. Clients once acquired become attached to „their" caretakers, resistant to the activities of the cooperative's competition, and less motivated by price in their choices. Loyalty and empathy are the only price paid by the cooperative's employees.

Channels

What channels can the social cooperative use to deliver their product or services to customers?

Buzz marketing and the reputation that the cooperative has earned for itself work well in care services. They are the most effective and do not require any financial outlay. In the first year of activity, in order to secure a continuity of services, the cooperative submitted a tender in response to a competition announced by the local authorities. This way, they obtained clients from a list provided by Municipal Social Welfare Center. The business approach adopted by the cooperative involved conducting the price calculation that allowed them to provide care services both through the municipal welfare centre and in the open market.

Revenue Streams

In return for what, how much and in what way does the social cooperative receive financial means from customers?

The cooperative's clients, i.e. individuals using care services, give their payments in cash to the caretaker, and she submits the money to the cooperative's teller desk. The rate per hour charged for care services in the open market in the city where the cooperative operates is now at the level of PLN 10-15. Payment due dates depend on when clients receive their disability/retirement pensions. The Municipality Office pays for distribution services via a bank transfer within 14 days from service provision, as agreed in the contract. Public funding, i.e. the State Fund for Rehabilitation of Disabled Persons, is an important revenue source for the cooperative. This is a partial refund of salaries, as the cooperative employs people with a severe or moderate level of disability. The funding results from the regulations of the Law on Social and Vocational Rehabilitation and Employment of Disabled, and will continue until the law remains in force. This is ca. 40% of the cooperative's revenue. If treated as a reimbursement of salaries or increased costs of employment of persons with disabilities, this has an impact on the level of prices for care services the cooperative provides. This supports them in being competitive and enables financial independence.

Cost Structure

What are the fixed and variable costs of the social cooperative's business model?

Due to the service nature of activities, 80% of the costs are salaries and benefits associated with them (social security contributions). It is the work of caretakers and disabled people assistants that generates the highest costs, but these are direct costs (salary for 1 hour is 10-15 PLN). It is the positive side of this business model because the funding is not allocated in stock, frozen in misguided investments, or in products for which there is no demand.

The presented social cooperative is aimed at social integration of people with a severe or moderate level of disability (aged 50+) through work. It pursues its objectives by creating employment and providing care services as well as distributing leaflets. Apart from creating *economic value* for customers such as high quality service, it also creates *social value* (creates employment for hard-to-employ individuals) and positively influences changes in customers' families. The social cooperative's model is targeted both at individual persons and ordering parties of care services such as local government and the commune. Given the fact that nowadays aging, the migration of the younger generation and prolonged retirement age are not fads but reality, such clear and definite *customer segments* seems to be reasonably set out. With respect to *key resources*, apart from tangible resources and equipment present at clients' houses, the cooperative relies heavily on knowledge to deliver value proposition. The primary sources of knowledge for the analysed social cooperative are life experiences of its members, practise gained during volunteering at a local social care institution, and free trainings (such as i.e. assistance to a disabled person) provided by social welfare centres, Red Cross and other professional training institutions. In terms of *human resources* the ultimate and imperative resource is people with great empathy who constitute the social cooperative. To separate the functions of a manager and an accountant, and to ensure the proper functioning of the cooperative, part-time employment (1/2 FTE) is devoted for each function. The additional 1/2 of working time is dedicated to the substantive work as an assistant to a disabled person. Such a model, i.e. combining the work of a manager and a disabled person's assistant, an accountant and a caretaker works very well, and reduces unnecessary administrative costs. A part-time manager and a part-time accountant (1/2 FTE per each) support 6 FTEs of caretakers/assistants. The number of staff devoted to care/assistance only could still be increased. *The key activities* in the social cooperative are management, organisation and promotion. These activities can be divided further into supervision of financial resources, setting work schedules for employees, organization of work in the cooperative, promotion and managing relationships with stakeholders. No activity is outsourced. *The key partners* strongly influence proper functioning of the social cooperative. Therefore, it is necessary to manage relationships with stakeholders such as municipal authorities, local administration, social welfare centres and customers themselves. Trust and competence are the key *customer relationship* features characterising analysed social cooperative. As the local government might be perceived as key ordering party in the future, good working relations are also of crucial importance. *The channels* of value delivery differ depending on the customer segments. Face to face contact or conversations over the phone are used as a channel to reach or update families of clients to whom social cooperative provides care services. In terms of providing care services through the municipal welfare centre more formal channels of communication are requisite, resulting from the fact that as at first public entities are required to issue requests for proposals. Generally, marketing is based on customer recommendations and promotion by befriended organisations, housing cooperatives, parishes, family members and friends, which makes maintaining professional relations with these actors vital. Social cooperative has various *streams of revenues* and financial support. Firstly, individuals using care services or their families pay caretakers in cash. The second stream of revenue comes from Municipality Office who pays for distribution services after the service is provided. In order to balance revenues it is necessary to ascertain that arrangements concluded with social welfare centre are secured, making sure to verify which price per hour of care ultimately wins the tender. An important and relatively stable source of revenue is also the State Fund for Rehabilitation of Disabled Persons. Possibility to use this source is a consequence of the fact that social cooperative employs people with a severe or moderate level of disability. *The costs structure* of the analysed social cooperative covers the operational expenses for managing it, including remunerations. Distribution services for the city generate material costs associated with the use of a private car for business purposes, but the price calculation of the service is very favourable, and the work done by one of employees is very profitable. However, due to rather low profitability of the business model, it is necessary to minimize costs.

SOLUTIONS AND RECOMMENDATIONS

The objective of the case presented above was to illustrate how business models for profit enterprises could be applied for social enterprises, which use both customary and innovative business models to achieve a social and environmental impact and/or meet sustainability goals. Below the author deepens the subject and uses the Business Model Canvas to present the analysis of 16 business models, prepared in collaboration between managers of social enterprises and experts within PI-PWP Replikator project co-financed by the European Union (Brzóska, Durczok, Jerzok, Zarębski & Żbik, 2014). The analysis of the aforementioned business model leads to the conclusion that in the area of (1) *economic and social value proposition* the managers indicated what specific value is targeted at a selected customer segment. This part of the business model focuses on proving that the activities taken by a social enterprise have real significance for the local community and contribute to the achievement of social cohesion. In the area of (2) *customer segments*, a manager specified who the customer of a social enterprise is, and for whom value is created. The emphasis was on properly identifying the groups of the beneficiaries of social enterprises. What appears significant in the created models is holistic thinking, consisting of broadening the field of results, and not only relating the results to those who work in a social enterprise, but also showing how and what activities impact the quality of life of the local community in which the organisation operates. The area concerning (3) *customer relationships* enables to define how to acquire, enhance and retain the value of customers in the long term. The main activities that managers in social enterprises indicated are related to establishing prestige and trust, and contribute to the networking and cooperation in local communities and among local authorities. In the area of (4) *channels*, it was identified which channels can be used to provide customers with products or services offered by social enterprises. In this area, it is crucial to take into account the means of communication that social enterprises employ to communicate with the potential recipients of services and/or products. The proper preparation of content for the area concerning (5) *the revenue streams* allows the estimation of for what, how much and how a social enterprises will receive funds from their customers, as well as support from their stakeholders operating in the task environment. The area of (6) *key activities* allows for the instantiation of measures that should be taken by a social enterprise in order for the planned business model to operate well. It is also necessary to determine (7) *key resources* (tangible, intangible and human resources), which will contribute to this. The fact that social enterprises are unique players in the market means that it is extremely important for these entities to correctly identify (8) *partners*, i.e. all organisations that support and assist in the implementation of planned activities and proceeding towards achieving the objectives. In the area of defining the business model, a manager has to consider who are the key partners and suppliers owing to whom a social enterprise creates value for its customers. The last area in the business model specifies (9) *the cost structure*, and mainly shows the breakdown into fixed and variable costs.

The study revealed that social enterprises operating as a social cooperatives could use the concepts and elements of business models characteristic to for-profit enterprises. However, both presented case and the abovementioned analysis of 16 business models reveals the need to expand the Business Model Canvas framework with additional components, as it overlooks some aspects in the context of social enterprises. Therefore, it needs to be modified to be more adjustable to social enterprises reality such as creating social value (primarily aiming at addressing and satisfying unmet social needs). In this vein, the author suggests that the social mission – being its reason for existence – should be reflected in the business model of a social enterprise. Taking into consideration the distinctive features of social enterprises, the social mission should be the integrating driver of the business model. Furthermore, BMC framework should also be extended with component addressing the benefits for social enterprise customers, being the social/environmental impact that social enterprise wants to achieve.

FUTURE RESEARCH DIRECTIONS

Although the work accomplished by social entrepreneurs is not new, the current momentum behind the movement provides significant base and data for further observations. As theoretical consensus on the business model concept either in profit or in non profit contexts has not been reached yet, future research is needed. It may for example seek to empirically explore differences in the nature of business models and their components depending on the legal and organisational form of social enterprise (for instance how the components of a business model differs between work integration social enterprises and non-profit organisations, as well as the implications these differences have for managers). Given the fact that social enterprises are culture-specific, socially embedded, and community-tied, future research in other countries should also be undertaken. Conducting more empirical research such as case studies could help understand the research subject in other situations or contexts. Such studies can also offer important insights for social entrepreneurs, who consider social franchising and replicating business models of successful social enterprises. Another interesting research avenue is examining whether social franchise and replicating business models help social enterprises in achieving market success and reduces risk of market failure. Due to the special role that social enterprises play in economies, the adoption of the engaged scholarship perspective would be beneficiary as it is committed to utilizing the resources of academic inquiry for practical concerns and useful ends. As Boyer (1996, p. 21) argues, the scholarship of engagement is a mode of inquiry "connecting the rich resources of the university to our most pressing social, civil, civic and ethical problems".

CONCLUSION

The development of social entrepreneurship is exceptionally important from the point of view of every national economy because social economy subjects, including social enterprises, constitute a reservoir of resources inaccessible to other forms of activity. These resources can be used to the advantage of local communities, as they play an important role in the implementation and provision of public and social services, and they can also be used in the process of building a democratic and civil society (Hausner 2008, Wygnański 2007). It is worthwhile to invoke here Ducker's words and vision formulated far back in the 1980's, when he pointed out that in the 21st century the condition of the organisations operating in the public and social sector would determine the condition of the economy as a whole. Therefore, nowadays it is crucial to search for the rules and routines of management, which transplanted from the private sector to the public and social sectors (especially to social enterprises), will stimulate the increase in the efficiency of the latter organisation.

Social enterprises are a unique form of entrepreneurship and business, because they meet both social and economic objectives, and in such a way that the economic results are largely channelled to help the organisation's employees and achieve a social mission. In the case of self-financing of social enterprises, the fulfilment of these goals is possible only if they are competitive. In some countries, for example in Poland, only scarce external funds are available to support social entrepreneurship development (apart from public funds and some additional support given to social enterprises). This leads to the situation in which social enterprises are not only grant-dependent, but also self-insufficient. To support their social value, social enterprises should pursue their activities in a thoroughly entrepreneurial way (Chell, 2007). It can be achieved through the development and application of appropriate business models, which is especially important if an organisation operates in an open market.

Previous studies investigated business models from a rather static point of view portraying them as a set of coherent interrelated core business model components. However, Demil and Lecocq (2010) propose a more dynamic transformational approach, which portrays a business model as a tool to manage change and innovation in the organisation. This is in line with Teece (2010), who points out that organisations should actively seek and consider improvements to their business models constantly, which means it should be an ongoing process enabling the adaptation to the turbulent and ever-changing environment. The ability to do that may actually be crucial to creating competitive advantage in the future. The author is convinced that applying the business model perspective would also have a positive influence on stronger legitimacy of social enterprises in the society and economy. As Margretta (2002, p. 87) points out, a business model is the story of the business you tell if you are asked what your company is doing. The skillful and deliberate use of business models could help social entrepreneurs tell this story in a more engaging and convincing manner.

The business model itself could also be considered as a framework of educational nature, as it shows business activities in a clear way and facilitate effective decision making with regard to operational and strategic activities in which social entrepreneurs engage. As Lee emphasises "through the development of a business model, social enterprises will be able to understand their business better, find ways to enhance their competitive advantage, and lay a ground work for implementation-oriented business planning" (2015, p. 295). Accordingly, the participation of social enterprises' employees in the development of such models could enable them to better understand the current or planned activities, but also unleash their creativity and innovation, as reflected in the value proposition.

REFERENCES

Afuah, A., & Tucci, C. L. (2001). *Internet Business Models and Strategies: Text and Cases*. New York, NY: Irwin/McGraw-Hill.

Amit, R., & Zott, Ch. (2012). Creating Value Through Business Model Innovation. *MIT Sloan Management Review*, *53*(3), 40–50.

Austin, J., Stevenson, H., & Wei-Skillern, J. (2006). Social Entrepreneurship and Commercial Entrepreneurship: Same, Different, or Both? *Entrepreneurship Theory into Practice*, *30*(1), 1–22.

Austin, J., Stevenson, H., & Wei-Skillern, J. (2006). Social and commercial entrepreneurship: Same, different or both? *Entrepreneurship Theory and Practice*, *1*(1), 1–22. doi:10.1111/j.1540-6520.2006.00107.x

Battilana, J., & Lee, M. (2014). Advancing Research on Hybrid Organizing: Insights form the Study of Social Enterprises. *The Academy of Management Annals*, *8*(1), 397–441. doi:10.1080/19416520.2014.893615

Boschee, J. (1998). *What does it take to be a Social Entrepreneur?* National Centre for Social Entrepreneurs. Retrieved May 10, 2016, from www.socialentrepreneurs.org/whatdoes.html

Bosehee, J., & Mc Clurg, J. (2003). *Toward a better understanding of social entrepreneurship: Some important distinctions*. Retrieved May 10, 2016, from http://www.se-alliance.org/better_understanding.pdf

Boyer, E. (1996). The scholarship of engagement. *Journal of Public Service & Outreach*, *1*(1), 11–20.

Brzóska, J., Durczok, M., Jerzok, I., Zarębski, A., & Żbik, W. (Eds.). (2015). *Formaty biznesowe*. Katowice: Stowarzyszenie Współpracy Regionalnej.

Brzóska, J., & Jerzok, I. (2014). *Koncepcje i aplikacje modeli biznesu w organizacjach ekonomii społecznej*. Gliwice: Wydawnictwo Politechniki Śląskiej w Gliwicach.

Casadesus-Masanell, R., & Zhu, F. (2013). Business Model Innovation and Competitive Imitation: The Case of Sponsor-Based Business Models. *Strategic Management Journal*, *34*(4), 464–482. doi:10.1002/smj.2022

Chell, E. (2007). Social enterprise and entrepreneurship: Towards a convergent theory of the entrepreneurial process. *International Small Business Journal*, *25*(1), 5–26. doi:10.1177/0266242607071779

Chesbrough, H., & Rosenbloom, R. S. (2002). The role of the business model in capturing value from innovation: Evidence from Xerox Corporations technology spin-off companies. *Industrial and Corporate Change*, *11*(3), 529–555. doi:10.1093/icc/11.3.529

Ciepielewska-Kowalik, A., Pieliński, B., Starnawska, M., & Szymańska, A. (2015). *Looking for Social Enterprise Models in Poland: Institutional and Historical Context*. ICSEM Working Papers, No. 2015-11. Retrieved May 19, 2016, from http://www.iap-socent.be/sites/default/files/Poland%20-%20Ciepielewska-Kowalik%20et%20al.pdf

Cooney, K. (2011). An exploratory study of social purpose business models in the United States. *Nonprofit and Voluntary Sector Quarterly*, *40*(1), 185–196. doi:10.1177/0899764009351591

Czemiel-Grzybowska, W. (2010). *Zarządzanie przedsiębiorstwem społecznym – raport z badań*. Białystok: Regionalny Instytut Społeczno-Ekonomiczny.

Dacin, M. T., Dacin, P. S., & Tracey, P. (2011). Social entrepreneurship: A critique and future directions. *Organization Science*, *22*(5), 1203–1213. doi:10.1287/orsc.1100.0620

Dart, R. (2004). Legitimacy of Social Enterprise. *Nonprofit Management & Leadership*, *14*(4), 411–424. doi:10.1002/nml.43

Dees, J. G., Elias, J., & Bowie, N. E. (1998). The challenges of combining social and commercial enterprise. *Business Ethics Quarterly*, *8*(1), 165–178. doi:10.2307/3857527

Defourny, J., & Nyssens, M. (2010). Conceptions of Social Enterprise and Social Entrepreneurship in Europe and the United States: Convergences and Divergences. *Journal of Social Entrepreneurship*, *1*(1), 32–53. doi:10.1080/19420670903442053

Defourny, J., & Nyssens, M. (2013). Social Coops: When Social Enterprises meet the Cooperative tradition. *Journal of Entrepreneurial and Organizational Diversity*, *2*(2), 11–33.

Deloitte. (2014). *Dual Purpose Corporate Structure Legislation: Stakeholder Engagement Report*. Ontario: Author.

Demil, B., & Lecocq, X. (2010). Business model evolution: In search of dynamic consistency. *Long Range Planning*, *43*(2/3), 227–246. doi:10.1016/j.lrp.2010.02.004

Dempsey, S. E., & Barge, J. K. (2013). Engaged Scholarship and Democracy. In D. K. Mumby & L. L. Putnam (Eds.), *Handbook of Organizational Communication: Advances in Theory, Research, and Methods* (pp. 665–688). Thousand Oaks, CA: Sage.

Doherty, B., Haugh, H., & Lyon, F. (2014). Social Enterprises as Hybrid Organizations: A Review and Research Agenda. *International Journal of Management Reviews*, *16*(4), 417–436. doi:10.1111/ijmr.12028

Drucker, P. (1954). *The Practice of Management*. Harper and Row Publishers.

DTI. (2002). *Social Enterprise: A Strategy for Success*. Retrieved May 10, 2016, from www.seeewiki.co.uk/~wiki/images/5/5a/SE_Strategy_for_success.pdf

Eisenhardt, K. M. (1989). Building Theories from Case Study Research. *Academy of Management Review*, *14*(4), 532–550.

European Commission. (2010). *A European strategy for smart, sustainable and inclusive growth*. Brussels: Author.

European Commission. (2014). *A map of social enterprises and their eco-systems in Europe. Country Reports: Poland*. Brussels: European Commission Directorate General for Employment, Social Affairs and Inclusion. Retrieved May 30, 2016, from http://ec.europa.eu/social/main.jsp?langId=en&catId=89&newsId=2149

Gibson, E., & Jetter, A. (2014). *Towards a Dynamic Process for Business Model Innovation: A Review of the State-of-the-Art. Proceeedings of PICMET '14: Infrastructure and Service Integration* (pp. 1230–1238). IEEE.

Głowacki, J., & Jelonek, M. (2013). *Badanie społecznej wartości dodanej podmiotów ekonomii społecznej w Polsce za pomocą narzędzia ESometr*. Kraków: MSAP.

GoGreen Portland. (n.d.). *The Social Enterprise Business Model Canvas*. Retrieved May 27, 2016, from http://portland.gogreenconference.net/downloads/GGPDX_Guided_Learning_Entrepreneurship(FRONT).pdf

Haigh, N., & Hoffman, A. J. (2014). The New Heretics: Hybrid Organizations and the Challenges They Present to Corporate Sustainability. *Organization & Environment*, *16*(4), 417–436.

Haugh, H. (2006). Social enterprise: Beyond economic outcomes and individual returns. In J. Mair, J. Robinson, & K. Hockerts (Eds.), *Social entrepreneurship* (pp. 180–205). Basingstoke, UK: Palgrave Macmillan. doi:10.1057/9780230625655_12

Hausner, J. (2007). Ekonomia społeczna jako kategoria rozwoju. In J. Hausner (Ed.), *Ekonomia Społeczna a rozwój* (pp. 11–28). Kraków: Małopolska Szkoła Administracji Publicznej.

Hausner, J., Laurisz, N., & Mazur, S. (2007). Przedsiębiorstwo społeczne – konceptualizacja. In J. Hausner (Ed.), *Zarządzanie podmiotami ekonomii społecznej* (pp. 9–22). Kraków: Małopolska Szkoła Administracji Publicznej.

Hedman, J., & Kalling, T. (2003). The business model concept: Theoretical underpinnings and empirical illustrations. *European Journal of Information Systems*, *12*(1), 49–59. doi:10.1057/palgrave.ejis.3000446

Herman, R. D., & Renz, D. O. (2008). Advancing nonprofit organizational effectiveness research and theory: Nine theses. *Nonprofit Management & Leadership*, *18*(4), 399–415. doi:10.1002/nml.195

Hockerts, K. (2005). Bootstrapping Social Change, Towards an Evolutionary Theory of Social Entrepreneurship.*1st International Social Entrepreneurship Research Conference (ISERC)*. Barcelona: IESE.

Johnson, M. W. (2010). *Seizing the white space - Business model innovation for growth and renewal*. Harvard Business Press.

Koźmiński, A. K., & Jemielniak, D. (2008). *Zarządzanie od podstaw. Podręcznik akademicki*. Warszawa: Wydawnictwa Akademickie i Profesjonalne.

KPRES. (2014). *The National Programme for the Development of Social Economy, Ministry of Labour and Social Policy, Warsaw*. Retrieved May 30, 2016, from http://www.ekonomiaspoleczna.pl/files/wiadomosci.ngo.pl/public/kor espondenci/portal_ekonomiaspoleczna/KPRES_23.06.2014.pdf

Leadbeater, C. (1998). *The rise of Social Entrepreneur*. London: Demos.

Lee, I. (2015). A Social Enterprise Business Model for Social Entrepreneurs: Theoretical Foundations and Model Development. *Social Entrepreneurship and Innovation*, *3*(4), 269–301. doi:10.1504/IJSEI.2015.069351

Leś, E. (2007). Rola trzeciego sektora w polityce społecznej. In G. Firlit-Fesnak & M. Szylko-Skoczny (Eds.), *Polityka społeczna* (pp. 361–372). Warszawa: PWN.

Magretta, J. (2002). Why Business Models Matter. *Harvard Business Review*, *80*(5), 86–92. PMID:12024761

Mair, J., & Martí, I. (2009). Social entrepreneurship in and around institutional voids. *Journal of Business Venturing*, *24*(5), 419–435. doi:10.1016/j.jbusvent.2008.04.006

Markides, C. (2006). Disruptive innovation: In need of better theory. *Journal of Product Innovation Management*, *23*(1), 19–25. doi:10.1111/j.1540-5885.2005.00177.x

Michelini, L., & Fiorentino, D. (2012). New business models for creating shared value. *Social Responsibility Journal*, *8*(4), 561–577. doi:10.1108/17471111211272129

Michelini, L., & Fiorentino, D. (2012). New Business Models for Creating Shared Value. *Social Responsibility Journal*, *8*(4), 561–577. doi:10.1108/17471111211272129

Molyneaux, D. (2004). Accountability and volunteers at social businesses: A role for ethical checklists. *Business Ethics (Oxford, England)*, *13*(1), 14–25. doi:10.1111/j.1467-8608.2004.00345.x

Moore, M., & Khagram, S. (2004). *On Creating Public Value: What Business Might Learn from Government About Strategic Management*. Working paper of the Corporate Social Responsibility Initiative. Kennedy School of Government, Harvard University, Cambridge.

Mort, G. S., Weerawardena, J., & Carnegie, K. (2003). Social entrepreneurship: Towards Conceptualisation. *International Journal of Nonprofit and Voluntary Sector Marketing*, *8*(1), 76–88. doi:10.1002/nvsm.202

Newth, F. (2013). *Business Models and Strategic Management: A New Integration*. Business Expert Press.

Nicholls, A. (2006). Playing the Field: A New Approach to the Meaning of Social Entrepreneurship. *Social Enterprise Journal*, *2*(1), 1–5.

Ospina, S., Diaz, W., & OSullivan, J. F. (2002). Negotiating accountability: Managerial lessons from identity-based nonprofit organizations. *Nonprofit and Voluntary Sector Quarterly*, *31*(1), 5–31. doi:10.1177/0899764002311001

Osterwalder, A., & Pigneur, Y. (2004). An ontology for e-Business Models. In W. Currie (Ed.), *Value Creation from E-Business Models* (pp. 65–97). Oxford, UK: Elsevier Butterworth- Heinemann. doi:10.1016/B978-075066140-9/50006-0

Osterwalder, A., & Pigneur, Y. (2010). *Business Model Generation*. John Wiley & Sons, Inc.

Palo, T., & Tähtinen, J. (2011). A network perspective on business models for emerging technology based services. *Journal of Business and Industrial Marketing*, *26*(5), 377–388. doi:10.1108/08858621111144433

Pearce, J. (2003). *Social Enterprise in Anytown*. London: Calouste Gulbenkian Foundation.

Perrini, F., & Vurro, C. (2006). Social entrepreneurship: Innovation and social change across theory and practice. In J. Mair, J. Robinson, & K. Hockerts (Eds.), *Social entrepreneurship* (pp. 57–85). Basingstoke, UK: Palgrave Macmillan. doi:10.1057/9780230625655_5

Rymsza, M. (2003). Aktywna polityka społeczna w teorii i praktyce. In T. Kaźmierczak & M. Rymsza (Eds.), *W stronę aktywnej polityki społecznej* (pp. 19–32). Warszawa: Instytut Spraw Publicznych.

Rymsza, M. (2007). Druga fala ekonomii społecznej w Polsce a koncepcja aktywnej polityki społecznej. In T. Kaźmierczak & M. Rymsza (Eds.), *Kapitał społeczny. Ekonomia społeczna* (pp. 175–192). Warszawa: FISP.

Rymsza, M. (2013). *Aktywizacja w polityce społecznej W stronę rekonstrukcji europejskich Welfare States?* Warszawa: Wydawnictwo Instytutu Filozofii i Socjologii PAN.

Sharir, M., & Lerner, M. (2006). Gauging the success of social ventures initiated by individual social entrepreneurs. *Journal of World Business*, *41*(1), 6–20. doi:10.1016/j.jwb.2005.09.004

Shaw, E., & Carter, S. (2007). Social entrepreneurship: Theoretical antecedents and empirical analysis of entrepreneurial processes and outcomes. *Journal of Small Business and Enterprise Development*, *14*(3), 418–434. doi:10.1108/14626000710773529

Short, C., Moss, T. W., & Lumpkin, G. T. (2009). Research in social entrepreneurship: Past contributions and future opportunities. *Strategic Entrepreneurship Journal*, *3*(2), 161–194. doi:10.1002/sej.69

Smith, G. (2012). *The Nonprofit Business Model Canvas | Innovative Nonprofit*. Retrieved May 27, 2016, from http://www.innovativenonprofit.com

Social Innovation Lab. (2013). *Social Business Model Canvas*. Retrieved May 27, 2016, from http://www.socialbusinessmodelcanvas.com

Starnawska, M. (2014). Behavior through networking in social entrepreneurship in response to challenging institutional framework – case of 5 social cooperatives. *Problemy Zarzadzania*, *12*(4), 97–116. doi:10.7172/1644-9584.49.6

Starnawska, M. (2014). Zachowanie poprzez sieciowanie w przedsiębiorczości społecznej w odpowiedzi na trudne otoczenie instytucjonalne – przypadek pięciu spółdzielni socjalnych. *Problemy Zarzadzania*, *12*(4), 97–116.

Szarfenberg, R. (2007). Rodzaje i formy aktywnej polityki społecznej. In G. Firlit-Fesnak & M. Szylko-Skoczny (Eds.), *Polityka społeczna* (pp. 401–416). Warszawa: PWN.

Teece, D. J. (2010). Business models, business strategy and innovation. *Long Range Planning*, *43*(2/3), 172–194. doi:10.1016/j.lrp.2009.07.003

Thompson, J., Alvy, G., & Lees, A. (2000). Social Entrepreneurship – A New Look at The People and The Potential. *Management Decision*, *5*(5), 328–338. doi:10.1108/00251740010340517

Thompson, J., & Doherty, B. (2006). The diverse world of social enterprise: A collection of social enterprise stories. *International Journal of Social Economics*, *33*(5/6), 399–410. doi:10.1108/03068290610660643

Timmers, P. (1998). Business Models for Electronic Markets. *Electronic Markets*, *8*(2), 3–8. doi:10.1080/10196789800000016

Tracey, P., Phillips, N., & Haugh, H. (2005). Beyond philanthropy: Community enterprise as a basis for corporate citizenship. *Journal of Business Ethics*, *58*(4), 327–344. doi:10.1007/s10551-004-6944-x

Ustawa z dnia 27 kwietnia 2006 roku o spółdzielniach socjalnych (Dziennik Ustaw Nr 94, poz. 651 z późn. Zm).

Van Berkel, R., & Hornemann Møller, I. (2002). *Active social policies in the EU. Inclusion through participation?* Bristol: Policy Press.

Van de Ven, A. H. (2013). *Engaged Scholarship: A guide for organizational and social research*. Oxford, UK: Oxford University Press.

Vis, B., van Kersbergen, K., & Hylands, T. (2011). To What Extent Did the Financial Crisis Intensify the Pressure to Reform the Welfare State? *Social Policy and Administration*, *45*(4), 338–353. doi:10.1111/j.1467-9515.2011.00778.x

Wilson, F., & Post, J. E. (2011). Business Models for People, Planet (& Profits): Exploring the Phenomena of Social Business, a Market-based Approach to Social Value Creation. *Small Business Economics*, *40*(3), 715–737. doi:10.1007/s11187-011-9401-0

Wronka-Pośpiech, M., & Frączkiewicz-Wronka, A. (2014). The use of ICT for achieving the objectives of the business model – social enterprise perspective. *Polish Journal of Management*, *10*(2), 33–42.

Wygnański, J. (2007). Ekonomizacja organizacji pozarządowych. In J. Hausner (Ed.), *Ekonomia społeczna a rozwój* (pp. 127–156). Kraków: Małopolska Szkoła Administracji Publicznej.

Yin, R. K. (2003). *Case Study Research* (3rd ed.). Thousand Oaks, CA: Sage Publications Inc.

Yunus, M., Moingeon, B., & Lehmann-Ortega, L. (2010). Building Social Business Models: Lessons from the Grameen Experience. *Long Range Planning*, *43*(2-3), 308–325. doi:10.1016/j.lrp.2009.12.005

Zott, C., Amit, R., & Massa, L. (2011). The business model: Recent developments and future research. *Journal of Management*, *37*(4), 1019–1042. doi:10.1177/0149206311406265

KEY TERMS AND DEFINITIONS

Business Model: The logic behind the enterprise, the way it functions and how it creates value for its stakeholders.

Business Model Canvas: A common language for describing, visualizing, assessing, and changing business models.

Business Model for Social Enterprise: Decision-making framework necessary for developing consistent activities to support social mission-driven value creation and formulating strategies to sustain competitive advantage.

Social Enterprise: An operator in the social economy whose main objective is to have social impact rather than make a profit for their owners or stakeholders. It operated by providing goods and services for the market in an entrepreneurial and innovative fashion and uses profits to achieve social objectives. It is managed in an open and responsible manner and, in particular, involves employees, consumers, and stakeholders affected by its commercial activities.

Social Cooperative: A specific form of cooperative work. It is an association of people at risk of social exclusion (including: long-term unemployed, the disabled, addicts or the homeless). Within its activities it combines both economic and social functions. The main objective of social cooperative is social and professional reintegration of its members.

Stakeholders: Those groups without whose support the organization would cease to exist and who has an interest in the value created by the organisation.

Social Impact: The effects (improvements) an organization has on society.

Chapter 7
Support of Social Innovations:
Case of the Czech Republic

Jarmila Šebestová
Silesian University in Opava, Czech Republic

Zuzana Palová
Silesian University in Opava, Czech Republic

ABSTRACT

The aim of this chapter is to summarize the theoretical knowledge from the field of social entrepreneurship and the creation of social innovation and highlight the impact EU funds have on the development of social innovation in selected regions of the Czech Republic. The authors assumed that there could be a positive link between the amount of financial support and the number of created social innovations within the chosen EU programmes. Classification of created social innovation according to type, creator, priority axis in relation to beneficiary etc. came under other objectives. Social innovations are created as a positive externality from other social projects. Finally, recommendation for sustainable support evaluation is provided.

INTRODUCTION

Social innovations are considered as "new ideas/products, services and models) that simultaneously meet social needs and create new social relationships or collaborations. In other words, they are innovations that are both good for society and enhance society's capacity to act "(Murray et al 2010) or "new and better solutions which meet urgent social (or societal) needs, and, at the same time, they create new social interactions and cooperation" (Kadeřábková, Moghadam Saman, 2013). Dobele (2015) added, that social innovations should bring newness, create new changes in society, solve social problems, create benefits for the whole of society, offer social value, and should be based on social motives. Social motives and other factors were also mentioned as determinant of social innovations by Murray et al. (2010). The impact of social innovations, stemming from those definitions, is crucial for economic and social development, and thus for social welfare. However, their implementation still depends largely on public support, especially in less developed countries. The main problems, which have an influence on

DOI: 10.4018/978-1-5225-2215-7.ch007

the practice of social innovation, are a lack of legal recognition of social innovation (the current definition is available as a "project" call of each Operational program), insufficient social innovation policy coordination as well as a lack of data for the measurement of output.

Regional socioeconomic analysis, which is related to the problem of output measurement is mostly based on three pillars, which assess the impact of actions on regional policy. These pillars are economic (dynamics of economic development), social (investments in social capital, the main indicator for societal development) and finally, environmental (eco-friendly approach, recycling). The main condition of social development is the maintenance of a balance between these pillars (Minařík et al., 2013). In this context, Kapstein and Kim (2011) presented a fully developed matrix of socioeconomic impacts on the local community and their model includes four dimensions: macroeconomic impacts (GDP contribution and other macroeconomic indicators), socioeconomic impacts and linkages (education, employment), community impacts (changes in settlement, social structure, migration), and environmental impacts (pollution, investment in environmental protection). Despite the differences, the inputs and outputs of the social innovation process clearly remain as a multidimensional phenomenon.

Earlier attempts to capture the core logic of the public sector business model highlighted the multidimensional and multi-level exchange of resources, capitals, power etc. (Alford, 1993), in contrast to the traditional business model framework for business companies (Osterwalder & Pigneur, 2010). Thus, the main goal of this chapter is to gain a better understanding of the multi-level specificity of the public sector business model and its innovation from the perspective of social innovation. Thus, this chapter evaluates the relation between the creation of social innovation and financial support from the Human Resources and Employment Operational Program (HREOP) and the Operational Program Education for Competitiveness (OPEC) in 2007-2013. The authors assumed that there could be a positive link between the amount of financial support and the number of created social innovations when social innovations were created as a positive externality from other social projects because of their form as a "voluntary output" of those projects (Dohnalová et al, 2015, Kadeřábková, Moghadam Saman, 2013). The undertaken study aims to explore the immediate effects of using European Structural Funds in the current context of regional development, highlighting the influences on socioeconomic indicators and the social return on investment ratio (SROI). The selected area is complex and is represented by the impact of structural funds finance on the socioeconomic results obtained at the level of Kapstein and Kim (2011) indicators and of social innovations, which express the new regional capacity to be dynamic at both local and regional levels. It is in such context, that this chapter firstly aims to measure, through marketing quantitative research, using the survey research method (based on an evaluation questionnaire directed at the beneficiaries), the level of impact of projects financed through the Human Resources and Employment Operational Program (HREOP) and the Operational Program Education for Competitiveness (OPEC) in 2007-2013 on the target audience in selected regions of the Czech Republic. Secondly, the effectiveness of support allocation is measured by a modified ratio of social return ratio (SROI), suggested by the authors. Finally, an evaluation tool based on performance evaluation is presented. The chapter deals with performance management on regional beneficiaries using the principles of the "3E's", which represent the areas of economy, effectiveness and efficiency, namely in the example of the selected regions in the Czech Republic. The authors also applied a classification of social innovation as used by Caulier-Grice et al., 2012) in their research.

Background

Social innovations might differ from innovations in the private sector, when the essence of innovations comes from social or public need and brings a new organization for the provision of social services to society (Mulgan, 2006, Bekkers et al., 2013). As has been mentioned, these innovations are mostly community-need oriented and serve a specific group. Feedback from the end-users of these services is needed (Lee et al., 2012). The special phenomena of social innovations can be seen in their non-significant relationship with technology driven innovations (Howaldt and Schwarz, 2010). On the other hand, social innovations support the more effective allocation of publicly supported services such as home care, health services, social care or education. The support of social innovation depends primarily on previous traditions of the public sector. (See Table 1).

The EUROPE 2020 strategy provides a framework of social policy interventions to support social innovation creation in three main areas (Phills et al., 2008, SSE, 2012):

- Novelty, which means a new approach based on community needs.
- Improvement, when social innovation must bring more effectiveness in the solution to new problems.
- Sustainability, where all realized innovations should be environmentally, organizationally and economically sustainable from the long-term point of view.

Social innovation faces two main problems within society, to be sustainable, such as a lack of coordination between actors involved in that process (local government, non-profit organizations, business entities) and financial resources due to poor methods of real societal impact evaluation (SSE, 2012). The process of creating social innovation across society can be demonstrated by a cycle of motivation, where at the start the pressure of business sustainability was identified as a push factor (Figure 1).

Each innovation must be able to cover its costs and resources in each phase (Figure 1). Three areas were mentioned as crucial for the support of sustainability (Tucker, 2014):

- Social innovation market development, where socially oriented entrepreneurs are supported and accordingly may create their own business model with which to reinvest financial returns for so-

Table 1. Dependence of tradition and social innovation environment

Type of State	Basic Characteristics	Socially Innovative Environment	Country Examples
Strongly centralized	• Strong rules and regulations • Protection of public interest • Innovations led from the top	• Political and administrative problems • Strong central and unitary government	UK, France, Romania, Slovakia, Czech Republic
Decentralized structure with strong local government	• Mostly federal structure of country • Collaborative and consensus-seeking environment • Environment prepared to meet local needs	• Strong civil society • Bottom-up innovations	Belgium, Denmark, The Netherlands

(Bekkers, 2013, Author's comments)

Figure 1. Process of social innovation development
Source: Author's illustration

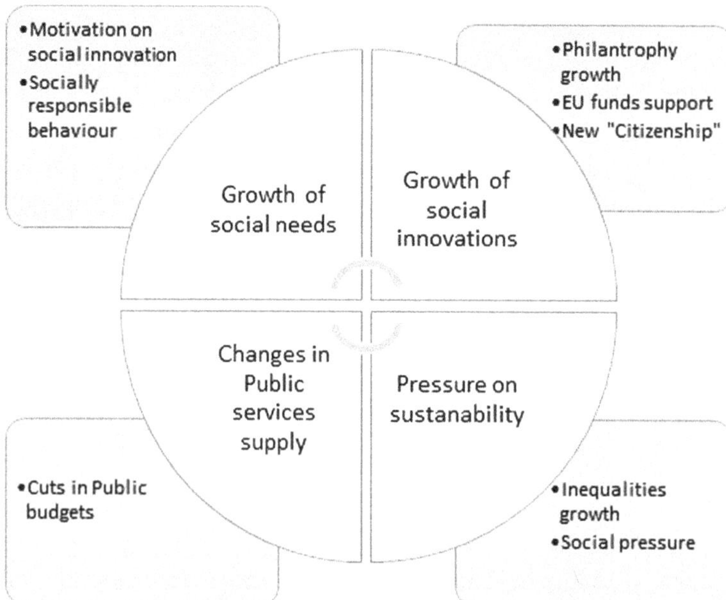

cial purpose. The encouragement of citizens is needed in order to identify opportunities for social innovation and to discover better solutions for their needs.

- Changes in roles in the relationship between public bodies and social entrepreneurs in order to be more effective.
- Changes in management and leadership of current business entities, where the primary emphasis is placed on socially oriented business plans. By using this approach, they minimalize the risk of failure by delivering a sustainable business model.

The following previous findings for a model of national social innovation system was proposed by Tucker (2014, p.28, Figure 2) to describe the relationship between key players on the social innovation market.

Cooperation between these actors should support the more effective allocation of financial support. Effectiveness, efficiency and economy within EU funded programs are the common terms used in assessing and measuring the performance of organizations (Mouzas, 2006). Efficiency is associated with "doing things right" and effectiveness with "doing the right things" (Drucker 1977). A measurement of effectiveness relates to the achievement of main project goals (Asmild et al., 2007). The efficiency of public expenses and support describes the relationship between the economic and social effects as a result of program implementation and the effort made to finance the program (economy side). The three "E's" were expanded by Norman-Major (2012) into four "E's", where economy means managing resources to provide the desired effect for the least amount of money, efficiency has a tie to the provision of services, other public goods and performance indicators as well (i.e. quantity of services, time schedule). This approach includes effectiveness in area of evaluation of goal achievement and equity, which means fairness and the complex treatment of social needs for disadvantaged persons.

Figure 2. System of social innovation development
Source: Author's illustration

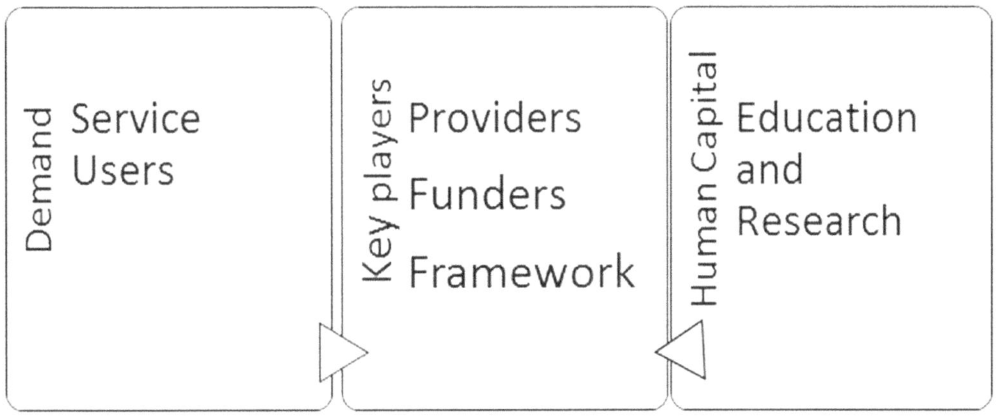

Basic items could alternately be interpreted as also meaning that economy signifies performance in relation to price, where the main criterion for that item is the inexpensive achievement of maximum output. The efficiency is described as a relationship between inputs and outputs based on the rule of achieving outputs with less money. Finally, the area of effectiveness expresses the degree of progress towards objectives (Otrusinova, Pastuszkova, 2012).

INFRASTRUCTURE FOR SOCIAL INNOVATIONS IN THE CZECH REPUBLIC

The largest financial support for social innovation in the Czech Republic comes from European Union funds. Social innovations in the Czech Republic are defined using a definition similar to that employed by EU Funds, as well as to that of TEPSIE project outputs (Eriksson, Einarsson and Wijkström, 2014). The concept is close to that of Gerometta, Haussermann and Longo (2005) who described the process of social innovation as "the satisfaction of human needs (content dimension); changes in social relations especially regarding governance (process dimension); and an increase in the socio-political capability and access to resources (empowerment dimension)". This approach is used by entrepreneurs who very often utilize support from operational programs in the Czech Republic, and it is clearly understandable for the Ministry of Labor and Social Affairs. They focus on social needs and social care more than other significant factors, which determine social innovations (such as civic demand). The main problem is the technical explanation of social innovation by public bodies to organizations interested in providing and developing social innovations. The supporting infrastructure (Figure 3.) for social innovation consists of three layers (Dohnalová et al., 2015):

1. Political and legal framework, when special conditions for social businesses are still missing;
2. Institutional platforms and support, which include ministries and associations connected with social issues;
3. Financial support, based on resources, which are used to create social innovation.

Figure 3. Current infrastructure for social innovations
Source: Author's illustration

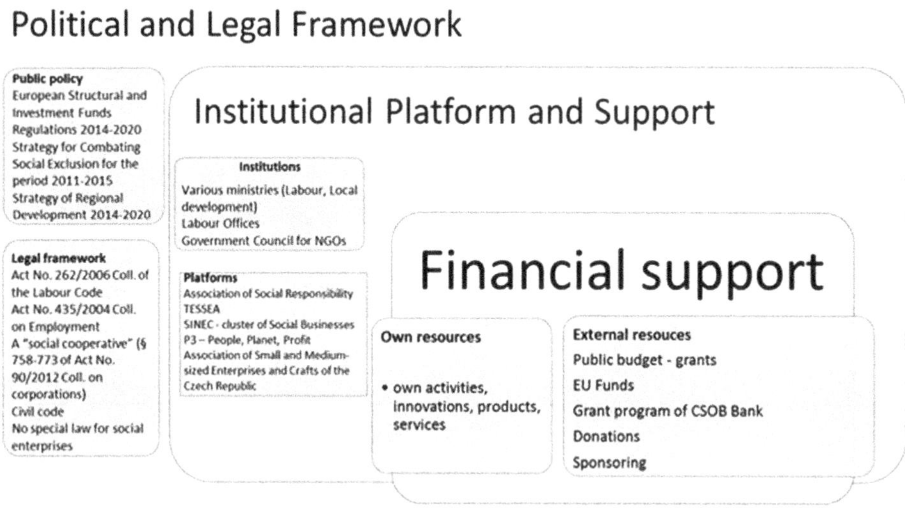

On the national level, governmental bodies are interested in corporate social responsibility actions and at present, the Czech Republic participates in the SozialMarie prize for awarding socially innovative projects. Social innovations are also connected with community plans in municipalities (analysis of demand for social care, social needs, and target groups) and local action groups (LAGs), who prepare a local strategy of development. They include social innovations and the applicant's requirement to meet the needs of the place where the project will be realized. Public administration does not reveal any significant potential to become a source of social innovations because of cooperation with the non-profit sector or other entities in the market (Eriksson, Einarsson and Wijkström, 2014).

Historical Background of Social Innovations

The background of the development of social enterprise and social innovation in the Czech Republic has a long and rich tradition of solidarity, foundations and cooperatives; such organizations have indeed existed under different forms in the country for over 100 years. The First Republic of Czechoslovakia, under the presidency of Tomáš Garrigue Masaryk (1850-1937) created a friendly environment for the development of civic organizations and philanthropy activities, supporting education, research and social and health services (Dohnalová et al, 2015). The shift started under the German occupation in 1939 when many non-profit organizations were banned and others were reorganized to serve the purposes of the state. After the end of World War II, non-profit organizations renewed their activities. Many organizations, however, became influenced by political parties and the further evolution of the non-profit sector was blocked by the communist regime (1948). The activities of churches were restricted to worship and education. The state patronized the provision of education, health and social care. These services were provided by budgetary and subsidiary organizations that created the so-called state non-profit sector. The political pressure of the state slowly faded in the 1960's. Citizens became more active in public affairs and several non-profit organizations were re-activated (Brhlíková, 2004).

Four decades of communist rule restricted and disfigured the organic development of the non-profit sector. The state administration was adapted to fulfil this monopolistic role. For this purpose, during the communist era the state built a relatively extensive network of state owned 'non-profit' organizations in the domain of social welfare. The relevant organizations were essentially 'budgetary and contributory' organizations, directly reporting to the appropriate ministries. State ownership and control in this domain was to stay remarkably constant, despite the change of regime in 1989. The immediate impact of the political changes after 1989, however, did not only mean the quantitative expansion of the third sector. The sector also started to consolidate its position in society, although the move to recover and build the third sector was limited by many factors such as limited financial resources, number of professionals and finally, the number of volunteers (Fric, 2005). The real beginnings of the social enterprise and social innovation phenomenon in the Czech Republic are mainly connected with the implementation of projects supported by the European Structural Funds.

Evaluation Criteria of Financial Support from EU funds

A significant macroeconomic impact on the Czech economy is exerted by the Cohesion Policy of the EU, especially EU funds. These funds should strengthen competitiveness and reduce economic and social disparities between regions, thereby helping to fulfil the goals of the Lisbon Strategy and Strategy Europe 2020.

In the operational period from 2007-2013 there were 26 operational programs in the Czech Republic. The objective was to support the economic and social development of less developed regions and member states. In the Czech Republic, this objective includes all the cohesion regions except for Prague, the capital of Czech Republic, and it was implemented through eight sectoral operational programs and seven regional operational programs. A total of EUR 25.89 billion (approximately CZK 730 billion) was allocated to the Convergence Objective in the Czech Republic (Table 2.), eight sectoral operational programs (OP) in value of EUR 21.23 billion, (approximately CZK 598.62 billion) and seven regional

Table 2. Financial support structure in percentage shares

<table>
<tr><th>Sectoral Operational Programs (OP)</th><th>Percentage Share</th><th>Regional Operational Programs (ROP)</th><th>Percentage Share</th></tr>
<tr><td>OP Transport</td><td>21.58%</td><td>ROP NUTS Northwest (Jihozápad)</td><td rowspan="9">17.23%</td></tr>
<tr><td>OP Technical Assistance</td><td>0.97%</td><td>ROP NUTS Northwest (Jihozápad)</td></tr>
<tr><td>OP Research and Development for Innovation</td><td>7.76%</td><td>ROP NUTS Northeast (Severovýchod)</td></tr>
<tr><td>OP Prague</td><td>1.54%</td><td>ROP NUTS Central Moravia (Střední Morava)</td></tr>
<tr><td>OP Human Resources and Employment</td><td>6.79%</td><td>ROP NUTS Moravia-Silesia (Moravskoslezsko)</td></tr>
<tr><td>OP Environment</td><td>18.43%</td><td>ROP NUTS Southeast (Jihovýchod)</td></tr>
<tr><td>OP Enterprise and Innovation</td><td>11.40%</td><td>ROP NUTS Southwest (Severozápad)</td></tr>
<tr><td>OP Education for Competitiveness</td><td>6.79%</td><td>ROP NUTS Central Bohemia (Střední Čechy)</td></tr>
<tr><td>Integrated support</td><td></td><td>Integrated support</td></tr>
<tr><td>European Territorial Cooperation</td><td>2.75%</td><td>Integrated Operational Program</td><td>5.82%</td></tr>
</table>

(Office of the Regional Council South-East, 2014)

operational programs (ROP) at the level of NUTS 2 cohesion regions (EUR 4.66 billion, approximately CZK 131.38 billion).

Firstly, the ESF funds support social innovation offering financial sources to strengthen the institutional capacity of public bodies to increase their abilities, promote socio-economic development and good governance as a key element of the Lisbon Strategy. There were three types of ESF interventions regarding increase in institutional capacity and three types of recipients of (public employment services, educational training systems and other public administrations (Tarnawska, Ćwiklicki, 2012). Secondly, European Regional Development Fund (ERDF) financed direct aid to investments in companies - small and medium sized enterprises (SMEs) - to create sustainable jobs as well as infrastructures linked notably to research and innovation, telecommunications, environment, energy and transport, but also social infrastructures (EU Commission, 2013). Viturka (2010) confirmed, that each priority axis is determined by quantifiable indicators of input, output, outcome and impact, the part with qualitative performance indicators is still missing. This is the reason why within the upcoming period a system of indicators for local action groups (LAGs) and LEADER principles for community led development will be applied in the evaluation of projects. The number of people supported by the project or number of innovations generated by the project does not answer the question about the real contribution to the development of local communities, increasing local prosperity and having security in life. On the other hand, Richmond, Mook, and Quarter (2003) suggested social accounting, but for many organizations it was still difficult to evaluate their work through a true quantitative approach. Finally, the New Economics Foundation presented a social ratio, called social return on investment (SROI) based on the return on investment in business analysis (Clifford, List, Theobald, 2010). The European Union recommended the use of the SROI and encouraged member countries to develop their own versions (Moody, Littlepage, Paydar, 2015), but this approach is still not widely used. On the other hand, effectiveness is the main indicator given by the ratio of the result of the obtained EU funds program to achieve and that area is widely expressed. Within the framework of the Human Capital Operational Program and Operational Program Education for Competitiveness, we can distinguish the following types of performance indicators, which we can find in other EU countries (Poland, Romania and others) in their national reports and research analyses covering the program above (Jałocha, 2012, Pădurean, et al., 2015, Wolińska, et al., 2010):

1. **Impact Indicators:** Describing immediate effects on beneficiaries and they are applicable in the medium term. The Managing Authority is responsible for measuring the impact indicators.
2. **Product Indicators:** Describing the product of activities of the persons as beneficiaries of the program. They differ per priorities of the program.
3. **Results Indicators:** Present information about the changes taking place regarding the direct beneficiaries (Šebestová, Palová, 2016, Potluka, Liddle, 2014).

Efficiency cannot be directly measured. Therefore, different approaches concerning data and methodological framework have been used. Generally expressed, one would expect that public funds are directed to essential policy areas, which support the overall objectives of the state, others prefer the regional development or disparity measurement, where indexes and performance indicators are used by themselves to measure efficiency.

ASSESSMENT OF EU PROGRAMS IN CONTEXT OF SOCIAL INNOVATION CREATION

The key role in the support of social innovation in the Czech Republic has been played by those structural funds. Many social innovation projects already received Structural Fund support from 2007-2013, also they are integrated into the program in 2014-2020. Social Innovations were supported from the European Social Fund (ESF) by co-funding national, regional and local projects that improve the levels of employment, the quality of jobs, and the inclusiveness of the labor market in the member states and their regions.

The undertaken study aims to explore the immediate effects of using European Structural Funds in the current context of regional development, highlighting the influences of socioeconomic indicators and social return on investment ratio (SROI) in the context of the "3E's" methodology. The objective of following study is to capture sufficient data and information to assess the economy, efficiency and effectiveness principle of the Human Resources and Employment Operational Program (HREOP) and the Operational Program Education for Competitiveness (OPEC) programs in the context of social innovation creation. Especially, Social Innovation Call (SIC) was opened in 2013 under the HREOP by the Czech Ministry of Labor and Social Affairs to foster the societal impact of the program, but only eight projects were realized.

According to the definition in that project call (SIC), "social innovations are new and better solutions which meet urgent social (or societal) needs, and, at the same time, create new social interactions and cooperation." (Kadeřábková, Moghadam Saman, 2013).

The detailed secondary information is based on data acquired from the monitoring system (IS Monit7+) up to September 2015. The research included all businesses except municipalities, regions and public administrations in these four regions of the Czech Republic (Figure 4.).

The authors chose the following regions: the Moravian-Silesian Region, the Usti Region, the Hradec Kralove Region and the Plzen Region (NUTS III). These regions were chosen because these regions met the conditions of (1) not sharing a border, (2) they differ in regional competitiveness indicators (Viturka, 2007) as Hradec Kralove (HK) and Plzen Regions (PR) are both classified in B group (in good competitive position) and the Moravian-Silesian Region (MSR) and the Usti Region (UR) are in contrast to them ("C" group – in less favorable position). This classification was supported by Melecký and Staníčková (2011) when they divided the NUTS II regions (8) as follows MSR (8th place), UR (part of the Severozápad region, 7th place), HK as part of Severovýchod (4th place) and PR as part of Jihozápad (3rd place).

Types of Social Innovation Created through Support from the HREOP and OPEC EU Programs

At the time of examination, the database contained data on 1,820 applications in the HREOP program and 1,665 applications in the OPEC program. A questionnaire survey was distributed to all the 3,485 program applicants in the selected regions at the beginning of 2016. Some of the applicants applied for EU funding several times. The survey obtained 158 valid responses from the OPEC and 165 valid answers from the HREOP program (Table 3). It was a statistically representative sample at a confidence level of 95% with a 5% margin of error.

Figure 4. NUTS III regions in the Czech Republic
Source: http://www.strukturalni-fondy.cz/getmedia/73d4d51b-39f4-4b60-8e70-1a8e355a4cdf/Regiony-NUTS.png

Table 3. Structure of research data in percentage shares

	HREOP		OPEC	
	Percentage Share in Economic Entities	**Percentage Share in of Social Innovations**	**Percentage Share in Economic Entities**	**Percentage Share in of Social Innovations**
Nonprofit organizations	46%	43.66%	81.8%	77.6%
Business entities	52%	54.54%	8.8%	12%
Universities	2%	1.8%	9.4%	10.4%
Total	100%	100%	100%	100%

(Own research, 2016)

Economic entities were divided into five groups per size according to the EU definition of small and medium sized companies (28% up to 9 employees; 28% up to 49 employees; 26% till 249 employees and 18% in size of 250+ employees). The most active segments in the area of social innovation are entities with up to 49 employees (56% of the sample). Based on the questionnaire survey, the minority of projects (30.3% in HREOP and 37.3% in OPEC) at the time of realization declared some type of social innovation to contribute to regional development as sufficient output. Social innovations were classified according to end-user groups (Table 4). Other types of innovations were identified as a work-life-balance program, gender statistics preparation, consultations, methodological manuals, strengthening the capacity of the social partners and employment.

Where the HREOP program focused primarily on Adaptability of SMEs, most innovations were dedicated to human capital development to support competitiveness. A small share of financial support went to social enterprise creation to support a future network of social innovation providers. In comparison

Table 4. Main types of social innovations

	HREOP	OPEC
Type of social innovation	Percentage share	Percentage share
Creation of social enterprise	14.55%	--
Innovation in education	43.64%	84.70%
Innovation in environment	1.81%	5.10%
Innovation in health care	3.62%	--
Innovation in social care	18.18%	--
Technical and technological innovation	7.30%	10.20%
Other	10.90%	--
Total number of social innovations	100%	100%

(Own research, 2016)

with the results from Kadeřábková and Moghadam-Saman (2013) in terms of need for innovation, the most frequent factor was the improvement of quality of life of the target group (37%, in the case of the HREOP program - 10.9%), followed by urgent or critical needs of the target group (e.g. health services, sanitation, food) and development of civic society, e.g. for the resolution of community issues (25% and 24% respectively; in HREOP in total, social and health care 21.8%), and capacity development (e.g. education) for employment or financial independence (14%, in case study HREOP 43.64%, OPEC 84.70%). In our case (Table 4), the problem of education was mentioned in the first place due to socioeconomic problems in the four analyzed regions (HREOP 43.64% and OPEC 84.70%) in combination with social enterprise creation (HREOP 14.55%) and technical innovation, especially IT support (HREOP 7.30%, OPEC 10.20%). To get widely comparable data for future research a classification of social innovations by Caulier-Grice et al. (2012) was used (Table 5), where significant differences between both operational programs were discovered. The social innovation types to some extent match the components of a classic business model framework (Osterwalder & Pigneur, 2010) or the business model innovation in general (Chesbrough, 2007).

The most important innovation for beneficiaries of the HREOP program were new platforms and cooperation to share knowledge from and to support their business ideas. Opposite to that, the OPEC

Table 5. Unified classification of social innovations by type

Type of Social Innovations	HREOP	OPEC
New products	20%	38%
New services	5%	17%
New processes	22%	40.8%
New markets	0%	0%
New platforms	38%	4.2%
New organizational forms	0%	0%
New business models	15%	0%

(Own research, 2016)

program focused on the development of new processes, especially new methods of education and support for the education of disadvantaged persons. Those connections (the relationship between the focus of the program and social innovations) were evaluated by correlation analysis. A very strong, direct correlation (correlation coefficient was 0.69, Sig. 0.00, $\alpha = 0.05$) between the amount of financial support (divided by priority) and number of social innovation (divided by priority axis) was found.

Evaluation of Real Economic Impact of Social Innovations Created through Support from the HREOP and OPEC EU Programs

All business model frameworks encompass a component related to a revenue model, profit formula etc. which explain how an organization captures value in exchange of the value proposition delivered to target customers (Osterwalder & Pigneur, 2010). Social innovations also allow the capture of value, yet its evaluation is somewhat problematic due to several dimensions of analysis.

Economic issues and ratios could be measured by standard financial ratios covered in cost-benefit analysis. EBIT was used as standard indicator of revenue flow; costs were defined as the amount of allocated money.

The SROI indicator was used to improve the efficiency evaluation of the program. Several modifications should be made in order to provide a more effective evaluation within operational programs. The basic SROI ratio covers EBIT and program costs (Cooney and Lynch-Cerullo, 2014). This approach had to be modified according to priority axes and each operational program (OP) as summarized in Table 6 (Šebestová, Palová, 2016). A formula modification for the HREOP program was calculated as follows: Priority axis 1[1] showed through outcome indicators a change in earnings of participants of the program. Accordingly, a five-year change in average wages and salaries were used separately in each region. The two-year effect of the project was accounted after its completion (due to "n+ 2" rules).

To define the efficiency of expenditures in education (OPEC program), the calculation started from the purpose of the education activity, following the way in which it may be achieved (completion of elementary education or university degree in tertiary education). The efficiency (Table 7) must be ana-

Table 6. SROI ratio modification

OP	SROI Formula Type	EBIT	Focus Group Size (Based on Survey)
--	Basic type	Number of clients enrolled * earning difference (ED)	--
HREOP	Axis 1 Adaptability	Number of successful participants * ED per participant for two years	50
	Axis 2 – Active labor market policy Axis 3 – Social Integration and Equal Opportunities	Number of newly employed person * savings from compensations in unemployment (5 months)	100
OPEC	Priority Axis 1 – Initial Education	Number of supported pupils*ED for two years (to complete elementary education)	100
	Priority Axis 2 – Tertiary education, research policy	Number of supported students*ED for three or five years (to complete academic degree)	100

(Own research, 2016)

Table 7. Economy and Efficiency evaluation (Own calculations)

	Priority Axis	Indicators	Moravian-Silesian Region	Usti Region	Plzen Region	Hradec Kralove Region
HREOP	Priority Axis 1 - Adaptability	Number of projects	112	58	41	41
		Total support (Mil. CZK)	322	55	137	16
		EBIT (Mil. CZK) per project	1.62	2.02	2.33	1.5
		SROI	0.56	2.13	0.7	3.84
	Priority Axis 2 – Active labor market policy	Number of projects	56	50	6	5
		Total support (Mil. CZK)	276	259	26	25
		EBIT (Mil. CZK) per project	6.57	6.24	5.58	6.347
		SROI	1.33	1.2	1.29	1.27
	Priority Axis 3 – Social integration and equal opportunities	Number of projects	162	136	37	26
		Total support (Mil. CZK)	612	604	135	65
		EBIT (Mil. CZK) per project	6.57	6.24	5.58	6.347
		SROI	1.74	1.41	1.53	2.54
OPEC	Priority Axis 1 – Initial education	Number of projects	446	131	150	260
		Total support (Mil. CZK)	648.5	496.55	387.44	688.13
		EBIT (Mil. CZK) per project	39.38	43.86	36.3	41.17
		SROI	27.08	11.57	14.06	15.56
	Priority Axis 2 – Tertiary education, research policy	Number of projects	76	20	35	23
		Total support (Mil. CZK)	721.69	165.53	465.17	377.09
		EBIT (Mil. CZK) per project	130.76	134.22	125.96	125.20
		SROI	13.77	16.22	9.48	7.64

lyzed and calculated per the rational production and use of resources allocated to training the individuals (Gherghina, Văduva, Postole, 2009, Duca, 2013). The economic efficiency of subsidies allocation in that priority axis was evaluated by the SROI index. (based on alternative accounting costs per completed education). The social rate of return summarizes the costs and benefits of the educational investment from the state's point of view, i.e., it includes the full resource cost of education, rather than only the portion that is paid by the recipient of education (Psacharopoulos,1995). EBIT in that program was calculated from salaries dependent on education where the outcome indicator within that program was persons, who "completed that level of education".

In economic contribution, the highest amount of EBIT was produced in Priority axis 3 – Social Integration (about 6 mil. CZK, approx. € 200 000), in the HREOP program it was Axis 2 – Tertiary education, where EBIT oscillated between 125 to 135 mil. CZK. (approx. € 4.5 mil.). There were significant differences in efficiency results. The maximum of SROI in the HREOP program is 3.84 in the Hradec Kralove Region (Axis 1), in the OPEC program the minimum is 7.64 in the same region (Priority Axis 2). It seems that it is more significant for regional development to develop human resources, which, as showed in the results, brings more than 7% in social innovations than the HREOP projects. Finally, an ordinal scale was used to compare efficiency in the examined regions (1 - the best performance in support, projects, EBIT or SROI; 4 - last in efficiency, Table 8).

Table 8. Efficiency evaluation (Own calculations)

		Moravian-Silesian Region	Usti Region	Plzen Region	Hradec Kralove Region
HREOP	Priority Axis 1 – Adaptability	2.25	2.25	2.25	3
	Priority Axis 2 – Active labor market policy	1	2.75	3	3.25
	Priority Axis 3 – Social integration and equal opportunities	1.25	2.75	3.25	2.75
OPEC	Priority Axis 1 – Initial education	1.75	3	3.5	1.75
	Priority Axis 2 – Tertiary education, research policy	1.5	2.5	2.5	3.5
	Average rating	1.55	2.65	2.9	2.85

The best efficiency was achieved in the Moravian-Silesian Region in most priority axes (except Adaptability). The worst efficiency was surprisingly found in Plzen and the Hradec Kralove Region. The efficiency relates to regional problems, that were confirmed the in Moravian-Silesian region (1.55), and not in the Ústí region (2.65).

Effectiveness was evaluated as a linkage between financial support from the Operational Program and the creation of social innovation, when correlation analysis was used. The interpretation of correlation analysis results was based on the Liebetrau scale (1983). Other assumptions were tested in a partial dependency with cross tables, depending on the region and implementation of the priority axis. Table 9 summarizes the various factors of relationships. The evaluation was based on the value of Cramer V for nominal values (Sig. 0.00, $\alpha = 0.05$) If a relationship between values has been statistically significant, the sign of "+ "was used. In the opposite case, when the significance level was above 0.05 (the factor wasn't statistically significant, $\alpha > 0.05$) a sign of "0" was used.

In summary, better effectiveness in outcomes were measured in the OPEC program, because there is an equal relationship between regional development and program priority. On the opposite end, the HREOP cared only in the effectiveness axis, not in regional development, this was confirmed in the efficiency part as well. On the other hand, the HREOP program showed more suitable ties for the sustainable support of social innovation, not only in relationship with the priority axis, but also in connection with the region.

Real Regional Macroeconomic Impact

Kapstein and Kim (2011) methodology was followed and accordingly their indicators were divided into four groups (Macroeconomic impacts - GDP contribution and other macroeconomic indicators; Socioeconomic impacts and linkages such as education, employment; Community impacts - changes in settlement, social structure, migration and Environmental impacts), where results from the survey were added into social impact as the value of social innovation, socially excluded locations and EU project share. A baseline for macroeconomic indicators a year of 2007 (Figure 5, Table 10 in Appendix) was chosen. An ordinal scale describes a cross regional comparison, where the lowest score represents the best performance.

Table 9. Effectiveness evaluation (Own calculations, Cramer V coefficients, when >0.5 high association; 0.3 to 0.5 moderate association; 0.1 to 0.3 low association; 0 to 0.1 little if any association)

Variable	HREOP				OPEC			
	Relationship to the Region		Relationship to Priority Axis		Relationship to the Region		Relationship to Priority Axis	
	Cramer V	Sig.	Cramer V	Sig.	Cramer V	Sig.	Cramer V	Sig.
Legal form of beneficiary	0.271	0	0.444	+	0.390	+	0.851	+
Number of employees	0.243	0	0.328	+	0.535	+	0.327	+
Main business activity	0.297	0	0.486	+	0.571	+	0.261	0
Type of innovation	0.221	+	0.380	+	0.188	0	0.605	+
Amount of innovations per project	0.229	0	0.159	0	0.526	+	0.517	+
Willingness to continue in 2014-2020	0.220	0	0.187	+	0.283	0	0.218	+
Networking	0.325	+	0.471	+	0.590	+	0.181	0
Total community impact (total "+")		**2**		**6**		**5**		**5**

Figure 5. Regional evaluation model 2007-2013
Source: Author's calculations

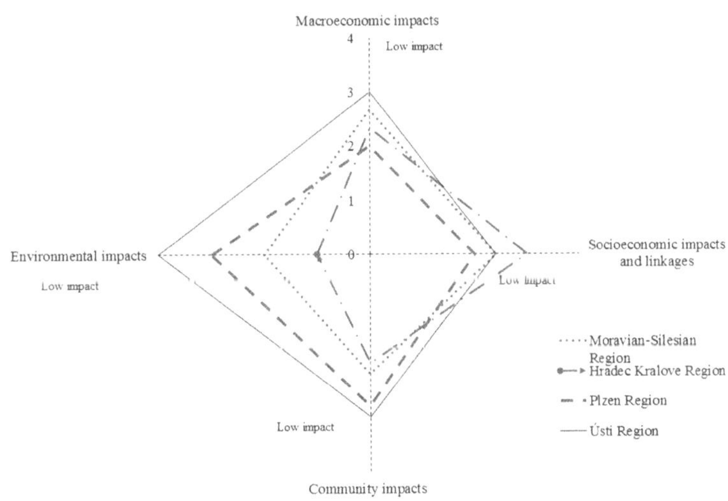

The best results in most of the four dimensions was obtained by the Plzen region, the worst performance under the operational program could be seen in the Usti region. When we focused on the socio-economic dimension, we could see an imbalanced relationship with macroeconomic indicators. When the impact on society was high, the result on macroeconomic indicators was lower (comparison of the Hradec Kralove and Moravian-Silesian regions). When we compare our findings with socioeconomic data, social innovation is found in regions, where the migration rate is highly positive, and therefore it reduces the movement of people. When the innovation activity is low and the motivation to remain is also low, migration and unemployment increase (low social impact of innovations).

SOLUTIONS AND RECOMMENDATION

The efficiency of EU operational programs causes a problem, which most local governments should face (as being analyzed in four selected regions), and which is determined mainly by the existence of some other external factors, which influence the final performance objective (Mihaiu et al., 2010). In comparison with the strategy of regional development 2014-2020 (MMR, 2013) it could be ascertained that both programs did not bring about the expected economic effect, such as the situation of the Ústí region, which mostly worsened and this region was classified as a region without any competitive advantage together with the region of Karlovy Vary (efficiency of Ústí region was 2.65). The other three regions were classified as regions with an unclassified competitive advantage, which describes the worsened situation in Plzeň and Hradec Králové Regions (confirmed as in last place for efficiency). Together they have two problematic districts, which still must be supported by government (Stříbro, Broumov). The situation in the Moravian-Silesian region is stable. The Ministry of regional development suggested two types of indicators, mostly based on macroeconomic data to evaluate the efficiency of their support as GDP per capita, migration, unemployment rate and others (MMR, 2015) that are insufficient for the sustainability of social innovation support. It seems that in the EU-funded projects discussed in this chapter, it is not always obvious and accepted that success is a multi-dimensional construct (which has been described above), because the criteria for success are very rigid and based mostly on indicators imposed by the financing bodies, not by local governments. Several local initiatives have been supported by the ESF within the OP Human resources and employment (HREOP) and OPEC program which stresses, at least on the programming level (much less in the implementation itself), the horizontal innovation principle in the projects. Other problems came from the definition of social innovation, when many beneficiaries are not able to define their contribution to society, it is still difficult to classify and evaluate sustainable social innovations.

To support more reactive behavior of citizens, government and social innovators, an increase of cooperation between local government and entrepreneurs is suggested. Primarily these activities are developed in an entrepreneurial environment (Valová, 2016) but also by non-profit organizations (Binek, 2011) in the form of the three circles of cooperation (Figure 6).

These activities that support local government could provide data about SROI, demand of citizens and could support a system of social innovation activities, which do not exist in the Czech Republic (Figure 2.). The first option for small business development and the associated development of the sustainable and social innovation friendly environment is a circle of local production in the form of a catalogue of service providers and traders. There are so many local social service providers, small craftsmen and traders, but there is lack of promotional activities. This circle could help them enforce their satisfaction with marketing activities and could reinforce connection with regional and service users. This circle should be one of the activities to support small business development, as the issue of local production is often overlooked in the development area and the potential of local tradesmen and artisans is insufficiently exploited.

The second circle is the creation and implementation of a social business circle to promote social business within the region to support creation in a number of business units. This circle represents a platform for building relationships with the private and public sector to support satisfaction with business environment. Generally, it is the creation of an economic and a social network in the region in a more informal way. As for the benefits of realization of the business circle, we see the following points: sharing workers, mutual promotion, stronger linkages to region, and an exchange of experience in socially

Figure 6. Circles of cooperation
Source: Author's illustration

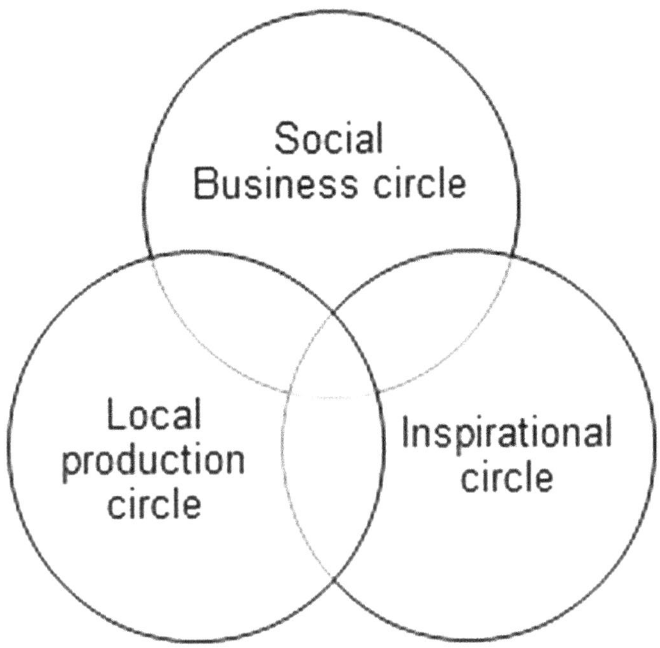

innovative projects. Another part could be a special grant system for innovative projects, which does not exist in the Czech Republic. This activity is closely connected with Tucker' s (2014) three areas of sustainable social innovations.

The third circle, called the Inspirational circle, will bring an innovative spirit into regions in not only a technical way. The main goal is seeking new proposals and solutions for the development of the local area. This idea is closely connected with the process of social innovation development, when many socially innovative activities reached the third stage of the process of social innovation (as illustrated in Figure 1.). They can involve people from the business community, students and officials and create an active group, which aims to improve the conditions and quality of life in the region. It could bring a creative platform for solving municipal and societal problems, and support of public and private projects.

Future Directions of Research

It would seem useful to continue in the development of an optimal method, which could help beneficiaries, supporting agencies and policy makers to evaluate the real social impact of given operational programs. It is still very hard to attain suitable information about the real impact and work within EU projects, as the survey has shown. Most beneficiaries cannot describe the real social value of their project and how it relates to the analyzed program and how they contributed to society i.e. via social innovation. The systematic support of social innovations is to start from 2016 within the Operational Program Employment and its priority axis "*Social innovation and international cooperation*". More systematic research and evaluation is needed, not only on a central level, but on a local level to support bottom-up innovations, not only centralized activities. Aging population, people with disabilities and changes in regions affected by industry reorganization are all challenges for systemic social innovation support.

CONCLUSION

There is opportunity for local governments to improve efficiency evaluation with the application of real societal impact index (SROI) at the regional level. Those findings were illustrated with significant social innovation supporting factors, based on field study, such as networking between organizations and the willingness of organizations to realize social innovations. By studying these indicators, new directions of action were identified along the lines of improving the use of European funds in the field of human resources support. In the presented case study, most of the beneficiaries created only one social innovation as a positive externality or due to the operational program requirement (Social innovation Call). Only 10% created two and more social innovations in their projects. Although non-profit organizations (according to business form) came in second place in a number of realized social innovations, it should not be neglected as an important aspect of the creation of social innovation. The conducted survey divided social innovation into innovation in the environment, education, social care, health care, creation of social enterprise and other areas. Social innovation in education was the most often innovation in the research sample, along with new business entity creation. The Authors confirmed a positive link between the amount of financial support (divided according to priority axes) and the number of created social innovations by correlation analysis. That means, when the amount of financial support changes, the number of created social innovations (divided according priority axes) will change positively. (correlation coefficient 0.69, sig. =0.00).

Finally, the 3E evaluation method was used to evaluate the economic contribution of supported social innovation. The model of three circles was suggested to support a sustainable social innovation environment in the Czech Republic. The mentioned analyses and ratios (3E, SROI) are fully transferable between other regions in the Czech Republic and for also, for other EU countries, because of the same system of data collection (e.g. list of beneficiaries, amount of financial support, socio-economic data) and a quite similar structure of Operational programs, especially within the Visegrad group countries (Czech Republic, Slovakia, Poland and Hungary). On a more general level, social innovation could be perceived to innovate public sector business models in the area of publicly supported services and goods.

ACKNOWLEDGMENT

This chapter was supported within the project SGS SU: The support social innovation from funds of European Union. Project registration number 16/2015.

REFERENCES

Alford, J. (1993). *Towards a New Pubuc Management Model: Beyond "Managerialism" and Its Critics*. Australian Journal of Public Administration; doi:10.1111/j.1467-8500.1993.tb00263.x

Asmild, M., Paradi, J. C., Reese, D. N., & Tam, F. (2007). Measuring Overall Efficiency and Effectiveness using DEA. *European Journal of Operational Research*, *178*(1), 305–321. doi:10.1016/j.ejor.2006.01.014

Bekkers, V. J., Tummers, J. M., Stuijfzand, L. G., & Voorberg, W. (2013). *Social Innovation in the Public Sector: An integrative framework.* LIPSE Working papers (no. 1). Rotterdam: Erasmus University Rotterdam.

Binek, J. et al., (2011). *Obce a regiony pro podnikatele; Podnikatelé pro obce a regiony.* Brno: GaREP.

Brhlíková, P. (2004). *The Nonprofit Sector in the Czech Republic.* CERGE-EI Discussion Paper, no. 128.

Caulier-Grice, D. J. A., Patrick, R., & Norman, W. (2012) Social Innovation Overview: A deliverable of the project: "The theoretical, empirical and policy foundations for building social innovation in Europe" (TEPSIE), European Commission – 7th Framework Programme. Brussels: European Commission, DG Research. The Young Foundation.

Chesbrough, H. (2007). Business model innovation: Its not just about technology anymore. *Strategy and Leadership, 35*(6), 12–17. doi:10.1108/10878570710833714

Clifford, A., & List, M. & Theobald, C. (2010). *Evaluating Economic.* London: Alliance of Sector Skills Councils.

Cooney, K., & Lynch-Cerullo, K. (2014). Measuring the Social Returns of Nonprofits and Social Enterprises: The Promise and Perils of the SROI. *Nonprofit Policy Forum, 5*(2), 367–393.

CSO. (2015). *Veřejná database.* Retrieved May 03, 2016, from: https://vdb.czso.cz/vdbvo2/faces/cs/index.jsf?page=statistiky&filtr=G~F_M~F_Z~F_R~F_P~_S~_null_null_

Dobele, L. (2015). Factors Which Influence the Development of Social Innovation in Latvia. *Proceedings of the 2015 International Conference "Economic Science for Rural Development".* Jelgava: LLU ESAF.

Dohnalová, M., Guri, D., Hrabětová, J., Legnerová, K., & Šlechtová, V. (2015). *Social Enterprise in the Czech Republic.* ICSEM Working Papers, No. 24. Liege: The International Comparative Social Enterprise Models (ICSEM) Project.

Drucker, P. (1977). *An Introductory View of Management.* New York, NY: Harper College Press.

Duca, I. (2013). Efficiency, Effectiveness and Profitability – Concepts Used in Assessing Public Expenditures in Education. *Journal of Knowledge Management Economics and Information Technology, 3*(6), 63–75.

Eriksson, M., Einarsson, T., & Wijkström, F. (2014). *Report on the European social innovation policy framework in light of third sector and civil society actors. Deliverable 1.2 of the project: 'Impact of the Third Sector as Social Innovation' (ITSSOIN), European Commission – 7th Framework Programme.* Brussels: European Commission, DG Research.

European Commission. (2013). *Guide to Social Innovation.* Author.

Fric, P. (2005). The third sector and the policy process in the Czech Republic. *Third Sector European Policy Working Paper, 6*(1), 1-55.

Gerometta, J., Haussermann, H., & Longo, G. (2005). Social Innovation and Civil Society in Urban Governance: Strategies for an Inclusive City. *Urban Studies (Edinburgh, Scotland), 42*(11), 2007–2021. doi:10.1080/00420980500279851

Gherghina, R., Văduva, F., & Postole, M. (2009). The Performance Management in Public institutions of Higher Education and the Economic Crisis. *Annales Universitatis Apulensis Series Oeconomica*, *11*(2), 639–645.

Gupta, S. (2014). Modeling Regional Disparities for a Balanced Quality of life and Apportioning Public Funding – a Graph Theoretical Approach. *Applied Research in Quality of Life*, *10*(3), 473–493. doi:10.1007/s11482-014-9327-1

Howaldt, J., & Schwarz, M. (2010). *Social innovation: concepts, research fields and international trends*. Dortmund: Sozialforschungsstelle Dortmund.

Jałocha, B. (2012). Projectification of the European Union and its implications for public labour market organisations in Poland. *Journal of Project Program & Portfolio Management*, *3*(2), 1–16.

Kadeřábková, M., & Moghadam-Saman, S. (2013). *Evaluation of social innovations –their characteristics and impacts, cross-country comparison and implications for policy support. Social Frontiers: The next edge of social innovation research Evaluation*. Prague: Centre for Innovation Studies.

Kapstein, E., & Kim, R. (2011). *The Socio-Economic Impact of Newmont Ghana Gold Limited*. Retrieved May 03, 2016, from http://www.newmont.com/files/doc_downloads/africa/ahafo/environmental/Socio_Economic_Impact_of_Newmont_Ghana_Gold_July_2011_0_0.pdf

Lauritzen, J. K. R. (2012). *Social Innovation in Local Government – experiences from Denmark*. Retrieved October 23, 2015, from: http://www.euclidnetwork.eu/news-and-events/sector-news/990-social-innovation-in-local-government-experiences-from-denmark.html

Lee, S. M., Hwang, T., & Choi, D. (2012). Open innovation in the public sector of leading countries. *Management Decision*, *50*(1), 147–162. doi:10.1108/00251741211194921

Liebetrau, A. M. (1983). *Measures of association*. Newbury Park, CA: Sage Publications. doi:10.4135/9781412984942

Melecký, L. & Staníčková, M. (2011). Evaluation of the Czech Republic regions' competitiveness in the context of the Lisbon strategy. *Ekonomická revue - Central European Review of Economic Issues*, *14*(3), 183-200.

Mihaiu, D. M., Opreana, A., & Cristescu Pompiliu, M. (2010). Efficiency, Effectiveness and Performance of the Public Sector. *Journal for Economic Forecasting Institute for Economic Forecasting*, *13*(4), 132–147.

Minařík, B., Borůvková, J., & Vystrčil, M. (2013). *Analýzy v regionálním rozvoji*. Praha: Professional Publishing.

MMR. (2013). *Strategie regionálního rozvoje ČR na období 2014–2020*. Praha: MMR.

Moody, M., Littlepage, L., & Paydar, N. (2015). Measuring Social Return on Investment. *Nonprofit Management & Leadership*, *26*(1), 19–37. doi:10.1002/nml.21145

Mouzas, S. (2006). Efficiency Versus Effectiveness in Business Networks. *Journal of Business Research*, *59*(10/11), 1124–1132. doi:10.1016/j.jbusres.2006.09.018

Mulgan, J. (2006). *Social innovation: what it is, why it matters and how it can be accelerated*. Oxford, UK: Oxford Said Business School.

Murray, R., Caulier-Grice, J., & Mulgan, G. (2010). *The open book of social innovation*. National Endowment for Science, Technology and the Art.

Norman-Major, K. (2011). Balancing the Four Es; or Can We Achieve Equity for Social Equity in Public Administration?. *Journal of Public Affairs Education*, *17*(2), 233–252.

Office of the Regional Council South-East. (2014). *Operational Programmes Allocations*. Retrieved October 23, 2015, from: http://www.jihovychod.cz/en/eu-funds/operational-programmes-allocations

Osterwalder, A., & Pigneur, Y. (2010). *Business Model Generation: A Handbook for Visionaries, Game Changers, and Challengers*. Hoboken, NJ: John Wiley and Sons.

Otrusinova, M., & Pastuszkova, E. (2012). Concept of 3 E's and Public Administration Performance. International Journal of Systems Applications. *Engineering & Development*, *6*(2), 171–178.

Pădurean, M. A., Nica, A. M., & Nistoreanu, P. (2015). Entrepreneurship in Tourism and Financing through the Regional Operational Programme. *Amfiteatru Economic*, *17*(38), 180–194.

Phills, J., Deiglmeier, K., & Miller, T. (2008). Rediscovering Social Innovation. *Stanford Social Innovation Review*, *6*(4), 34–43.

Potluka, O., & Liddle, J. (2014). Managing European Union Structural Funds: Using a Multilevel Governance Framework to Examine the Application of the Partnership Principle at the Project Level. *Regional Studies*, *48*(8), 1434–1447. doi:10.1080/00343404.2014.898837

Psacharopoulos, G. (1995). *The Profitability of Investment in Education: Concepts and Methods*. New York: World Bank. Retrieved June 13, 2015, from https://www.commdev.org/userfiles/files/1863_file_Profitability_Investment_Edu.pdf

Richmond, B. J., Mook, L., & Quarter, J. (2003). Social accounting for nonprofits: Two models. *Nonprofit Management & Leadership*, *13*(4), 308–324. doi:10.1002/nml.2

Šebestová, J., & Palová, Z. (2016). Real Societal Impact of Projects Supported within Human Resources and Employment Operational Programme in Selected Czech Regions. In *Proceedings of the 3rd International Conference on European Integration 2016*. Ostrava: VŠB - Technical University of Ostrava.

Social Service Europe (SSE). (2012). *Social Innovation: The Role of Social Service Providers*. Retrieved May 20, 2016, from: http://www.socialserviceseurope.eu/#!publications/c1muj

Tarnawska, K., & Cwiklicki, M. (2012). Evaluation of The European Social Fund as a Measure To Support Social Innovation in The Public Sector. *Economics and Management*, *17*(1), 237–243. doi:10.5755/j01.em.17.1.2273

Tucker, S. (2014). *Social Innovation for Public Service Excellence*. UNDP Global Centre for Public Service Excellence. Retrieved May 20, 2016, from: http://www.undp.org/content/dam/uspc/docs/GPCSE_Social%20Innovation.pdf

Tvrdoň, M. (2015). Moravskoslezský kraj z pohledu makroekonomických ukazatelů. In *18th International Colloquium on Regional Sciences. Conference Proceedings*. Brno: Masarykova univerzita. doi:10.5817/CZ.MUNI.P210-7861-2015-8

Valová, G. (2016). *Development of Business Environment and Small Businesses in the Microregion of Valašskomeziříčsko-Kelečsko* (Unpublished master's thesis). Karviná: SU OPF.

Viturka, M. (2007). Konkurenceschopnost regionů a možnosti jejího hodnocení. *Politicka Ekonomie*, *55*(5), 637–658. doi:10.18267/j.polek.617

Wokoun, R. (2007). Regionální a strukturální politika (politika soudržnosti) Evropské unie v programovém období 2007-2013. *Urbanismus a územní rozvoj*, (10), 1, 3-7.

Wolińska, I., & Rokicki, B., &Zub, M., &Niedoszewska, A. &Piekutowski, J. (2010). Raport końcowy. *Ocena systemu monitorowania i wartości wskaźnikow PO KL 2007-2013*. Warszawa: PSDB grupa WYG.

KEY TERMS AND DEFINITIONS

3 E Method: A method based on three pillars: economy, efficiency and effectiveness.

Hradec Králové Region: A geographic region of the North-Eastern part of the Czech Republic. Well known for motor production, agriculture and Krkonoše mountains.

Moravian-Silesian Region: A geographic region of the Eastern part of the Czech Republic. Well known by heavy industry production, automotive industry and high number of socially excluded locations.

Operational Programme: An initial strategic document that define goals, priority axes and other conditions for each area of development (Education, innovation, Environment, transport). This document is prepared by each EU country member according its National strategy of development.

Plzeň Region: A geographic region of the Western part of the Czech Republic. Well known for food industry, especially beer production, also ceramics and engineering industry is important.

Social Innovation: An innovation which is mostly inspired by community need and it brings new solution of community problems.

Social/Community Impact: An effect of organizational action on well-being of the local community.

SROI: A ratio, which measures Social return on investment, especially in social area.

Three Circles of Cooperation: A concept of cooperation which covers three different activities to support sustainable social innovations. They are named as Circle of local production, Social Business circle and Inspirational circle.

Ústí Region: A geographic region of the North-western part of the Czech Republic.Well known for chemical industry, brown coal, mechanical engineering and Bohemia glass production.

ENDNOTES

1 A priority axis corresponds to a thematic objective and comprises one or more investment priorities of that thematic objective in the Operational program.

APPENDIX

Data used for Regional Evaluation model 2007-2014. Secondary data was filtered from Czech statistical Office (CSO,2015). An index method was used to be able to compare economic development in years 2007-2014. Index (I): indicator data were compared to year 2007 as a baseline:

$$Index_{2007\,/\,2014} = {Indicator_{2014} - Indicator_{2007}} \Big/ {Indicator_{2007}} * 100[\%]$$

An ordinal scale was used for final evaluation in form of Points (P), when 1 is the best result within 5 years, 4 is the worst result within 5 years in chosen group of regions.

Table 10. Regional Evaluation model 2007-2014

	Moravian-Silesian Region		Hradec Kralove Region		Plzen Region		Ústi Region	
	P	I	P	I	P	I	P	I
Macroeconomic impacts (GDP contribution and other macroeconomic indicators)								
GDP contribution	3	-3.0%	2	2.2%	1	4.0%	4	-6.3%
Disposable income of households	3	-1.9%	2	0.0%	1	1.9%	4	-5.6%
Gross fixed capital	2	2.1%	3	-8.8%	4	-25.8%	1	27.3%
Average score	*8*		*7*		*6*		*9*	
Socioeconomic impacts and linkages (education, employment)								
Average monthly wages and salaries	4	-1.3%	2	2.0%	1	2.5%	3	0.1%
Participation rate of people aged 15 and more	1	1.3%	4	-1.7%	1	1.3%	3	-0.7%
General unemployment rate	2	-14.1%	4	8.7%	3	0.6%	1	18.5%
Tertiary education rate	3	-1.7%	4	-2.3%	2	0.1%	1	52.8%
Net increase of businesses (units)	2	-46.4%	1	-14.3%	3	-50.8%	4	-59.5%
Average score	*12*		*15*		*10*		*12*	
Community impacts (changes in settlement, social structure, migration)								
Net migration	4	-755.2%	2	-146.5%	1	-9.8%	3	-153.1%
Mid-year population growth	2	-46.4%	1	-14.3%	3	-50.8%	4	-59.5%
Social innovations	1	72	3	36	4	42	2	89
EU Projects share	1	341	3	77	4	84	2	249
Socially excluded localities	3	19	1	4	2	3	4	11
Average score	*11*		*10*		*14*		*15*	
Environmental impacts								
Environmental protection expenditure	2	28.7%	1	31.5%	3	4.0%	4	-40.6%
Average score	2		1		3		4	

(Sebestova,Palova,2016)

Chapter 8
Local Public Enterprise Business Model as Multiple Value Creation System

Beata Maria Staszewska
Kozminski University, Poland

ABSTRACT

This chapter demonstrates that the local public business model is a phenomenon connected with the multiple value creation system. The primary objective of this study is to gain a more comprehensive understanding of how local public companies may build their business models and simultaneously create new values for customers, stakeholders, and society. The study examines the links between the business model, multiple value creation, business ecosystems with regards to public enterprise. The present study analyses the business model of the local public enterprise in Poland. The findings describe the multiple value creation system from different perspectives and allow for delineating the public business model framework in the tourism industry. Profound understanding of values perception from many perspectives may bring the local public enterprise business model as complex phenomenon closer to recognition.

INTRODUCTION

Local public enterprise represents a popular approach to the communal service delivery in many European countries in the 21st century. The activities of local government companies generate a large part of a gross domestic product, employment, and investments. Public enterprise has a long tradition, and there are continuous discussions and contrary tendencies about privatization or further development of the local public sector. After World War II, multiple public enterprises were created in both democratic and communist countries to address market deficits, reduce mass unemployment and promote economic development. However, from the eighties rising corruption, management inefficiencies, overstaffing and current account deficits exposed serious government business failures which consequently increased tendencies toward privatization and limitation of the quantity of the public enterprises worldwide (Khan, 2008). Moreover, globalization and marketization of the international economy contributed to

DOI: 10.4018/978-1-5225-2215-7.ch008

the development of the private sector and limitation of the public sector including local enterprises. Performance goals such as increasing efficiency are among the primary motives for local privatization and restriction of the public companies. Private firms outperform public ones in several aspects, like productivity, investment, return and access to these services (Perrone Pinheiro et al., 2016). Nonetheless, the general feeling regarding privatization is now more balanced than in the past, due to unsatisfactory results obtained by the private enterprises and difficulties with the control of local economies in turbulent times (Bognetti, & Obermann 20108). From the last economic crisis, the transparency of the bank policy and other global firms are under discussion. Private sector companies have not produced clear, definite improvements in recent years. On the other hand, the public sector in many countries has had difficulty in resolving local economic and social problems due to limited financial resources and insufficient organizational capacities. Therefore, governance alternatives to directly managed local public enterprise have become very familiar nowadays with local governments in Poland, and other countries like United Kingdom, Canada, Japan.

Municipal enterprises are the businesses owned by local governments that provide services and generate new local revenues, which can be used for the purpose of benefiting the community. While national governments have adopted many strategies to promote the transfer of activities from the public to the private sector, there is a widely held belief that basic services like health, education, water, electricity and transportation should remain as public enterprises (Khan, 2008). Local public enterprises play a significant role in building sustainable local economy nowadays for several key reasons: create stable jobs for community members, increase local economic stability by reducing a community's reliance on private businesses, and provide goods and services to important but overlooked areas by for-profit enterprises. Moreover, local public companies often provide more accountability, transparency, and democratic control than the for-profit company. Public entrepreneurs do, in many ways, act like private ones, although there are important differences related to the difficulty in measuring performance, defining objectives, softer budget constraints, and the problems of the underspecified legal status of some public activities (Klein, et all 2010). Despite the fact that rapid increase in the number of municipal enterprises are significant across many countries; there is a limited understanding of what a municipal enterprise is and what it means from an academic perspective (Kallio & Kuoppakangas, 2013). Compared with the widespread debate on corporate governance in the private sector, the international research on the public sector organizations needs more attention from an academic perspective. Moreover, local public enterprises that combine the organizational forms of both business and public mission values at their cores are a common type of hybrid organization, thus making the public enterprise a particularly attractive setting to study.

The local public company is the suitable place where we may observe coexistence of two different logics: the business management model and public management model. In acknowledging the potentially conflicting goals that institutional environments impose on public enterprises, there is an increasing research interest in the organizational and strategically level characteristic for the hybrid organization (Crucke et al., 2015). Research has acknowledged the importance of resolving tensions that possibly emerge from the conflicting demands associated with the social and economic missions, and the key role of decision makers in shaping organizational level goals and strategies (Tracey, Phillips, and Jarvis, 2011). The main objective of this study is to gain the more comprehensive understanding of how a public company builds its hybrid business models in adverse conditions and simultaneously creates new values. The theoretical background consists of three sections. The first part of the theoretical study is focused on business model theory in the context of the value creation and definition of the business model as a research phenomenon. The second part is dedicated to the characteristic of the local public enterprise as a hybrid organization. The third part of the theoretical consideration is aimed at extending the understanding of the business model of the local public enterprise. Afterward, the main subject local public enterprise as a multiple value creation system was considered. This section contains two parts and presents: value creation system for customers and

value creation system for stakeholders and local community. The research section consists of basic information about the method and the recreation park business model as a case example. The empirical part of the research was focused on value phenomenon from various perspectives. Moreover, the full chapter includes solutions and recommendations, proposals for further research directions and conclusion.

BACKGROUND

Business Model and Value Creation

The business model is a theme that connects perfectly the two worlds of business practice and theory. There is no question that interest in business models has grown exponentially in the past few years. Cycles of the firm's existence are shorter and shorter, and traditional business models have evolved to many multidimensional and complex forms, in recent years. New technological possibilities and globalization gives us enormous plurality and diversity of new ways of doing business. The emergence of new modern markets opens sources of the business model that require a shift in thinking about strategy towards more integrative, dynamic, adaptive, and entrepreneurial strategies (Amit, & Zott, 2001). New business models develop every day and over the years, they have become much more sophisticated and complex. According to Priem et al. (2013), there were only ten scholarly articles on business models in leading management journals. Surprisingly, for the same time frame (from 1975 to 2009), sixty papers focused on the business model have been found in significant and well-known practitioner journals. This lack of scholarly concentration on the business model concept appeared striking, especially given the close association of innovative business models with strategic success and the outstanding level of interest in business models among practitioners. In particular, the business model of the public enterprise is underexamined. Why is it so difficult to expand business model theory simultaneously with model business practice? Firstly, little research proposes standard languages, conceptual frameworks, and methods that could be used to understand business models in practice. Secondly, designing new business models is closer to an art than science (Casadesus-Masanell & Ricart, 2010). Most successful and innovative business models are often unique, unrepeatable and difficult to describe. The business model idea is not new. The first signals of that concept in the management literature come from Lang (1947). There is always widespread agreement among business model researchers that core description category for the business model is valued transformation system, however relatively rarely this category is the foundation for the empirical research. Academics have approached the model of a business phenomenon from different points of view. Particularly common is the holistic approach that used system theory to explain the business model, a value-centered theory, and business model framework, that describes the components of the business model. These three areas of analysis are significant for the business model phenomena description and its proper definition. The business model construct creates a holistic perspective on business. According to Timmers (1998) a business model relates to an integrated system consisting of products, services and information flows, including roles of the participants, their potential benefits, and the profit sources. Chesborough and Rosenbloom (2002) confirms that a business model provides a holistic perspective of the business, which helps to understand its internal functions and structures, as well as its interconnectivity and interaction dynamics with the external world. Kindström (2010) claims that nowadays, companies particularly need to focus on their business models in a holistic way, and not just change selected elements.

The business model concept is a systemic and analytical framework. However, business model implementation contains its translation into concrete things, such as a business structure, processes, and infrastructure. Amit and Zott (2010) claim that the business model is a system of activities. Casadesus-Masanell and Ricard (2010) argue that the business model is made of real strategic choices and the consequences of these choices. Morris et al. (2005) similarly define a business model as a concise representation of an interrelated set of decisions. These decisions are addressed to create a sustainable competitive advantage in the specific market. According to Morris et al. (2005), there are three levels of business model definitions: economic, operational, and strategic and there are six subcomponents of the model: factors related to the offering, market factors, internal capability factors, competitive strategy factors, economic factors and personal investor factors. Johnson et al. (2008) argue that the business model consists of four fundamental elements: customer value proposition, profit formula (revenue model, cost structure, margin model and resource velocity), key resources, and key processes. Osterwalder and Pigneur (2010) propose a business model canvas which consists of customer segments, value proposition, channels, customer relationships, key resources, key activities, key partnerships, revenue streams, and cost structure.

There are many various and valuable definitions of the business model phenomena and there is no agreement about the scope of the basic business model components in the academic environment. However, there is a considerable consensus among business model researchers that critical matter for every business model is the value creation. A business model is often seen as a structured presentation of the value creation process and describes how the firm appropriates that value. According to Magretta (2002) a business model answers the following questions: Who is the customer? What does the customer value? How do we make money in this business? And what is the underlying economic logic that explains how we can deliver value to customers at an appropriate cost? Björkdahl (2009) describes the business model as the logic and the activities that create an appropriate economic value. Casadesus-Masanell and Ricart (2010) define the business model as a logic of the firm, the way it operates, and how it creates value for its stakeholders. Amit and Zott (2010) describes a firm's business model as a system of interdependent activities that transcends the firm and spans its boundaries. That activity system enables the firm, with its partners, to create value and also to appropriate its share of value. According to Cao, et al., (2011), a business model should include at least three essential value-based premises: value proposition; value creation and delivery; and value appropriation. Due to this categorization, value proposition concerns the identification of target clients and an answer to the question of why they should buy a product or service. Value creation and delivery are about the design of the retail value chain. Value appropriation describes the retail profit formula. Similarly, Moingeon and Lehmann-Ortega (2010) defines the business model as a description of the mechanisms enabling it to create value through the value proposition made to the clients, its value architecture, and to harness this value and transform it into profits. According to Johannes Sang Un and Hedman (2015) the business model includes five main dimensions: value service, value network, value architecture, value finance, and threats. Value structure is the least understood dimension, despite the fact that value is a cornerstone of a business model (Nag et al., 2007). Therefore, value is the core category for the business model analysis and the main research topic in this study.

The companies that manage to create value over extended periods of time in the same time successfully shape, adapt and renew their business models to fuel such value creation (Achtenhagen et al., 2013). Anderson et. al. (2009) describe value as the sum of the economic, technical, service and social benefits that a customer receives in exchange for a price he or she pays for an offer. Value creation is directly associated with the customer and the company. The concept of shareholder value has been used to evaluate business performance. Shareholder value represents the economic return arising from the implemented decisions and strategies. A business model also involves many non-financial performance categories that determine the creation of value. The value creation index represents the sum of all main intangible assets that the company can evaluate. Low (2000) describes the nine intangible categories of the value innovation: quality, customer relations, management capabilities, alliances, technology, brand value, employee relations, and environmental and community aspects. These categories enable the measurement of the firm's intangible assets, and provide information on the company's performance (Aranha et al., 2015).

In conclusion, this study proposes the definition of the business model as an interactive system of internal and external components of the business that enables value creation. This definition considers the following assumptions: the business model is a system (holistic approach), value creation is a core aspect of the business model, and the business model consists of external and internal components. The attributes and relations among business components make the content for the business model description. This study assumes that the business model consists of subsystems: internal subsystem (resources, activities and offer) and external subsystem – business eco-system (stakeholders-investors, customers, providers, competitors, legislators). Value is created through relationships among business model components, and the sustainability of the business requires that the business model components demonstrate consistency. Morris et al. (2005), argue that consistency of the business model can be described regarding both an internal and external ''fit,'' concerned with a configuration of key activities within the firm and the appropriateness of the configuration given external environmental conditions. Value creation possibilities are the results of changes in the relationships among business model components in order to increase the consistency of the system. Therefore, the business model description concerns not only internal or external components characteristics but especially should include their relations analysis and influences assessment.

Public Enterprise as a Hybrid Organization and Its Business Ecosystem

Public enterprise is most frequently defined as an organization established with a legal personality which is autonomous or semi-autonomous, provides goods and services on a full or partial self-financing basis, and in which the government participates by way of having shares or representation in its decision-making structure (Saussier & Klien, 2014). According to Wilson and Post (2013), public business as a hybrid organization refers to two different sectoral standards: logics, and value systems. The hybrid organizational form requires the coexistence of values from two or more categories. The hybrid company synthesizes activities, structures, processes and meanings by which organizations make sense of and combine multiple organizational forms (Battilana and Lee, 2014). The term hybrid organization has also been employed to describe organizations that span institutional boundaries (Brandsen & Karré, 2011). The implication for practice is the desirability of achieving congruence between the internal environment (decision-making and control) and entrepreneurial activities (Kearney, 2012). However, there are many problems that characterize hybrid organization and influence their results. In public organizations, those difficulties are much more explicit than in the private ones.

Firstly, unlike private entrepreneurs, public entrepreneurs cannot use privately appropriated benefits as criteria for success, and the selection mechanism for allocating resources over time towards more successful public entrepreneurs is complex and poorly understood. Competitive forces are typically weaker in the public domain than in the open market. Moreover, the objectives of the public enterprise are often not clear, performance is more difficult to manage, and political business cycles have its influence on the system. Assent to political power can also entrench individuals and organizations leading to duopolistic political structures with similar policies and collusive attitudes, and even political dynasties and clans – not unknown even in mature democracies (Mueller, 2003). Importantly, political entrepreneurs are likely to undertake actions that foster economic value if they also personally benefit from these actions in ways other than the mere private appropriation of value create (Klein, et al., 2010). If public enterprises are prohibited from distributing all their profits to their shareholders, they will have a direct incentive to take account of their social purposes, both in operating the undertaking and in the subsequent disposal of profits. On the other hand, this may lead to the risk of the owners trying to extract money from the company by other means. Rules to prevent this are also needed if restrictions on the payment of dividends are to be effective (Sørensen, 2014). Greiling and Grüb (2014) suggest that public enterprise should offer a better reporting of their results regarding different dimensions beyond the standard financial accounts. The government would need to require public enterprise at the national and local level to provide both transparent evidence of public mission performance regarding perceptions of users, and factual evidence along with a set of predefined yardsticks. However, the more the external governance model is overloaded, the greater the risk of gaps in the incentive structure for managers, as different branches of government may think that it is not their responsibility to define a public mission or to establish a monitoring and evaluation system for the public enterprise. This, in turn, may favor uncertainty or opportunism in managerial behavior (Florio, 2014). Therefore, the key challenge is to guarantee the competent public authorities' freedom of choice of management modes (Bauby, 2012). However, the public sector generates a lower rate of entrepreneurship than the private sector, and does not appear to be the destination for those with entrepreneurial qualities (Özcan & Reichstein, 2009). Modern public entrepreneurship promotes non-bureaucratic mechanisms to the limitation of the main problem of traditional bureaucracy through changing processes, and cultures through the dimensions of risk-taking, innovation, and pro-activeness: inclined to shared governance, promotion of customer satisfaction, more responsive to its stakeholders, and promotion of cost-efficient performance (Malik & Mahmood, 2012).

Local Public Enterprise Business Model

The study proposes the definition of the business model as an interactive system of internal and external components of the business that enables value creation. The business model consists of subsystems: internal subsystem (resources, activities and offer) and external subsystem – business eco-system (stakeholders-investors, customers, providers, competitors, legislators). This definition is appropriate for private and public enterprise as well. However, the meaning of the main components is different in accordance with the specificity of the local public enterprise.

Firstly, value for stakeholders has wider and more complex meaning in the public enterprise than in the private company. For public business, profit is not the most important indicator of success. The difference between commercial enterprises and public companies lies not in the form of ownership or source of the control but concerns the relative priority that the financial and the social objectives have. Compared to the private sector, accountability expectations are higher for public companies due to public ownership and goals associated with the public mission. (Greiling & Grüb, 2014) There are differences in performance between a local utility firm with varying degrees of public ownership. In particular, public–private utilities show better economic performance than publicly owned firms, especially regarding profitability. The majority of private ownership is not necessary for better performance (Monteduro, 2014). Public businesses need to recover their full costs, so as to achieve self-sustainability, while at the same time, achieving other objectives significant for the local community. Managers of the public company are responsible for overseeing the achievement of financial sustainability at the same time as meeting the different demands of a wider variety of stakeholders (Wilson & Post, 2013). While commercial enterprises might pay some attention to the social values, financial performance is the definitive goal for them. Financial performance is necessary for a public enterprise to survive in the long term, but not sufficient to realize its mission connected with local community requirements. The pursuit of financial sustainability and other significant public objectives requires sufficient revenue (Moizer & Tracey, 2010). Public businesses need to recover their full costs, so as to achieve self-sustainability, while at the same time, achieving their social objective. The significant negative relation between the social and economic mission suggests that the constructs are at two ends of a continuum. Higher levels of social mission imply lower levels of economic mission and vice versa. This result may have various explanations. The strong inverse relation could be the result of the idea that both the social mission and the economic mission in a social enterprise compete for scarce organizational resources.

Secondly, the owner and customer in the public enterprise are partially composed of the same group of people – community members. Public enterprise managers are often challenged in managing the identity of a hybrid organization, responding to pressures from customers, stakeholders, and legislative authorities. The community as a stakeholder is not the homogeneous entity, and contains many different groups of interests, each of them with their own, often diverging, priorities. In practice, this means that the strategy development process will involve networking, communicating, and negotiating with stakeholders to achieve a consensus on main issues and to build and retain legitimacy (Doherty et al., 2014). Local governments often contract out some services. Contracting out may provide immediate electoral advantages as awarding public service contracts allows governments to cater to important voter groups, such as private sector contractors and business associations. Contracting also serves to make the general public more market-friendly as the number of voters employed in the private sector increases. Furthermore, contracting often serves as an argument for privatization of local public enterprises, however, providing immediate short-term electoral support for elected politicians (Sundell & Lapuente, 2012). A managerial approach in the public enterprise is crucial to focus efforts on targeted actions, on services rather than on financial incentives, on a few businesses with the greatest growth potential, rather than on a widespread distribution of funds (Vecchi, et al., 2014). Public sector organizations are often motivated by social and political goals, which furthermore are often numerous, varied and challenging to prioritize. These companies are less visible to the market and its strict application of cost-cutting measures or financial efficiency (Mohanty, 2013). Instead, resources are sometimes allocated with political pressures, and equity is taken into consideration. An organization needs to create space for negotiation to provide a mechanism to resolve the tension that structurally differentiated staff with potentially different interests will face. The combination of structural differentiation and spaces of negotiation facilitate on-going coordination. According to Battilana and Lee (2014), there are two critical conditions that enable the effective functioning of spaces of negotiation: staff members' common understanding of the organization's superordinate goal, and awareness of their mutual interdependencies. Achieving congruence between the internal environment and entrepreneurial activities is crucial for the sustainability of the local public enterprise. Without a fine level of congruence, these activities will lack focus and productivity.

Thirdly, the activities of the local public company and consequently the value proposition for consumers are regulated by the law more strictly, than in the case of private enterprise. Public enterprise represents different characteristics from country to country. Communal enterprises show signs of distinctive national characteristics unique from a historical perspective, or social and economic structure. Therefore, the main differences are noticeable from the point of view of the location of public and private commerce, ownership, the centralization and decentralization of public economy, and the variety of legal forms for public companies (Cox, 2008). Local public enterprises are present in a variety of different sectors and provide a wide range of services. On the one hand, there are classic utility services like water, sewage, waste, electricity distribution and local public transport. On the other hand, local governments create various types of local public enterprises for economic, regional, and business development and planning or even health and care services (Saussier & Klien 2014). What has happened in many governments, it could be argued, has been that central agencies have had a tendency to maintain the old controls and to add new ones, making it even more difficult to manage the established rules or to think outside the box (Meynhardt & Diefenbach, 2012). Regulation of the competition is not enough; regulations must also have as its purpose fulfilment of the public service. The regulation theory has neglected public service regulation in many countries; As a consequence, the particular interest in competition has no doubt led to the devaluation of general interest services (Cox, 2008). The fundamental problem for local business enterprise as a hybrid organization is an inconsistent law system. On the one hand, the local enterprise is a business entity and should perform according to a strong business model. On the other hand, the area of activities is restricted to the communal scope for action. Therefore, possibilities of business model diversification are significantly limited. The development of the local public enterprises and the level of their strategic and operational performance depend on the core arrangement at the highest level of governance and afterward, the policy of the local community. For example, Poland has progressed through a fundamental transformation of its social, economic and political structures since the fall of communism. The economy has been growing steadily. However, the unemployment rate remains relatively high, and living standards of the population are lower than in western Europe. The public sector in Poland has had difficulty in resolving local problems due to limited financial resources and insufficient organizational capacities. Local public enterprises and their activities are often a significant issue for municipal authorities and have an impact on the local development. Related to the lack of financial resources, and in response to the problems with the insufficient financial performance of local public enterprises, the ideological debate about the right size and scope of the public sector started to be tracked with rising attention in Poland. A public service company's activity is strictly regulated by the law. Obligation concerns require public supervision by the regulation body. Competition regulation is already highly developed, although public service regulation is still in its infancy. The situation of the local public enterprises in Poland is unstable. The final legal act of local public entrepreneurship is inconsistent, and becomes a significant source of problems for a local business enterprise. Public enterprise has indeed encountered various degrees of their legal obligations. During the last two decades, the regulation of local public service provision has been changed several times. There are inconsistencies as regards the present organization and its legal form. The decided majority of local public enterprises fulfil public service tasks. However, the legislator gave the ability to establish local public enterprises conducting business activities to a greater extent, in certain cases (Nowak, 2011).

Fourthly, community members as stakeholders and as local suppliers have a greater influence on the financial politics of the enterprise and value proposition. Public companies are in competition with private companies and have to follow the same general market and competition rules unless there is the legitimate exception of competition (Cox, 2008). The public organizations had similar aims: to operate as real market-based organizations, although they were mostly operating in quasi-markets. Also, they wanted to achieve cost and operational efficiency as well as to streamline organizational functions regarding decision making. To do so, they were hovering between a public limited company and a municipal enterprise form. Municipal enterprises are under tight public governance that does not allow for flexible and efficient decision making as in the private sector. Municipal enterprises belong to the municipality and are subject to political control. Too much independence in decision making was seen as a threat to citizen trust (Kuoppakangas, 2014). Local governments in various countries have gradually abandoned the direct management form for indirect forms, including corporatisation and public-private collaboration. These forms of management have comprised, respectively, the transformation of units within local governments into autonomous organizations with their private law legal status and with considerable managerial freedom (Peda et al., 2013). Corporatization is certainly one of the major trends in public and local public enterprises policy. Corporatization was, however, not only restricted to public enterprises but may also be seen as a part of a general trend towards organizational autonomy in public organizations (Saussier & Klien, 2014). Evolution has taken place where the local public service is characterized by the rethinking of the role played by local government. With the objective of providing better quality services to citizens in a more competitive environment, local governments have been involved in new forms of externalization. The choice of an externalized, private law form of organization, especially to manage services of economic significance, has been driven by the need to establish more business-like organizations. That kind of organization demands some degree of autonomy. Mainly, it implies a significant change in the ownership rights and the introduction of incentives to managers' performance. Managers are more keen to hold additional risk and to face increased responsibility. By reallocating control rights to managers, corporatization thus acts as a potentially effective instrument for providing incentives to improve performance and increase efficiency. The results suggest that the transformation of a local firm into an autonomous company – corresponding to the first stage of the corporatization process into a limited liability company can exert a reducing impact on production costs (Cambini, et al., 2011). In recent years, an intense debate has developed regarding the introduction of new tools and control systems in local businesses. Particular attention has been paid to control systems, human resources management, and performance management. It is crucial to develop the analysis of corporate governance mechanisms with regard to their relationships with the external actors (stakeholders) and the internal ones (management) (Gnan et al., 2013). However, this chapter assumes that the core problem with the poor performance of local public companies is connected with the hybrid characteristic of the organization and insufficient awareness of the business model.

LOCAL PUBLIC ENTERPRISE AS A MULTIPLE VALUE CREATION SYSTEM

Value Creation System for Customers

The study analyses the recreation park as a local public enterprise business model that is based on the experience. Experience is a focal point for current tourism management (Tung and Ritchie, 2011a), and the main factor for the value proposition in theme park and leisure industry. Companies innovate by recognizing trends in customer needs and aspirations that, in consequence, provide opportunities to design and develop business models that offer high-value experiences (Pine II and Gilmore, 2016). Ritchie and Hudson (2009) noted that one of the main challenges facing tourism planners is designing the tourism experience. Tourism providers must, therefore, develop a clear understanding of the experience tourists seek and the qualities of the experience that tourists value to ensure they can meet or exceed expectations. Three fundamentals in tourism experience design are suggested: interactive designing process, human-centeredness, and a holistic experience concept as an outcome of designing business model. The value creation process for the customers is connected with the quality of the customer experience. Individual and collective consumer experiences of value creation are multiple, dynamic, individually and socially constructed, and the value of customer experiences emerge in different consumer contexts over time (Kelleher & Peppard, 2011). Both hedonic and utilitarian value dimensions are present for the different tourist's experiences. Functional values and value for money influence satisfaction as well as hedonic value (Prebensen & Rosengren, 2016). These theoretical considerations have influenced the assessment of the value focused on experience in this present study. To enhance memorable tourists, an experience analysis is necessary to understand how tourists produce positive feelings. In recent years, Dijk and Weiler (2009) used ten indicators, including tourist knowledge, attitude, emotional thought and behavioral intention to evaluate tourism experience in three aspects: cognitive, affective and behavioral constructs. Filep and Deery (2010) suggested that tourists' happiness can be evaluated in the three phases of a tour over time: anticipation, on-site experience, and reflections after the visit. Tourism experiences usually link leisure, culture, education and social interaction at the visited place and create a holistic immersion process. A memorable tourism experience is a crucial factor for enhancing tourists' satisfaction with the visited place. Tung and Ritchie (2011a) explored the essence of a memorable tourism experience and identified four dimensions concerning a tourism experience which was particularly memorable for tourists: Namely, affect, expectations, consequentiality, and recollection. The innovation of the experience is described as a collaborative and integrated process. The interpretation of a tourist experience is determined by individual attitudes and beliefs. The visitors exhibited substantial variation in emotions connected with the visited places, activities, and the people they met. It is concluded that the ability to measure a tourist's emotions in real time and natural settings lead to new knowledge, which, in turn, provides a useful foundation for designing and managing tourism experiences (Kim & Fesenmaier, 2015).

According to Özel and Kozak, (2012) tourists could be classified into five distinct groups, each with different travel motivations. The accuracy of the cluster classification was tested using discriminant analysis and delineated clusters were labeled: relaxation seekers, sports seekers, family oriented, escapists, achievement and autonomy seekers. The motive for travelling contains a wish to realise a great number of various human needs and desires such as the need to spread knowledge, for aesthetic experiences, encountering new areas and people, change, freedom and relaxation, adventure and play, as well as the need for travelling in search of one's own personality. According to Rageh et al., (2013) there are eight dimensions: comfort, educational, hedonic, novelty, recognition, relational, safety, and beauty, which are consistent with major studies on experience. Lache and Alexa (2011) claim that tourist consumption represents the expression of a very varied set of needs and motivations: rest, relaxation, escape from the every-day setting, recreation or the satisfaction of certain cultural, aesthetic, emotional aspirations, based on the principle of pleasure, of the integration of individuals in their social environment, and the relation with their personality and family or personal experiences. In this study, the framework for research is developed that contains categories for describing valuable tourist experiences such as uniqueness, diversification, activeness, relaxation, knowledge, entertainment, challenge, friendship as a social aspect and engaging the senses in a positive way.

Uniqueness is an important factor for a memorable experience. According to the Poulsson and Kale (2004) experience, value is connected with personal relevance, novelty, surprise, and engagement. According to Tung and Ritchie (2011b), if tourists find something beyond their expectation, a feeling of surprise will arise and enhance their emotional experience. The core experience generates new impressions, surreal feelings, incomparable and memorable evaluations of the whole experience (Maklan & Klaus, 2011). On the other hand, diversification of the tourist offer enables them to join many different activities, plurality of impressions, and experiences to satisfy clients. The process that identifies required experiences allows tourists to make a good choice between different alternative activities. Activeness in the physical and mental sense ensures a strong engagement in tourist offers. Tourists in some places were not a prerequisite to emotional or spiritual experiences, for example being in and engaging with the landscape but effectively becoming part of it – especially physical activity is fundamental to emotional responses (Jepson & Sharpley, 2015). The movement of people and their urge to travel are connected to biological needs which offer relaxation to tourists and result in greater cultural knowledge (Zbornik, 2012). Tourist knowledge experience relates to the free-choice learning, rather than the goals for factors affecting or outcomes of free-choice learning, has revealed that these learning experiences are often characterized as entertaining, fun, enjoyable, multi-sensory, fascinating, effortless and involve discovery and freedom of choice (Winkle and Lagay, 2012). Entertainment as a value of tourist experience category describes a full range of positive emotions. In the study of Filep and Deery (2010), tourism was considered as a powerful intervention to create happiness, because tourists can experience positive emotions such as joy, interest, and contentment during their activities. The entertainment factor is connected with hedonic enjoyment and refers to the hedonic pleasure and excitement, which is felt, perceived and experienced by participants at any time during the trip. The hedonic motives also express a deeper level of enjoyment resulting from skill mastery, risk normalization and a state of flow. In other words, hedonic enjoyment is distinguished from the enjoyment of extraordinary experiences. Personal progression in this research focuses on the skill acquisition experience. Personal progression does not include new identity development, safety, or survival concerns, which are parts of the efficacy motives (Klaus and Maklan, 2011). Many tourists today seek challenging and extreme or extraordinary experiences. Friendship as an experience factor relates to the social interaction during the tourist trip, this experiential factor is related to other people before, during, and after the experience. This dimension, the interaction with other tourists, staff, and coaches, facilitates the core experience. Moreover, the tourist experience is an offer that is consistent with the theme and engages the customers' senses. That kind of offer has to be personal, unique, memorable and sustainable over time (Pine II and Gilmore, 1998). For the tourist, no matter the motivation, everything sensorial matters. Sensory experience offers benefits to the tourist through the sensations and emotions that are felt and the memories that are created, as well as a perception of quality. Ditoiu et al., (2014) describes sensory dimension as: visual dimension (what the tourist sees), auditory dimension (what the tourist hears), olfactory dimension (what the tourist smells), gustatory dimension (what the tourist tastes), and tactile dimension (what the tourist feels).

Value Creation System for Stakeholders and Local Community

Municipal enterprises are the businesses owned by local governments that provide services and generate new local revenue that can be used for community benefiting purposes. Local public enterprises play a significant role in building a community sustainable local economy for several key reasons: create stable jobs for community members and increase local economic stability by reducing a community's reliance on private businesses, and provide goods and services to underserved areas overlooked by for-profit enterprises. International assessments show no consistent results on the performance of companies in the public sector and question social outcomes, as in many cases the private sector presents higher costs and tariffs than the public sector (Perrone Pinheiro et al., 2016). Moreover, local public companies often provide more accountability, transparency, and democratic control than the for-profit company.

Performance criteria for the public and private enterprise are different. For example, when issues of social affordability, environmental sustainability, security of supply, and other policy goals are to be evaluated (Florio, 2013). According to Gutu (2014), the administration of the local public property is obliged to the principles: efficiency, legalization, and transparency. McDonald and Ruiters (2012) identify several normative 'criteria for success' of a contemporary public enterprise. The performance criteria would go beyond economic efficiency to include equity, participation, accountability, transparency, sustainability, solidarity, public ethos. The municipal enterprise is an economic entity with juridical personality constituted exclusiveness by the municipal property. Public enterprise produces certain types of consignments, executes the works and carries out the services that are necessary for the satisfaction of the requirements of the founder and the realization of the social and economic interests of the work staff (Gutu, 2014). Networking and collaboration allow for overcoming the lack of resources faced by local public firms. Networking was beneficial for the local public company, for several reasons. An economic benefit in the initial phase of the firms and the network resulting from joint marketing. Moreover, many firms increased their market shares, and the benefits of friendship and fellowship become more important than the economic benefits of the network (Tolstad, 2014). Enterprises, which develop their abilities to build strong relationships with customers and other partners, develop an offer that is adaptive to changing customer needs and environmental conditions (Kindström, 2010). Through partnering with customers, managers are equipped with the ability to anticipate discontinuity in consumer preferences, and through customer knowledge, they can meet future demand expediently (Agarwal et al., 2014). The importance of business networks is that they have resources that can supplement internal skills and knowledge and build relationships with suppliers and customers that is beneficial for all sides (Macpherson et al., 2004).

Resource interactions and integrations are fundamental for value to be co-created since they give rise to new interpretations and meaning creations (Fyrberg Yngfalk, 2013). Tantalo and Priem (2016) introduced the stakeholder synergy perspective and offered the proposition for realizing broader value creation that is likely to produce a sustainable competitive advantage. A stakeholder synergy perspective identifies new value creation opportunities that are particularly effective strategically because a single strategic action increases different types of value for two or more essential stakeholder groups simultaneously, and does not reduce the value received by any other essential stakeholder group. Multiple sources of value creation exist for each important stakeholder group. Actions that meet particular criteria increase the size of the value available for important stakeholder groups and obtain their increasing effort and commitment.

ILLUSTRATIVE CASE STUDY OF A LOCAL PUBLIC ENTERPRISE

Research Approach

This study presents an exploratory research of how one of the local public companies creates its business model. The case study method was chosen for the research purpose. Case studies explore and investigate contemporary real-life phenomenon through a detailed contextual analysis of a limited number of events or conditions, and their relationships (Zainal, 2007). The emphasis of the classic case study approach is to highlight a construct by showing its operation in an on-going social context. Case study methods enable us to understand the behavioral conditions through the actor's perspective. The classic case study is a coherent, credible, and memorable story (Eisenhardt, 1989). Yin (1984) defines the case study method as an empirical inquiry that investigates a contemporary phenomenon within its real-life context; particularly when the boundaries between phenomenon and context are not clearly evident. Moreover, case study research allows the exploration and understanding of complex issues. It can be considered a proper research method, particularly when a holistic, in-depth examination is required. The business model in its true essence is a complex and required holistic and thoroughgoing examination, particularly when its context is not clear. Therefore, the case study has been chosen as the proper method for the empirical research.

There are several categories of the case study. Yin (2003) notes three categories, namely exploratory, descriptive and explanatory case studies. First, exploratory case studies set to explore any phenomenon in the data which serves as a point of interest to the researcher. Descriptive case studies set to describe the natural phenomena which occur within the data in question. Explanatory case studies examine the data closely both at a surface and deep level in order to explain the phenomena in the data (Zainal, 2007). This study has an explicit exploratory character Moreover, the case study with mixed research data, including both quantitative and qualitative data, helps explain both the process and outcome of a phenomenon through observation, reconstruction, and analysis of the cases under investigation (Tellis, 1997). This study was an exploratory character and was based on qualitative and quantitative date as well. The sources of information were: questionnaires, the firm's documents, reports, strategy and plans, and information from the website.

Three steps were adopted for the development of this study. In the first step, the main framework of the value proposition in the business model context analysis was designed, identifying the two subjects for the empirical research – exploring the value proposition for customers and exploration value proposition for stakeholders. In the second step, a questionnaire for the empirical research was developed with the constructs from the literature reviews, and documents analysis (annual reports and presentations). Two questionnaires were created and used to collect data from respondents. The respondents groups were customers for the first questionnaire, the council board members and managers for the second questionnaire. The first questionnaire for the customers contains research categories: overall value proposition (uniqueness, diversification, activeness, relaxation, knowledge, entertainment, challenge, friendship), sensory value categories (visual impressions, gustatory impressions, tactile impressions, auditory impressions), and functional categories (infrastructure size, equipment quality, area cleanness, objects cleanness, personnel knowledge, personnel professionalism, personnel attitudes). Overall value proposition and sensory value categories were developed on the basis of the literature review. Functional categories were added by pre-test on the group of department managers and core personnel. The second questionnaire for the stakeholders contains the following research categories: recreation park as leisure place for the residents, new employment places, broad network of local suppliers, residents qualifications development, quality of the management of the communal property, financial profit and activity self-funding, investments in the communal property, new local tourist products development, promotion of regional tourism brand, new opportunities for local tourism sector development, and popularization of the valuable ideas (education, ecology, modernity). These categories were developed on the basis of information from company presentation and annual reports. Constructs were measured using a five-point Likert-like scale and there were some open questions for the deeper exploration research phenomenon. Since there are tourist sector specific and company specific scales items, this questionnaire is adapted for the aims of this particular research. As a result, the specific method was created based on construct definitions, designed scales, and conducted qualitative and quantitative pre-tests. The questionnaires were pretested quantitatively on a sample group from the company (department managers and core personnel). Additional adjustments were made and discussed with managers before launching the survey. In the third step, information from the questionnaires was gathered and analyzed. The sample for the empirical research was three groups of respondents: customers (78 respondents), town council members (13 respondents), company managers (3 respondents). Attendance in the survey research was anonymous. The company name was changed to restrict identification of the organization and potential damage to a firm's good reputation.

Recreational Park Business Model as Multiple Value Creation System

Recreational Park is a limited liability company, with 100% local government ownership. Recreational Park was founded in 2011 with the primary purpose of the new recreational complex management (with European Regional Fund contribution). In the following five years, the enterprise grew into a regional tourism leader in the field of outdoor recreation and interactive education. The owner of the enterprise is a small community with limited co-finance possibilities, so financial independence from the communal budget was the main assumption for the company business model. The business model allows for self-funding (percentage revenue growth 2011-2015 – 170%, independence from the community budget from 2013 to 2015). Awards for the product quality in 2015 - best leisure infrastructure in Europe – winner 2015/2016, best educational complex in Poland – winner 2015.

The Recreational Park creates values for the customers and the local community. The business offer contains commercial products like educational laboratories, rope course, creative park, water track, coffee bar, memorabilia shop, accommodation complex. Moreover, the recreational park of the 10 ha is a public area, and everybody may benefit from the vast beach with secure bathing places, outdoor gym, gardens, observation tower, pathways and the central square with fountains (free offer). The value proposition of the Recreational Park combines many fields: nature, innovation, education, active recreation, ecology and represents a complex offer based on experience. Value for customers includes overall value proposition focused on the strong brand (uniqueness, diversification, activeness, relaxation, knowledge, entertainment, challenge, friendship), sensory value (visual impressions, gustatory impressions, tactile impressions, auditory impressions), and functional values (infrastructure size, equipment quality, area cleanness, objects cleanness, personnel knowledge, personnel professionalism, personnel attitudes).

Data in Table 1 shows that the respondents admit tremendous value to personnel attitudes, professionalism and knowledge. These three categories have the biggest value and the expected value is equal to the current value. Cleanness of the area is the next factor significant in current value proposition for customers. The current value proposition for customers is focused on knowledge, activeness, diversification and entertainment. Sensory visual impressions and auditory impressions are valuable factors of the whole value proposition. The weakest categories of the current offer are sensory gustatory impressions (in this category, the expected value is significantly above the current offer), and uniqueness of the whole offer (in this category, the expected value is equal to the current value. The difference between the current value and expected value also appeared in object cleanness. Object cleanness and gastronomy offer need some work to increase customer satisfaction in the future.

Recreational Park values for the local community include: a leisure place for the residents, new employment places, broad network of local suppliers, residents qualifications development, quality of the management of the communal property, financial profit and activity self-funding, investments in the communal property, new local tourist products development, promotion of regional tourist brand, new

Table 1. Customer opinions on the value of the Recreation Park (N=78)

Items - Customer Value Proposition	Average Satisfaction	Average Priority	Expected Value	Total Value	Value Rank
Overall Value Proposition					
Knowledge	1.83	1.85	-0.01	3.38	**5**
Activeness	1.79	1.82	-0.03	3.27	**6**
Diversification	1.74	1.79	-0.05	3.13	**9**
Entertainment	1.76	1.8	-0.01	3.11	**10**
Challenge	1.7	1.63	0.05	2.73	**14**
Friendship	1.65	1.64	0.01	2.71	**15**
Relaxation	1.56	1.67	-0.1	2.61	**17**
Uniqueness	1.61	1.59	0.03	2.57	**18**
Sensory Value					
Visual impressions	1.79	1.78	0.01	3.2	**7**
Auditory impressions	1.74	1.77	-0.03	3.08	**11**
Tactile impressions	1.65	1.6	0.05	2.65	**16**
Gustatory impressions	0.74	0.92	-0.18	0.69	**19**
Functional Value					
Personnel attitudes	1.9	1.89	0.01	3.58	**1**
Personnel professionalizm	1.88	1.9	-0.01	3.57	**2**
Personnel knowledge	1.88	1.88	0	3.55	**3**
Area cleaness	1.86	1.87	-0.01	3.48	**4**
Objects cleaness	1.72	1.83	-0.11	3.15	**8**
Infrastructure size	1.68	1.68	0	2.82	**12**
Equipment quality	1.68	1.67	0.01	2.8	**13**

Source: Study based on the own survey 2016

opportunities for local tourist sector development and popularization of the valuable ideas (education, ecology, modernity).

Data in Table 2 described town council members' opinions about stakeholders' value proposition. Value for the stakeholders is focused on the four value categories: 1. Promotion of the community area as an attractive tourist place, 2. Popularization of the valuable ideas- education, ecology, modernity, 3. Recreational Park as a leisure place for the residents and 4. Creation of the strong tourist brand. The weakest value categories are residents' qualifications development, development of the local supplier's network and investment in the local property. Differences between the current value and expected value show the categories of new employment places, development of the local supplier's network and residents qualifications development. These are categories that describe area for improvement. Especially worthy of note is that financial profit and activity self-funding took the last place as an average priority in the opinion of the municipal councilors.

Table 3 contains data that show management opinions about Recreational Park value creation system for the stakeholders. Significant values relate to activities focused on the strong tourist brand creation and popularization of valuable ideas like education, ecology, and modernity. Promotion of the community area as an attractive tourist place is an important source of value for the communal environments. Categories, which create weaker value for the local community, according to the management opinions, are new local tourism product development and financial profit and activity self-funding. Creation of new local tourism products is relatively unimportant (average priority -0,67). However, financial profit and activity self-funding is a priority significance factor, and low value is connected with the lowest satisfaction level (0,33). The expected value of this factor is lowest (-1,67) and indicates that according to the management opinion, the business model of the company should be focused on the financial performance more than thus far. The value creation system focused on the stakeholders' perspective significantly differ in the

Table 2. Town council members opinions about value proposition for the stakeholders (N=13)

Items - Stakeholders Value Proposition	Average Satisfaction	Average Priority	Expected Value	Total Value	Value Rank
Promotion of the community area as an attractive tourist place	1.46	1.62	-0.15	2.36	**1**
Popularization of the valuable ideas- education, ecology, modernity	1.23	1.23	0	1.51	**2**
Recreational Park as a leisure place for the residents	1.08	1.31	-0.23	1.4	**3**
Creation of the strong tourist brand	1.15	1.08	0.08	1.24	**4**
Quality of the management of the communal property	0.92	1.15	-0.23	1.06	**5.5**
New local tourist products development	0.92	1.15	-0.23	1.06	**5.5**
Tourism local sector development	0.85	1.23	-0.38	1.04	**7**
Financial profit and activity self-funding	0.92	1	-0.08	0.92	**8**
New local employment places	0.69	1.23	-0.54	0.85	**9**
Investments in the communal property	0.77	1.08	-0.31	0.83	**10**
Development of the local suppliers network	0.61	1.15	-0.54	0.71	**11**
Residents qualifications development	0.54	1.08	-0.54	0.58	**12**

Source: Study based on the own survey 2016

Table 3. Management opinions about the value proposition for the stakeholders (N=3)

Items - Stakeholders Value Proposition	Average Satisfaction	Average Priority	Expected Value	Total Value	Value Rank
Creation of the strong tourist brand	2	1.67	0.33	3.34	**1.5**
Popularization of the valuable ideas- education, ecology, modernity	2	1.67	0.34	3.34	**1,5**
Promotion of the community area as an attractive tourist place	1.67	1.33	0.34	2.22	**3**
Recreational Park as a leisure place for the residents	1.33	1.33	0	1.77	**4**
Quality of the management of the communal property	1.67	1	0.67	1.67	**5,5**
Investments in the communal property	1.67	1	0.67	1.67	**5,5**
New local employment places	1.67	1	0.67	1.67	**7**
Quality of the management of the communal property	1.67	1	0.67	1.67	**7**
Investments in the communal property	1.67	1	0.67	1.67	**7**
Development of the local suppliers network	1.33	0.67	0.66	0.89	**8.5**
Residents qualifications development	1.33	0.67	0.66	0.89	**8.5**
Tourism local sector development	1	0.67	0.33	0.67	**10**
Financial profit and activity self-funding	0.33	2	-1.67	0.66	**11**
New local tourist products development	0.67	0.67	0	0.45	**12**

Source: Study based on the own survey 2016

perspective of the council board members and the management of the company in their opinion about expected value in the category of financial profit and activity self-funding.

SOLUTIONS AND RECOMMENDATIONS

This study analyses the local public enterprise as a hybrid organization and a complex business system. According to Cox (2008), the local public entrepreneurship should be understood as a research subject covering many theoretical disciplines like business management theory, competition theory, institutional economics, juridical and socio-economic. Public enterprises combine social goals and revenue generating activities in a variety of organizational constellations and institutional contexts. It is a challenging and ambiguous research field. However, because of its importance for the local economy, it cannot be overestimated. This empirical research concerns the local public enterprise business model as a hybrid organization which is an appropriate sample for the analysis of potential values conflict. Companies adopting the hybrid business model engage in more collaboration, but at the same time also establish a great proportion of exploration (Greiner and Ang, 2012). Business models relate to how firm offerings are developed based on problem solving and interactions with customers and stakeholders, which implies that the focus of the business model creation should be set on the ground at which this interaction occurs (Bankvall et al., 2014). This study describes the business model as a system consisting of subsystems: internal subsystem (resources, activities and offer) and external subsystem (stakeholders-investors, customers, and legislators). Different kind of relationships in the business network often cross over organizational boundaries. In facing the dynamic of an environment, the firm maintains its strategies to look for new resources, and continually build, enhance and reconstruct the business networks. This research introduces multiple value creation system, which allows observing company activities through the perspective of the different interest groups of customers, stakeholders, and managers.

This study indicates that the local public business model is a phenomenon connected with the multiple value creation system. The value creation system for the customers in this case should be appropriate for the tourism sector and is connected with the experience economy. In this study, the value proposition for the tourism sector was described and main categories for the value evaluation were established. Value for customers includes overall value proposition focused on the strong brand, sensory values, and functional values. The value system for stakeholders include: leisure place for the residents, new employment places, broad network of local suppliers, residents qualifications development, quality of the management of the communal property, financial profit and activity self-funding, investments in the communal property, new local tourist products development, promotion of regional tourist brand, new opportunities for local tourist sector development and popularization of the valuable ideas (education, ecology, modernity). Implication of the findings of the empirical research is that a value creation system focused on the stakeholders' perspective may significantly differ from the perspective of the council board members and the management of the company. In particular, financial profit and activity focused on self-funding has a different approach to the management and town council members' perspectives. Another submission is that public sector organizations can and should be viewed as entrepreneurial entities, particularly because public-sector organizations, like private sector ones, can create value for society through the thoughtful business model.

This study leads to some recommendation for the practice. Congruence between company stakeholders, town council members and managers in the matter of the value creation system is particularly significant for the effective business model of the local public enterprise. Moreover, congruence between the internal environment and entrepreneurial activities is crucial for the sustainability of the local public enterprise. Without a fine level of congruence, these activities will lack focus and productivity. To achieve congruence between the internal and external environment, management may negotiate goals and values with the many groups of stakeholders. According to Crilly et al., (2016) cognitive and linguistic perspective explains how difficult is to gain consensus and commitment among different stakeholder groups. Moreover, stakeholders with conflicts of interest are unable to detect these linguistic nuances. Communication and collaboration are key aspects of the value congruence that constitute the base for the sustainable business model. However, the communication development among different groups of stakeholders is possible only when differences in value evaluation from many perspectives are adequately diagnosed. Business model as a multiple value creation framework allows for recognition some linguistic categories for discussion. Differences in perception relates to value have to be recognized and analysed during the process of the business model development. This area of consideration is crucial for the business model creation and implementation.

FUTURE RESEARCH DIRECTIONS

This research as a single case study has its limitations. The case study method has been criticized for its lack of academic rigor and the tendency for a researcher to have a biased interpretation of the data. The most common disadvantage of the case method research approach concerns the inter-related issues of methodological rigor, researcher subjectivity, and external validity. Reliability and generality are also subjected to scepticism when a small sampling is deployed (Zainal, 2007). Case studies provide a very little basis for scientific generalization since they use a small number of subjects. This research was conducted with only one subject. The case study research method and, in particular, intrinsic case studies which set to solve the specific problem of an individual case, do not allow for the generalization of findings to a bigger population. There are some limitations connected with data collection and analysis, also. In particular, the research sample was chosen from three groups: customers, stakeholders, and managers. The research did not investigate other partners, like co-operators or suppliers. Constructs for the questionnaires were chosen through literature research and qualitative research of the company documents like annual reports and presentation. There was some area for subjective interpretation in choice of particular constructs for the questionnaire. Hence, future research could comprise more comprehensive analysis of quantitative data. Moreover, there is a need for more research on the business model focused on the value creation process. Further research may be aimed at exploring the relations between the values system and other components of the business model. Another interesting avenue for future research is concerned with the business model development in the long-term perspective. That study has its limitations, therefore the business model focused on value creation should be investigated with a different methodological approach.

CONCLUSION

A key objective of this study was to gain a better understanding of the local public enterprise business model as a multiple value creation system. The findings reveal a worthwhile information, but particularly, three issues are worth mentioning in the first place. Value is created through relationships among public sector business model components. Such components must be consistent in order to make the model sustainable. Moreover, the public enterprise business model involves many financial and non-financial performance categories that determine the creation of value. Consistency of created values is difficult to accomplish, due to the plurality of values expected by many interest groups from internal and external environments, such as customers, managers, employees, suppliers and other stakeholders. The worth of the business model depends on the vast range of values for many groups and their sophistication. Profound understanding of values on various levels may bring the particular business model phenomenon closer to recognition.

The study provided a small step for better understanding of the local public enterprise business model phenomenon. Conducted research outlined some categories for value description from different perspectives and may be usable for the creation of the public enterprise sustainable business model in several sectors relevant for local and regional development, like tourism for instance. Nevertheless, a thoroughgoing analysis of the value creation system is necessary for establishing the sustainable business in every particular case and circumstances.

REFERENCES

Achtenhagen, L., Melin, L., & Naldi, L. (2013). Dynamics of business models – Strategizing, critical capabilities and activities for sustained value creation. *Long Range Planning*, *46*(6), 427–442. doi:10.1016/j.lrp.2013.04.002

Agarwal, R., Selen, W., Sajib, S., & Scerri, M. (2014). Dynamic capability building in service networks: An exploratory case study. *Journal of New Business Ideas and Trends*, *12*(1), 27–41.

Amit, R., & Zott, C. (2001). Value creation in E-business. *Strategic Management Journal*, *22*(6-7), 493–520. doi:10.1002/smj.187

Anderson, J., Narus, J., & Narayandas, D. (2009). *Business market management: Understanding, creating, and delivering value* (3rd ed.). Upper Saddle River, NJ: Prentice Hall.

Aranha, E. A., Garcia, N. P., & Corrêa, G. (2015). Open innovation and business model: A Brazilian company case study. *Journal of Technology Management & Innovation*, *10*(4), 91–98. doi:10.4067/S0718-27242015000400010

Bankvall, L., Dubois, A., & Lind, F. (2014). Towards a network-based business model concept and the case of a truck OEM considering its business model in India. *IMP Journal*, *8*(3), 44–50.

Battilana, J., & Lee, M. (2014). Advancing research on hybrid organizing – insights from the study of social enterprises. *The Academy of Management Annals*, *8*(1), 397–441. doi:10.1080/19416520.2014.893615

Bauby, P. (2012). Local services of general economic interest in Europe. Water services: What are the challenges? *Annals of Public and Cooperative Economics*, *83*(4), 561–583. doi:10.1111/j.1467-8292.2012.00477.x

Björkdahl, J. (2009). Technology cross-fertilization and the business model: The case of integrating ICTs in mechanical engineering products. *Research Policy*, *38*(9), 1468–1477. doi:10.1016/j.respol.2009.07.006

Bognetti, G., & Obermann, G. (2008). Liberalization and privatization of public utilities: Origins of the debate, current issues and challenges for the future. *Annals of Public and Cooperative Economics*, *79*(3/4), 461–485. doi:10.1111/j.1467-8292.2008.00367.x

Brandsen, T., & Karré, P. M. (2011). Hybrid organizations: No cause for concern? *International Journal of Public Administration*, *34*(13), 827–836. doi:10.1080/01900692.2011.605090

Cambini, C., Filippini, M., Piacenza, M., & Vannoni, D. V. (2011). Corporatization and firm performance: Evidence from publicly-provided local utilities. *Review of Law & Economics*, *7*(1), 195–217. doi:10.2202/1555-5879.1428

Cao, G., Wiengarten, F., & Humphreys, P. (2011). Towards a contingency resource-based view of IT business value. *Systemic Practice and Action Research*, *24*(1), 85–106. doi:10.1007/s11213-010-9178-0

Casadesus-Masanell, R., & Ricart, J. (2010). From strategy to business models and onto tactics. *Long Range Planning*, *43*(2/3), 195–215. doi:10.1016/j.lrp.2010.01.004

Chesborough, H., & Rosenbloom, R. (2002). The role of the business model in capturing value from innovation: Evidence from Xerox Corporations technology spin-off companies. *Industrial and Corporate Change*, *11*(3), 529–555. doi:10.1093/icc/11.3.529

Cox, H. (2008). Public enterprises and service providers in institutional competition and undergoing structural change. New challenges to the theory of public economics and public services in Germany. *Annals of Public and Cooperative Economics*, *79*(3/4), 527–547. doi:10.1111/j.1467-8292.2008.00369.x

Crilly, D., Hansen, M., & Zollo, M. (2016). The grammar of decoupling: A cognitive-linguistic perspective on firms sustainability claims and stakeholders. *Academy of Management Journal*, *59*(2), 705–729. doi:10.5465/amj.2015.0171

Crucke, S., Moray, N., & Vallet, N. (2015). Internal representation and factorial faultiness as antecedents for board performance in social enterprises. *Annals of Public and Cooperative Economics*, *86*(2), 385–400. doi:10.1111/apce.12076

Dijk, P. V., & Weiler, B. (2009). An assessment of the outcomes of a Chinese-language interpretive tour experience at a heritage tourism attraction. *Tourism Analysis*, *14*(1), 49–63. doi:10.3727/108354209788970144

Ditoiu, M., Stancioiu, A., Teodorescu, N., Onisor, L., & Radu, A. (2014). Sensory experience - between the tourist and the marketer. *Theoretical & Applied Economics*, *21*(12), 37–50.

Doherty, B., Haugh, H., & Lyon, F. (2014). Social enterprises as hybrid organizations: A review and research agenda. *International Journal of Management Reviews*, *16*(4), 417–436. doi:10.1111/ijmr.12028

Eisenhardt, K. M. (1989). Building theories from case study research. *Academy of Management Review*, *14*(4), 532–550.

Filep, S., & Deery, M. (2010). Towards a picture of tourists happiness: An insight from psychology. *Tourism Analysis*, *15*(4), 399–410. doi:10.3727/108354210X12864727453061

Florio, M. (2013). Rethinking on public enterprise: Editorial introduction and some personal remarks on the research agenda. *International Review of Applied Economics*, *27*(2), 135–149. doi:10.1080/02692171.2013.785664

Florio, M. (2014). Contemporary public enterprises: innovation, accountability, governance. *Journal of Economic Policy Reform*, 201-208.

Fyrberg Yngfalk, A. (2013). Its not us, its them! – Rethinking value co-creation among multiple actors. *Journal of Marketing Management*, *29*(9/10), 1163–1181. doi:10.1080/0267257X.2013.796318

Gilmore, J. H., & Pine, B. J. II. (1998). Welcome to the experience economy. *Harvard Business Review*, *76*(4), 97–105. PMID:10181589

Gnan, L., Hinna, A., Monteduro, F., & Scarozza, D. (2013). Corporate governance and management practices: Stakeholder involvement, quality and sustainability tools adoption. *Journal of Management & Governance*, *17*(4), 907–937. doi:10.1007/s10997-011-9201-6

Greiling, D., & Grüb, B. (2014). Sustainability reporting in Austrian and German local public enterprises. *Journal of Economic Policy Reform*, *17*(3), 209–223. doi:10.1080/17487870.2014.909315

Greiner, R., & Ang, S. (2012). Biotechnology collaborations: Does business model matter? *Journal of Management & Governance*, *16*(3), 377–392. doi:10.1007/s10997-010-9156-z

Gutu, G. (2014). General administration vs. right to manage of local community public property. *Juridical Current*, *17*(2), 54–59.

Jepson, D., & Sharpley, R. (2015). More than sense of place? Exploring the emotional dimension of rural tourism experiences. *Journal of Sustainable Tourism*, *23*(8/9), 1157–1178. doi:10.1080/09669582.2014.953543

Johannes Sang Un, C., & Hedman, J. (2015). Business models for NFC based mobile payments. *Journal of Business Models*, *3*(1), 29–48.

Johnson, M., Christensen, C., & Kagermann, H. (2008). Reinventing your business model. *Harvard Business Review*, *86*(12), 50–59. PMID:18681297

Kallio, T. J., & Kuoppakangas, P. (2013). Bandwagoning municipal enterprises: Institutional isomorphism and the search for the third way. *Policy Studies*, *34*(1), 19–35. doi:10.1080/01442872.2012.731842

Kearney, J. D. Sr. (2012). Strategic vs. financial buyer: Choosing the best option. *Financial Executive*, *28*(6), 30–33.

Kelleher, C., & Peppard, J. (2011). Consumer experience of value creation: A phenomenological perspective. *Advances In Consumer Research - European Conference Proceedings*.

Khan, M. A. (2008). Reinventing public enterprises in public enterprises. In *Public enterprises: Unresolved Challenges and New Opportunities*. New York: United Nations.

Kim, J., & Fesenmaier, D. R. (2015). Measuring emotions in real time: Implications for tourism experience design. *Journal of Travel Research*, *54*(4), 419–429. doi:10.1177/0047287514550100

Kindström, D. (2010). Towards a service-based business model – Key aspects for future competitive advantage. *European Management Journal*, *28*(6), 479–490. doi:10.1016/j.emj.2010.07.002

Klaus, P., & Maklan, S. (2011). Bridging the gap for destination extreme sports: A model of sports tourism customer experience. *Journal of Marketing Management*, *27*(13/14), 1341–1365. doi:10.1080/0267257X.2011.624534

Klein, P. G., Mahoney, J. T., McGahan, A. M., & Pitelis, C. N. (2010). Toward a theory of public entrepreneurship. *European Management Review*, *7*(1), 1–15. doi:10.1057/emr.2010.1

Kuoppakangas, P. (2014). Adopting a municipal enterprise form: Expectations and outcomes – Three cases and dilemmas. *Journal of Change Management*, *14*(1), 126–145. doi:10.1080/14697017.2013.778895

Lache, C., & Alexa, E. L. (2011). Particular aspects of the tourist products consumers' loyalty. *Agronomy Series Of Scientific Research*, *54*(1), 234–238.

Lang, F. (1947). Insurance research. *Journal of Marketing*, *12*(1), 66–71. doi:10.2307/1246301

Love, J. H., Roper, S., & Vahter, P. (2014). Learning from openness: The dynamics of breadth in external innovation linkages. *Strategic Management Journal*, *35*(11), 1703–1716. doi:10.1002/smj.2170

Low, J. (2000). The value creation index. *Journal of Intellectual Capital*, *1*(3), 252–262. doi:10.1108/14691930010377919

Macpherson, A., Jones, O., & Zhang, M. (2004). Evolution or revolution? Dynamic capabilities in a knowledge-dependent firm. *R & D Management*, *34*(2), 161–177. doi:10.1111/j.1467-9310.2004.00331.x

Magretta, J. (2002). Why business models matter. *Harvard Business Review*, *80*, 86–92. PMID:12024761

Maklan, S., & Klaus, P. (2011). Customer experience. *International Journal of Market Research*, *53*(6), 771–792. doi:10.2501/IJMR-53-6-771-792

Malik, N. J., & Mahmood, R. B. (2012). Facilitating corporate entrepreneurship in public sector higher education institutions: A conceptual model. *Issues In Social & Environmental Accounting*, *6*(1/2), 26–49.

McDonald, D. A., & Ruiters, G. (2012). *Alternatives to privatization. Public options for essential services in the global south*. New York: Routledge.

Meynhardt, T., & Diefenbach, F. E. (2012). What drives entrepreneurial orientation in the public sector? Evidence from Germanys Federal Labor Agency. *Journal of Public Administration: Research and Theory*, *22*(4), 761–792. doi:10.1093/jopart/mus013

Mohanty, J. (2013). Are government owned organizations deliberately demonized? The organizational evolution or revolution? Citizenship behavior indicators. *Advances in Management & Applied Economics*, *3*(4), 177–191.

Moingeon, B., & Lehmann-Ortega, L. (2010). Creation and implementation of a new business model: A disarming case study. *Management*, *13*(4), 266–297.

Moizer, J., & Tracey, P. (2010). Strategy making in social enterprise: The role of resource allocation and its effects on organisational sustainability. *Systems Research and Behavioral Science*, *27*(3), 252–266. doi:10.1002/sres.1006

Monteduro, F. (2014). Public-private versus public ownership and economic performance: Evidence from Italian local utilities. *Journal of Management & Governance*, *18*(1), 29–49. doi:10.1007/s10997-012-9235-4

Morris, M., Schindehutte, M., & Allen, J. (2005). The entrepreneurs business model: Toward a unified perspective. *Journal of Business Research*, *58*(6), 726–735. doi:10.1016/j.jbusres.2003.11.001

Mueller, D. C. (2003). *Public Choice*. Austria: Universitat Wien. doi:10.1017/CBO9780511813771

Nag, R., Hambrick, D. C., & Chen, M. J. (2007). What is strategic management, really? Inductive derivation of a consensus definition of the field. *Strategic Management Journal*, *28*(9), 935–955. doi:10.1002/smj.615

Nowak, M. J. (2011). Municipality companies in Poland. Formula objectives and scope of activities. *Samorząd Terytorialny*, *1*(2), 17–26.

Osterwalder, A., & Pigneur, Y. (2010). *Business model generation –A handbook for visionaries, game changers, and challengers*. John Wiley & Sons, Inc.

Özcan, S., & Reichstein, T. (2009). Transition to entrepreneurship from the public sector: Predispositional and contextual effects. *Management Science*, *55*(4), 604–618. doi:10.1287/mnsc.1080.0954

Özel, Ç. H., & Kozak, N. (2012). Motive based segmentation of the cultural tourism market: A study of Turkish domestic tourists. *Journal of Quality Assurance in Hospitality & Tourism*, *13*(3), 165–186. doi:10.1080/1528008X.2012.645199

Peda, P., Argento, D., & Grossi, G. (2013). Governance and performance of a mixed public-private enterprise: An assessment of a company in the Estonian water sector public. *Organization Review*, *13*(2), 185–196. doi:10.1007/s11115-013-0233-z

Perrone Pinheiro, F. A., Ferreira Savoia, J. R., & de Angelo, C. F. (2016). A comparative analysis of the public and private water supply and sanitation service providers' work in Brazil. *Brazilian. Business Review (Federal Reserve Bank of Philadelphia)*, *13*(1), 115–136.

Pine, B. J. II, & Gilmore, J. (2016). Integrating experiences into your business model: Five approaches. *Strategy and Leadership*, *44*(1), 3–10. doi:10.1108/SL-11-2015-0080

Poulsson, H. G., & Kale, H. S. (2004). The experience economy and commercial experience. *The Marketing Review*, *4*(3), 267–277. doi:10.1362/1469347042223445

Prebensen, N. K., & Rosengren, S. (2016). Experience value as a function of hedonic and utilitarian dominant services. *International Journal of Contemporary Hospitality Management*, *28*(1), 113–135. doi:10.1108/IJCHM-02-2014-0073

Priem, R. L., Butler, J. E., & Li, S. (2013). Toward reimagining strategy research: Retrospection and prospection on the 2011 Amr Decade Award Article. *Academy of Management Review*, *38*(4), 471–489. doi:10.5465/amr.2013.0097

Rageh, A., Melewar, T., & Woodside, A. (2013). Using an ethnography research method to reveal the underlying dimensions of thc customcr/tourist experience. *Qualitative Market Research: An International Journal*, *16*(2), 126–149. doi:10.1108/13522751311317558

Ritchie, J. B., & Hudson, S. (2009). Understanding and meeting the challenges of consumer/tourist experience research. *International Journal of Tourism Research*, *11*(2), 111–126. doi:10.1002/jtr.721

Saussier, S., & Klien, M. (2014). *Local public enterprises: A taxonomy*. Chaire Economie des Partenariats Public-Privé Institut d'Administration des Entreprises.

Sørensen, E. (2014). *The metagovernance of public innovation in governance networks*. Paper presented at the University of Bristol Policy & Politics Conference, Bristol, UK.

Sundell, A., & Lapuente, V. (2012). Adam Smith or Machiavelli? Political incentives for contracting out local public services. *Public Choice*, *153*(3/4), 469–485. doi:10.1007/s11127-011-9803-1

Tantalo, C., & Priem, R. L. (2016). Value creation through stakeholder synergy. *Strategic Management Journal*, *37*(2), 314–329. doi:10.1002/smj.2337

Tellis, W. M. (1997). Introduction to Case Study. *Qualitative Report*, *3*(2), 1–14. PMID:9035020

Timmers, P. (1998). *Business models for electronic markets*. European Commission, Directorate-General III.

Tolstad, H. K. (2014). Development of rural-tourism experiences through networking: An example from Gudbrandsdalen, Norway. *Norwegian Journal of Geography*, *68*(2), 111–120.

Tracey, P., Phillips, N., & Jarvis, O. (2011). Bridging institutional entrepreneurship and the creation of new organizational forms: A multilevel model. *Organization Science*, *22*(1), 60–80. doi:10.1287/orsc.1090.0522

Tung, V. W. S., & Ritchie, B. J. (2011a). Exploring the essence of memorable tourism experience. *Annals of Tourism Research*, *38*(4), 1367–1386. doi:10.1016/j.annals.2011.03.009

Tung, V. W. S., & Ritchie, J. B. (2011b). Investigating the memorable experiences of the senior travel market: An examination of the reminiscence bump. *Journal of Travel & Tourism Marketing*, *28*(3), 331–343. doi:10.1080/10548408.2011.563168

Vecchi, V., Brusoni, M., & Borgonovi, E. (2014). Public authorities for entrepreneurship: A management approach to execute competitiveness policies. *Public Management Review*, *16*(2), 256–273. doi:10.1080/14719037.2012.725759

Wilson, F., & Post, J. E. (2013). Business models for people, planet and profits: Exploring the phenomena of social business, a market-based approach to social value creation. *Small Business Economics*, *40*(3), 715–737. doi:10.1007/s11187-011-9401-0

Winkle, C. M., & Lagay, K. (2012). Learning during tourism: The experience of learning from the tourists perspective. *Studies in Continuing Education*, *34*(3), 339–355. doi:10.1080/0158037X.2011.646981

Yin, R. K. (2003). *Case study research: Design and methods*. Thousand Oaks, CA: Sage Publications Inc.

Zainal, Z. (2007). Case study as a research method. *Jurnal Kemanusiaan*, *9*, 1-6.

Zbornik, R. (2012). Tourist destination branding. *Ekonomskog Fakulteta u Istocnom Sarajevu*, *6*, 503–514.

Zott, C., & Amit, R. (2010). Business model design: An activity system perspective. *Long Range Planning*, *43*(2/3), 216–226. doi:10.1016/j.lrp.2009.07.004

KEY TERMS AND DEFINITIONS

Business Ecosystem: The complex of a business organization and its environment functioning as an interdependent system.

Business Model: An interactive system of internal and external components of the business that enables value creation.

Experience: The act or process of directly perceiving events or reality.

Hybrid Organization: An organization refers to two different sectoral standards: logics, and value systems.

Public Enterprise: An organization established with a legal personality, in which the government participates by way of having shares or representation in its decision-making structure.

Stakeholder: A person or business that has invested money in enterprise and has a stake in an enterprise.

Value Proposition: A sum of the economic, technical, service and social benefits that a customer receives in exchange for a price he or she pays for an offer.

Chapter 9
Public Sector Marketing in Poland

Robert Seliga
University of Social Sciences, Poland

ABSTRACT

Marketing is increasingly interdisciplinary area of high-scale diversity. Noticeable is the implement and adapt the tools and the concept of marketing in the different sectors of the economy such as the private sector, non-governmental sector or the public sector. Marketing is a concept in the area of management science and it plays a special role in the area of public management. For selected areas of public life such as: health services and higher education, it will be possible to use more sophisticated tools than in the case of administrative operations. The aim of the chapter is to create a comprehensive model of interdependence and cooperation of various marketing concepts in the public sector in Poland. The empirical part of the article was based on quantitative research conducted in 2015 – 2016.

INTRODUCTION

Year 1980 was a breakthrough not only for Polish history, but also for the broadly defined market behavior. After 1980, Poland has also seen a change in public services. The rate of change in the public service sector in Poland gave the Polish accession to the European Union and NATO. This situation resulted in the need to adapt public services to Western standards, including a shift towards New Public Management (Kożuch, 2004) The public sector in Poland has wide changes, there are adequate public entities responsible, among others, for marketing activities. Managers of public entities in Poland noted the need to implement some commercial management tools to the public services sector. Public marketing has become increasingly important in the management of public entities in Poland. Such concepts as social relationship marketing and public relations have earned popularity. This notion fits more general and worldwide research trends related with conceptualization of public sector marketing and therein a shift from transaction- to a relationship-orientation (Kaplan & Haenlein, 2009). Although New Public Management has been implemented variously around the world over the last few decades (Osborne &

DOI: 10.4018/978-1-5225-2215-7.ch009

McLaughlin, 2005; Ostrowska, Frączkiewicz-Wronka, & Bratnicki, 2013), the reform in post-communist countries, like Poland, has probably its own specificity.

Despite the differences and similarities between classic marketing in the corporate and commercial sectors and marketing in the public sector, marketing in general plays a key role in a business model. It is well depicted by the Business Model Canvas model, which contains nine building blocks necessary to create a complete business model, such as: partners, key activities and resources, customer relationships, customer segments, channels, value proposition, cost structure and revenue streams (Osterwalder & Pigneur, 2010). Hence, there is a need to gain better understanding of public marketing, also in post-communist countries, as a theoretical framework strongly underlying business models of public sector organizations.

The aim of the chapter is to create a framework of interdependence of various marketing concepts in the public sector in a post-communist country, in this case Poland. The model is characterized by a dynamic character, because it will combine the description of the individual marketing concepts applicable in the public sector in Poland. For the realization of this objective a few research problems have been outlined in a sequence reflected by the structure of this chapter. The first key issue is to try to identify the key factors determining the implementation of individual marketing concept in the area of public sector management in Poland. Next step pertains to the separation of model features of public marketing. The third problem is the design and marketing model diversity in public sector in Poland. Finally, on the basis of literature review and conducted research, a framework of Public Sector Marketing in Poland is provided.

BACKGROUND

The current state of knowledge in marketing allows the conclusion that marketing is increasingly interdisciplinary area of high-scale diversity. Noticeable is the implementation and adaption of tools one chooses and the concept of marketing in the different sectors of the economy such as the private sector, non-governmental sector or the public sector. Modern marketing is characterized by a high degree of practice and scientification.

The dynamics of changes in the turbulent environment of each type of organization makes significant barriers to the implementation of certain marketing concepts. There is big distraction at the same time, dovetailing the resulting theory of marketing. Increasingly, utilities, marketing concepts are directed towards the client, whose role in the processes of management in recent years has undergone very broad changes. The client has gone through a process of transformation of the attitudes from passive to active. However, it should be noted that this process has had different mileage depending on the sectors of the economy in which it was implemented. Modern marketing with very often clearly defined limits is continuous modification, it draws expertise from other areas of science. It can be predicted that, therefore, the marketing area will be subject to further changes, however, due to the changing environmental conditions in the long term, one cannot determine its direction.

Marketing is a concept in the area of management science (Flynn 2007, Ewing, Caruana 1999). It plays a special role in the area of public management. The evolution of the changes that have been initiated in the commercial sectors also has changed the perception of marketing in the public sector. The change entailed separation of the concept of public marketing (Walsh 1991; Kaplan & Haenlein, 2009).

Analysis of the contemporary literature tends to a certain reflection: there is extensive research in the area of public management, and part of the broader management is marketing, so the following questions can be asked:

1. Is it possible to extract marketing research field?
2. What specific factors should characterize public marketing?
3. What is the role of public marketing in the process of full public management?

Trying to find the answers to the above questions considerations should begin by defining the concept of "public management". Analysis of the literature does not indicate the standardized definition of public management. There is a variety of approaches and definitions. Kożuch (2004) notes that public governance can be treated in several ways:

- Involvement of the public service and from the selection of functional recognition of politicians;
- Management actions directed to the public of regulatory structures (along with all of the tasks carried out by the competent authorities);
- Management of public organizations or public bodies;
- Sub-discipline of the science of management.

In the literature there are more and more common statements justifying the treatment of public management as a sub-discipline of management science (Kożuch, 2004; Cyfert, Dyduch, Latusek-Jurczak, Niemczyk, & Sopińska, 2014; Sudoł, 2014). However, there are also opinions which suggest that public management should also be of interest to the realm of administration science (Wrzosek, 2012).

Szumowski identified three shots in the approach to the functioning of public administration: the political scene, the recognition of the legal and managerial approach (management) (Figure 1).

Considering the nature of modern public management it can be stated that its interpretation is multifaceted. Discerned in it can be both an instrument approach and subjective. Subjective approach takes place in the human, instrumental approach directs the management of the public in the area of business (managerial). Such an interpretation directs the management of the public in the two evolutionary approaches to public management: new public management and public governance. Two presented approaches are a response to the lack of flexibility in the structures of public administration and the criticisms of the classical approach to the area of public administration (Hausner, 2002).

The above-described approach to public management is similar. Public governance is a specific response to the theory of new public management. Such an approach guided by the concept of public management in the direction of the classical theory of management, which uses modern management methods and techniques, which should increase the efficiency of the public area (Walsh 1994).

Response to the demands of the modern sector of public management is the concept of public governance which, in its assumptions it implies cooperation between public organizations and external stakeholders. W. Rudolf claims that public administration should co-operate and serve its external stakeholders (Rudolf, 2010). This approach makes the public management public marketing area has been extracted, whose actions are synonymous with the modern concept of public management principles.

It seems that the answer to the modern requirements in the area of public management can become a concept of Nadler's ideal system (Figure 2). The essence of this method is a holistic approach to ef-

Figure 1. Approach to the analysis of the functioning of the public administration
Source: adapted from: Szumowski, 2014

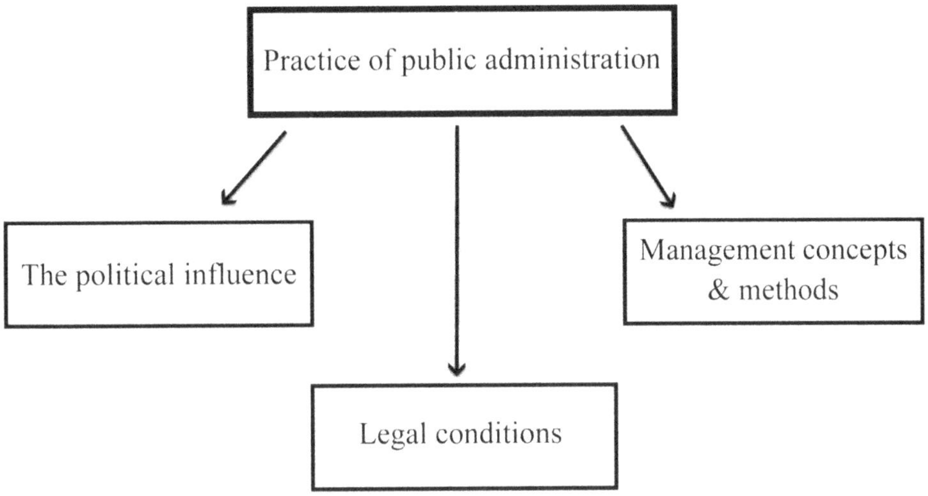

Figure 2. G. Nadler's model
Source: adapted from: (Burke, Lake, Paine, 2008)

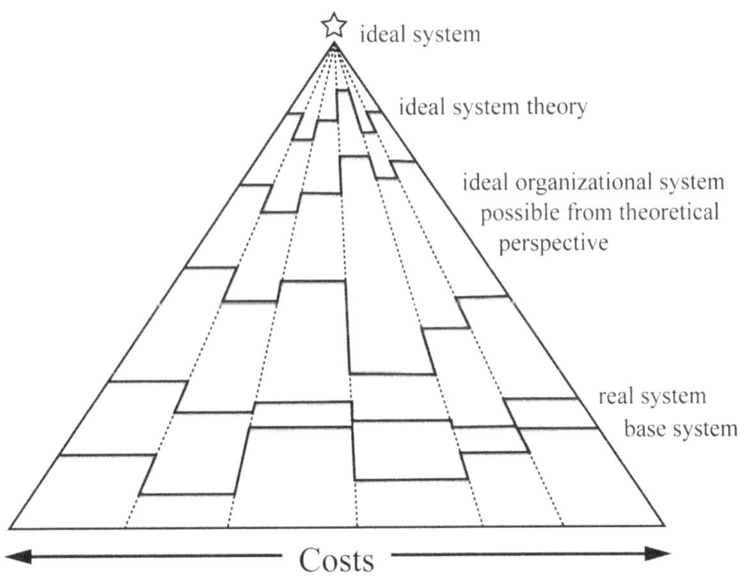

fective management of the overall process. This model assumes output from the ideal system and then gradually getting closer to the current system (Bińczycki, 2004).

Because the public governance by its nature includes a variety of management areas (sub-processes) in the following development attention will be paid to the area of marketing. From this point of view, the questions arise: what should the perfect system in the area of marketing look like?, do commercial tools have a direct impact on the marketing activities in the public sector?

According to a study of their own contemporary public marketing should be[1]:

- Cooperation of public institutions with the external environment (citizens, companies, etc.) on open, transparent and ethical principles,
- Popularization of pro-social attitudes among the public,
- Developing friendly image of public utilities
- Creating patriotism in the society.

It can be noted that public marketing by the respondents should meet the needs of society, to a large extent should be directed at the public should have a platform - a platform for cooperation between the public sector and society. Therefore, it can be stated that systematic research into the development of the concept of marketing public play a special role in the diagnosis of needs and social preferences. Proper placement during the marketing concept of the public allows for correction of irregularities which effectively reduce the communication process of social needs. Currently, the process of identifying social needs in the public sector occurs as a result of adequate support from the marketing concept and its impact on the market. To do this, increasingly in the public area of the marketing sophisticated approaches including the concept of relationship marketing, the concept of social marketing, etc. are implemented.

When analyzing the literature one can see the evolution that occurs in the area of defining public marketing. Under the definition of public marketing can also be found, such as the definition of marketing in public administration. Marketing in public administration is defined as a team processes and relationship marketing. Processes and marketing are adequately coordinated and organized and lead to meet the public interest (Florescu, Malcomete, Pop, 2003; Buurma 2001; Proctor 2007).

In turn, Petrescu and Muscalu (2003) believe that marketing in public administration is the process of "market", in which the research needs of citizens help or impede the marketing management in public administration. It can be concluded that the current marketing of public administration looking at a comprehensive definition of public marketing is only one of its areas. A. Laing believes that the reluctance of the public administration to implement marketing stems from the perception as a commercial concept (Laing, 2003).

Changing perceptions of the role of marketing in public administration forced introduction of market principles, tools for management of public administration. Today, public organizations have become "players" which is ever closer to the behavior of companies from the private sector, where the risk, uncertainty and turbulent environment determine their actions (Lisiecka, Papaj, 2009). This environment forces conducting creative activities, clear and, above all, relational in relation to society. Modern society expects from the government new forms of thinking and ways of doing things so that they correspond to their individual needs. It seems that such an approach on the one hand can lead to the development of the ideal model of the wider public communication with the public. On the other hand, taking into account the variability of the environment there can be observed an urgent need to introduce systematic changes to this model of communication. In response to the earlier question: "does a commercial tool have a direct impact on the marketing activities conducted in the public sector?" One can answer that it does. In this way, the concept of marketing public is a further evolution, because of this turbulent environment, it is difficult to develop the ideal model of marketing for the area of public action. However, there are marketing concepts, whose role in the process of marketing communication is unquestionable (social marketing, relationship marketing). It seems that implementing the principles of marketing in public services should be the basic decision of the management in the public sector. Argument for the above statement, is that adopted by the public administration marketing orientation largely strives to meet social needs, which in consequence have a direct impact on building a positive management's image. A study commissioned by the Chancellery of the President of Poland in 2011 on a sample of more than 2,000 people (survey) indicates that 81% of respondents would like to adopt an active attitude in the process of opinion-projects carried out by public administration (e.g. public consultation)[2].

RELATIONSHIP MARKETING: RANGE OF ISSUES

Today in Poland the fastest growing marketing concept in relation to the public area can be regarded as territorial marketing. The territorial marketing focuses on the involvement of stakeholders involved in the process of public management (Kożuch, 2004). To begin the considerations of relationship marketing, its origins, further development and its key elements, we will start from presenting definitions which should help us understand the essence of the topic. Since there were many scholars who made attempts to define the concept under study, a lot of definitions appeared as a result. Some of them are similar, but there are also definitions in which the authors concentrate on secondary aspects or present them in a more detailed way.

One of the first scholars to deal with the concept of relationship marketing was Leonard Berry. He was the first to define this new conception of marketing in 1983. According to Leonard Berry "“relationship marketing is creating, maintaining and enriching relationship with the customer. Acquiring a new customer is only a first step in the marketing process" (Leonard, 1983). The author of the above definition pays particular attention to the need of creating relationship with the customer, which, he believes, has a long-term character. Seven years later, in 1990, Grönroos proposed a new definition, which is, however, very much similar to the one given by Berry. It says: "Relationship marketing is creating, maintaining and enriching bonds with customers and their partners in such a way that objectives of both sides can be achieved by mutual exchange and realization of promises made" (Grönroos, 1990). In this definition Grönroos introduced a new marketing category – promises. Their purpose is to build a preliminary relationship with the customer. Then fulfilling of promises may strengthen the first bond. In this way both sides benefit. The author also pays attention to partnership relations between both sides. In the same year two scholars Capulsky and Wolf (1990) proposed another definition of relationship marketing. In their opinion relationship marketing is "a process consisting in creating a data base about existing and prospective customers and approaching them creating a bond with them using various information, characteristic for each customer. It is necessary to carry out an analysis of costs of acquiring and retaining each customer and long-term evaluation of the formed relationship". As we can notice, this definition is more extensive. Authors pay particular attention to getting detailed information about the customer. In the definition we have a concept of database, where information about the customer is stored. Furthermore, it is the first time when the cost aspect connected with acquisition and retention of customers appears.

Polish scholars also take part in discussion on defining relationship marketing. In 1995 Rydel and Ronkowski created and presented their own definition of the conception. According to Rydel and Ronkowski relationship marketing is: "the conception of managing and operating on the market according to which market effectiveness of firms depends on establishing good, equal terms relations with participants of the market. This conception includes developing relations with customers based on loyalty and strategic alliances with business partners" (Rydel, Ronkowski, 1995). Polish scholars emphasized the necessity of sustaining bonds not only with customers but also with other, co-operating firms. The term "partnership" has also been used. It is believed that it should be maintained, as finances of the firm may depend on it.

Analyzing the topic of relationship marketing, we can find a lot of definitions similar to that of Rydel and Ronkowski, e.g. the viewpoint of Fonfara (1999). Another additional issue that this author emphasizes is the time, or more precisely, long-term character of the firm's relationships. Thus Fonfara (1999) believes that "relationship marketing emphasizes the process of managing, that is creating, developing and maintaining bonds of the firm with other units. It focuses on direct, interactive, two- and multipleside character of contacts between the seller and the buyer and other subjects in the process of creating and supplying products and services. Additionally it stresses long-term character of bonds between partners of exchange. Formation and maintenance of these bonds takes time, requires effort and money" (Fonfara, 1999).

In the analysis of the subject we cannot miss the definition given by Philip Kotler, the author of numerous publications in the field of Marketing, whose contribution to the development of relationship marketing theory is considerable. According to Kotler (2005), relationship marketing is: "developing mutually beneficial, long-term relations with key participants of the market – customers, suppliers, distributors – in order to earn and maintain the firm".

As can be seen, there are a lot of definitions of relationship marketing. With so many of them it may be difficult to choose the best one, especially that differences between them are insignificant.

Criticism of Transactional Marketing and Origins of Relationship Marketing

Uppsala University is considered to be the source of the conception of relationship marketing. Beginnings of this conception date back to the 1960s, when scholars from this University analyzed the so-called interactive approach in industrial marketing. Is it, then, justified to believe that relationship marketing originates from industrial marketing? The answer is yes, because characteristics of relationship marketing are close to industrial marketing, in which relations with customers have always been important (Gummesson 2011; Shani, Chalasani 2013). However, it should be noticed that relationship marketing as an independent discipline was first introduced in the 1980s, that is almost 20 years after the first studies on the conception. A considerable contribution to these studies had scholars from the so-called "Nordic School". Their works helped to create the theory of relationship marketing. Growing interest in this new discipline of marketing in Finland and Sweden resulted in formation of solid "bases" for further studies. This important contribution to the development of relationship marketing justifies the opinion that Scandinavia is the real birth place of this conception. Another important reason why relationship marketing developed was frequent criticism of the main assumptions of transactional marketing and the marketing-mix. Scholars selected four main areas of criticism which refer to: theoretical bases of the traditional conception of marketing, excessive generalization, orientation to product and not to customer, and concentrating marketing operations on the transaction (Otto, 2004).

Figure 3 shows four main areas of criticism of the above two conceptions. In consideration of marketing one should go back to its roots in the theory of economics. Scholars have doubts as to the usefulness of the 4P theory as the main marketing theory. Grönroos (1990) believes that 4P model is today only a list of Ps, deprived of theoretical roots. Besides, it is also believed that nowadays the connection between theory of microeconomics and marketing-mix is much smaller. Grönroos (1990) demands changes in the traditional marketing conception, because in his opinion, it is very poor in theory. Excessive generalization is the second area of criticism referring to the classical marketing conception. It is though that 4P model does not reflect all its assumptions when it is used on various markets, because it was built on the basis of research conducted on a very well developed American market. And each market has its own,

Figure 3. Main areas of criticism of the classical marketing conception
Source: adapted from: (Otto, 2004)

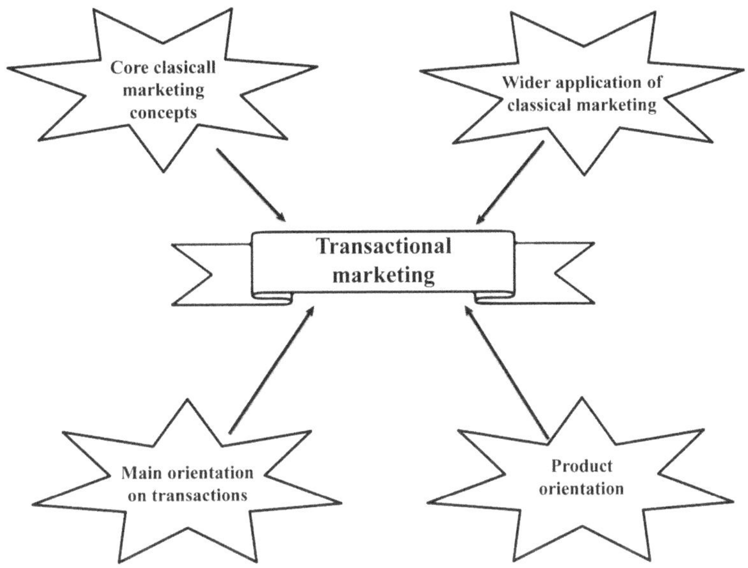

specific character, so it is believed that each of these markets should have its own, adequate and precise marketing instruments (Otto, 2004).

Moreover, Grönroos (1990) criticized the source of traditional marketing. He claims that marketing mix and its 4Ps consolidate the production oriented definition of marketing and not market - or consumer – oriented. On the basis of 4Ps various departments appeared in enterprises, which are not always customer friendly. Also the transactional approach of the traditional conception of marketing was subject to strong criticism, primarily due to its attitude to the customer. Customer's role is supposed to be passive, and the seller's role – active. It is believed that these roles should be changing during the transaction process (the customer is the one who takes decisions, it is for the customer to decide if he or she wants to be active or passive). In the traditional conception of marketing all efforts are concentrated on the transaction.

The commitment - trust model of relationship marketing was created by Morgan and Hunt (1994), who define relationship marketing as the one which "comprises all marketing activities oriented to building, developing and maintaining fruitful relational exchanges". This definition pays particular attention to creating bonds and trust between the seller and the customer. In their work Morgan and Hunt (1994) listed four areas of relational exchanges, such as: supplier partnerships, lateral partnerships, customer partnerships, and internal partnerships. Figure 4 depicts this relationship marketing model, in which a firm is placed in the centre of relational exchanges, and several types of relations exist.

Some of the above mentioned models are broader than others, but all of them contribute to development of knowledge on relations between the firm and its environment. According to Kotler relationship marketing consists in "long-term, based on mutual trust and beneficial for both sides relationships with customers, distributors, dealers and suppliers. It is achieved by keeping the promise of delivering high quality products, good service and good prices as well as by building strong economic, technical and social bonds" (Kotler, 1994). Kotler believes that relationship marketing reduces costs and time of the

Figure 4. Relational exchanges in relationship marketing
Source: adapted from: (Morgan, Hunt, 1994)

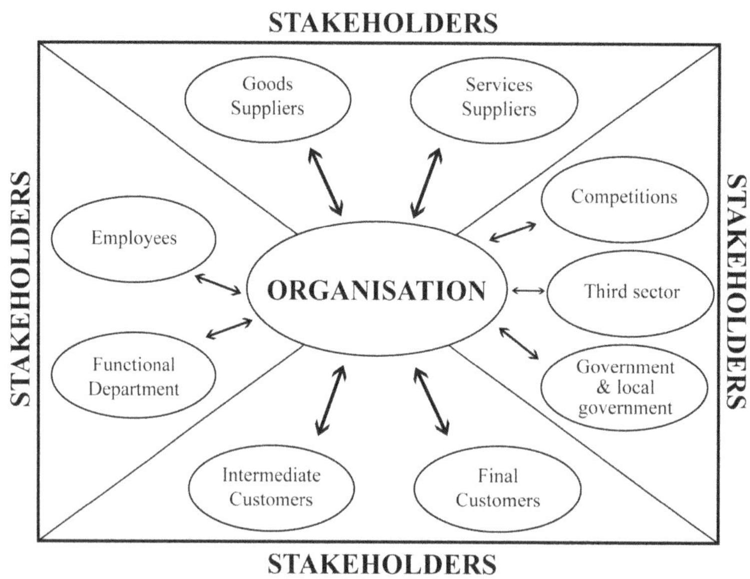

action. He also claims that in some cases the transaction process is not a subject of negotiations anymore, it becomes a routine process. In this way the final result of relationship marketing should be the creation of special kind of assets of the enterprise in a form of marketing network. Such a network includes the firm and its suppliers, distributors, customers and other co-operating entities. The firm should build long-term and reliable relationships with them. This framework is presented in Figure 5.

Additionally, Kotler (1994) defined and presented five potential levels of relationships:

Figure 5. From transaction marketing to relationship marketing
Source: adapted from: (Kotler,1994)

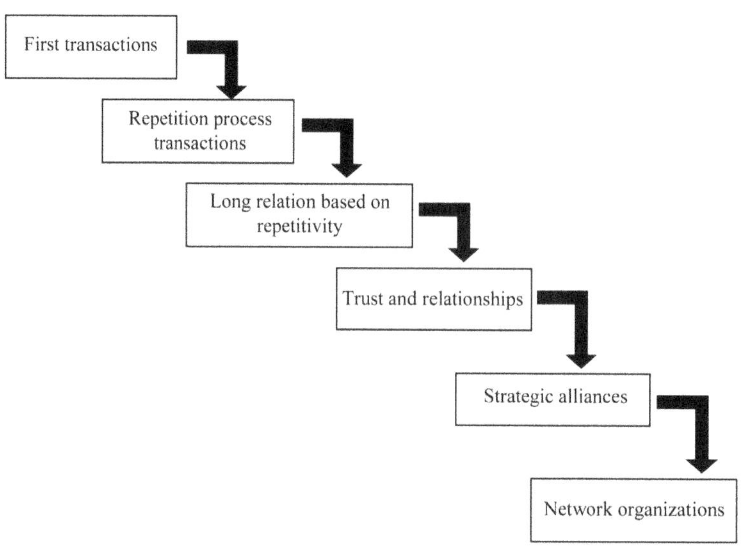

1. **Basic Way:** After the product is sold, the seller does not contact the customer any more. It is orientation to product only.
2. **Reactive Way:** During the transaction the seller asks the customer to call them in case of any questions or doubts.
3. **Accountable Way:** After the transaction the seller calls the customer to check if the purchased product meets their expectations. The seller also asks for suggestions about the product. Collecting such suggestions will let the firm continuously work on improving quality of their products.
4. **Proactive Way:** Contact with the customer is established in order to present them a new offer of products and encourage them to buy such products
5. **Partnership:** It consists in constant co-operation of the firm with the customer, offering the customer such purchasing solutions that would give them savings at purchase and let them optimally use the possibility of a given product.

This classification also shows involvement of the firm in relationship marketing.

The above presented models and positions referring to the conception of relationship marketing have a lot in common. The main idea common to all analyzed views is the long-term character of the conception. It is understood as building long-lasting relations between the firm and the co-operating environment. It is emphasized that positive relationship with the customer is not sufficient. The only way to success is to build good relationships with the customer and other stakeholders of the firm.

Relationship-Building Activities in Relationship Marketing

Building relationships with organization's stakeholders is at the core of relationship marketing. It comprises activities in the following three areas (Berry 2000):

1. **Defining the Goals of Activities:** Consists in increasing the value of each side of the relationship and in rationalizing of marketing activities.
2. **Defining Partners of Activities:** Consists in defining the target group (it can be a group of customers or a single person). Key customers of the firm should be defined.
3. **Defining Programmes of Activities:** There are three types of programmes, depending on the group they are directed to:
    a. Programmes oriented to durability of relationship – often called „loyalty" or „membership" programmes. They consist in promoting repeated purchases.
    b. Programmes oriented to individualization of relationship – their main aim is to satisfy the needs of key customers in a unique and individualized way. This programme requires frequent contacts and broader co-operation with the customer.
    c. Partnership programmes – which are to encourage other firms to create a common offer, unique on the market.

There are several subprocesses, that are necessary for appropriate application of relationship marketing (MacMillan, Money, Money &Downing 2005):

- **Defining Roles:** Refers to division of roles between the firm and the customer and also employees and co-operating firms.

- **Communication:** Realized by building open and interactive communication channels between all subjects involved in formation of the relationship.
- **Creating Bonds:** Consists in maintaining direct contacts with customers (Public Relations department is especially important in this process).
- **Planning Jointly:** It is participation of customers in planning of marketing activities. Customers' loyalty to the firm grows.
- **Ensuring Operational Coherence:** Integration of all processes in the firm.
- **Motivating and Training of Employees:** Consists in creating such an organizational culture that would enhance positive relations between employees.
- **Monitoring:** Consists in periodical monitoring of goals and the achieved results as well as solving of appearing problems.

Moreover, the evaluation of results is of key importance for further implementation of strategy. It enables corrective actions concerning goals, programmes and the way the whole relationship marketing process is managed. Nowadays the most frequently used method of measuring the relationship capital of an enterprise is the so-called consumer satisfaction indexes (Mitręga, 2005).

Evolution practically begins during the realization of relationship marketing programme (the only condition is that this programme must already be in progress for some time). The authors Sheth and Parvatiyar (2000) of stages claim that it is not possible to plan the evolution stage. Its development is influenced by factors independent of the programme, e.g. external factors (Palmatier, Dant, Grewal & Evans 2006), which play an important role in social marketing.

The Essence of Social Marketing

Social marketing has been known for over 40 years. In the literature we can find many definitions reflecting the idea of social marketing. The summary of the main ones is presented in Table 1.

The literature distinguishes three dimensions of social marketing: I - economic marketing pro-social (Ph. Kotler, G. Zaltman 1971), II - marketing in relation to the organizations providing social services

Table 1. The main definitions of social marketing

No.	Author	Definition
I	A. Andersen (1995)	Social marketing is the application of commercial marketing techniques to the analysis, planning, implementation and evolution of programs affecting the behavior of consumers in order to improve personal well-being of its members and society as a whole.
II	J. French, C Blair-Stivens (2005)	Social marketing is the systematic application of marketing concepts and techniques to achieve specific behavioral goals relevant to social good.
III	B. Smith (2006)	Social marketing is the process of creating, communicating and delivering the benefits that your target audience wants, in exchange for attitudes and behavior required by society without profits for broadcasters
IV	N. Lee, Ph. Kotler (2011)	Social marketing is the process of using the principles and techniques of commercial marketing to create currently affecting the behavior of the public health, safety, environment

Source: own study

(L. Hermel, P. Romagini 1990), and III - marketing as a tool for shaping opinions and social attitudes (A. Sergeant 1999). They are described in Table 2.

The dimensions of social marketing intertwine to create space for pro-social activities. Therefore, the concept of social marketing is part of the assumptions in the area of public marketing. Supplemented by the tools of relationship marketing is becoming a leading concept in the field of marketing in the public sector. Moreover, social marketing involves creating favorable attitudes to improve the standard of living of both society as a whole and the individual units, but requires systematic application of methods and marketing techniques to achieve specific goals of behavior for social good (Truss, Marshall, Blair-Stevens, 2010).

Social marketing has become a concept corresponding to modern requirements of market actors (French, Blair-Stevens, 2007). And there is an increasing need to approach marketing in a socially responsible way. Thus, social marketing is related also to the concept of corporate social responsibility (Czerw, 2010). The idea assumes that companies operating in the market are guided by ethical principles in relation to society (Mruk, 2008). CSR is a concept that builds a network of mutual relations between market participants and society members. Thus, the factors influencing the implementation of the idea of social marketing by companies comprise also the factors shaping stakeholders behavior, like social and cultural norms, social and economic condition, personal traits (e.g. education, skills, goals, aspirations), and influence of family, peers and local community (Hastings, 2010).

Moreover, social issues, which also social marketing struggles with, are related to the process of globalization. According to Lefebvre (2011) there are six global challenges for social marketing, such as: human rights, social networks as determinants of behavior, critical approach to marketing, durability of social programs, scale and dissemination (disseminating social programs will give greater capabilities), and introduction of Total Market Approach (TMA).

TMA is a system in which all sectors: public and commercial work together to give the public the possibility of enjoying the benefit of health for all social groups. Access to benefits would depend on the wealth of the target group. The poorest group would have free access to benefits, with increasing prosperity have gained the fee for the services. Higher performance TMA concept in the market strengthen subsidies for social programs. Despite the underlying model, social programs are at the heart of social marketing.

All outlined marketing concepts help to better understand the logic of public sector business models. The way value is created, delivered and captured is not the same in each of those approaches. Underlying marketing concept may possibly change the business models, and even innovate it.

Table 2. Three dimensions of social marketing

Dimension	Author	Definition
Dimension I	Ph. Kotler, G. Zaltman	Social marketing is designing, implementing and monitoring the program calculated to influence the acceptability of social ideas and development considerations relating to product planning, pricing, communication, distribution and marketing research.
Dimension II	L. Hermel, P. Romagini	Social marketing is the social activities of the organization
Dimension III	A. Sergeant	Social marketing is the design, implementation and control of programs designed to implement social ideas

Source: own study based on: (Kotler, Zaltman, 1971; Hermel, Romagni, 1990; Sargeant, 1999)

APPROACH TO PUBLIC MARKETING IN POLAND: SELECTED ISSUES

Methodological Note and Sample Structure

On the basis of the analysis of the literature three working hypotheses have been proposed in this pilot study.

H.1: Noticeable is the perception of increasing application of specialized marketing in the public sector in Poland.
H.2: Noticeable is the perception of growing importance of using the tools of social marketing and relational marketing communication in the public sector in Poland.
H.3: The differentiation of marketing activities depending on the type of public organization in Poland is increasing.

The results of hypothesis verification presented in this chapter is part of a wider pilot study carried out in 2015 and 2016, to answer the questions: how marketing in the area of public services in Poland is evaluated?, which tools in this area are most commonly used? what is the degree of implementation of marketing activities in public entities?

The preceding qualitative study involved individuals who constitute the primary source of information (primary research). Due to the nature of the information received, there followed a quantitative research (questionnaire)

In the sample of 843 people (Table 3) more than 64% percent of the respondents were women - 64.52%, while men 35,48%. In the sample they accounted for the vast majority of people aged between 26 to 35 years (approx. 47.67%), while the second group in terms of number were people aged 36 - 50 years (22.22%). Over 60% of people had higher education (62.01%). The largest group of respondents were people living in cities between 50 and 200 thousand inhabitants (40.14%).

Results

More than 57% of respondents see marketing efforts among public universities. Respondents also note marketing activities in public administration and in medical units. About 40% of respondents have no opinion as to the marketing efforts on the part of firefighters. More than 33% of respondents noted marketing activities of the police (Table 4).

According to the surveyed public universities focused their efforts around marketing tools such as advertising (68.82%) and customer orientation (72.76%). Public administration units and fire brigade lead marketing through social campaigns. Public medical units and police build their image through marketing efforts (Table 5).

Approximately 70% of the respondents believe that all institutions of the public services sector should engage in marketing activities. Respondents believe that public universities usually engage in marketing activities (approx. 98%) (Table 6).

According to the interviewed person in contact with the police, public universities, public authorities, public medical units should be treated like a customer. About 30% of respondents believe that the fire brigade not really have to treat the person contacting the customer (Table 7).

Table 3. Structure of the sample

		Number of responses	Percent
Sex	Women	180	64,52%
	Men	99	35,48%
Age	18-25 years	38	13,62%
	26-35 years	133	47,67%
	36 -50 years	62	22,22%
	above 50 years	46	16,49%
Education	Primary education	0	0%
	Vocational education	22	7,88%
	Secondary education	84	30,11%
	Higher education	173	62,01%
Place of residence	City below 50 thousand residents	76	27,24%
	The city of 50 thousand to 200 thousand residents	112	40,14%
	The city more than 200 thousand residents	91	32,62%

Source: own study based on the research.

Table 4. Question 1: Do you see the marketing activities in the activities of the various institutions of public services?

	Definitely not	No	Probably not	I have no opinion	Rather Yes	Yes	Definitely yes
Police	2,87%	8,24%	26,51%	22,60%	**33,33%**	6,45%	0%
Public universities	0%	0%	0%	0,72%	4,70%	37,28%	**57,30%**
Public administration	0%	6,87%	10,39%	11,05%	**44,09%**	27,60%	0%
Public Health Units	0%	18,64%	19,35%	18,67%	**30,80%**	12,54%	0%
Fire brigade	0%	27,24%	2,87%	**39,07%**	16,13%	14,69%	0%

Source: own study based on the research.

Table 5. Question 2: What kind of marketing you see in the various institutions of public services? (Can select multiple answers)

	Police	Public universities	Public administration	Public Health Units	Fire brigade
Advertisement	14,38%	**68,82%**	0,36%	16,44%	0%
Social communication (social campaigns)	12,77%	21,50%	**50,89%**	11,84%	**26,3%**
Image building	**37,28%**	31,86%	6,13%	**24,73%**	0%
Customer Orientation	24,73%	**72,76%**	0,36%	2,15%	3,4%

Source: own study based on the research.

Table 6. Question 3: Do you think the following institutions of public services should be involved in marketing activities?

	I definitely agree	I tend to agree	Rather I disagree	Strongly disagree	Hard to say
Police	32,97%	39,43%	12,19%	0,72%	14,69%
Public universities	64,52%	33,33%	0,36%	1,79%	0%
Public administration	41,30%	33,25%	9,68%	6,45%	9,32%
Public Health Units	50,90%	36,20%	10,39%	0%	2,51%
Fire brigade	42,97%	39,07%	0,72%	0%	17,24%

Source: own study based on the research.

Table 7. Question 4: Please determine the extent of 1 to 10, a person in contact with the following entities should be treated as a client?

	1 - a very low level	2	3	4	5	6	7	8	9	10 – a very high level
Police	0%	0%	8,49%	0%	1,79%	16,49%	9,44%	15,77%	**35,48%**	12,54%
Public universities	0%	0%	0%	0%	2,51%	0,72%	2,87%	32,62%	24,73%	**36,55%**
Public administration	0%	0%	0%	0%	0%	0%	4,30%	22,94%	**41,94%**	30,82%
Public Health Units	0%	0%	0%	0%	0%	0%	3,27%	20,79%	**39,43%**	36,51%
Fire brigade	0%	6,40%	10,84%	16,70%	**29,36%**	11,20%	6,40%	7,80%	6,10%	5,20%

Source: own study based on the research.

Over 80% of respondents believe that public awareness campaigns are needed to create good attitudes in society. Approximately 71% of respondents say they see a change in the perception of the role it plays in society contact with public entities (Table 8).

Over 70% of respondents believe that the conduct of marketing communication by the police is effective. Of which more than 50% of respondents believe it to be useful. For more than 35% of the responding marketing communication is fairly static and lacks the dynamism visible features. About 31% of

Table 8. Question 5: Please answer the sentence contained in the table

Questions	Yes	No	I have no opinion
Social campaigns are needed to create good attitudes in society	**80,04%**	7,17%	12,79%
Changing perception of the role of society by public institutions (applicant becomes a client)	**71,20%**	28,44%	0,36%
Growing importance of marketing communication between public institutions and society	**69,53%**	2,87%	27,60%

Source: own study based on the research.

respondents could not assess its completeness, while more than 98% of respondents believe marketing message to be credible, which reaches to the society (about 77%) (Table 9).

According to the respondents marketing communications conducted by public universities is effective (66%) and useful (97.85%). Respondents believe that the marketing communication of universities is dynamic and complete. About 55% of respondents believe it to be credible, coming into the society (Table 10).

The respondents believe that marketing communications conducted by public medical units is useful (about 35%), they cannot determine its effectiveness (44%) and completeness (29.77%). According to the respondents lacks dynamics. For more than 40% of respondents marketing communication public medical units it is reliable and reaches its message to the public (Table 11).

Approximately 60% of respondents assess marketing communications led by public administration units as effective. According to about 60% of communication it is useful. Approximately 31% of the respondents determines the dynamics of the campaign at the level of static. For 30% of respondents marketing communication carried out by the public administration it is incomplete and difficult to determine its reliability (37%). According to more than 65% of communication is not quite reach the public (Table 12).

Table 9. Question 6: Do you think that marketing communication in the following sectors are

POLICE								
	-3	**-2**	**-1**	**0**	**1**	**2**	**3**	
ineffective	0%	17,20%	0%	10,41%	**35,12%**	**37,27%**	0%	effective
useless	0%	0%	1,79%	10,39%	18,65%	17,92%	**51,25%**	useful
static	0%	0,36%	**35,48%**	12,55%	20,43%	31,18%	0%	dynamic
incomplete	18,64%	15,08%	11,87%	**31,03%**	13,47%	9,91%	0%	complete
unbelievable	0%	0%	0%	1,79%	**63,08%**	**35,13%**	0%	credible
does not reach the public	0%	11,70%	9,09%	1,79%	**43,73%**	**33,33%**	0,36%	goes to the public

Source: own study based on the research.

Table 10. Question 7: Do you think that marketing communication in the following sectors are

PUBLIC UNIVERSITIES								
	-3	**-2**	**-1**	**0**	**1**	**2**	**3**	
ineffective	0%	0%	2%	1,79%	**35,85%**	**31,18%**	29%	effective
useless	0%	0%	0%	2,15%	**34,77%**	**27,92%**	**35,16%**	useful
static	0%	0%	0%	3,58%	**50,89%**	**28,68%**	17%	dynamic
incomplete	0%	0%	1,79%	0,36%	**84,58%**	**12,56%**	1%	complete
unbelievable	0%	0%	0%	**45,16%**	**54,48%**	0%	0%	credible
does not reach the public	0%	0%	0%	15,41%	**28,68%**	**55,91%**	0%	goes to the public

Source: own study based on the research.

Table 11. Question 8: Do you think that marketing communication in the following sectors are

PUBLIC HEALTH UNITS								
	-3	**-2**	**-1**	**0**	**1**	**2**	**3**	
ineffective	0%	0%	20,43%	**44,44%**	17,20%	17,93%	0%	effective
useless	0%	0%	18,64%	18,28%	29,03%	**34,05%**	0%	useful
static	7,20%	**26,20%**	**24,44%**	21,96%	10,80%	8,20%	1,2%	dynamic
incomplete	0%	10,20%	20,78%	**29,77%**	24,20%	15,05%	0%	complete
unbelievable	0%	0%	0%	39,06%	**43,73%**	17,21%	0%	credible
does not reach the public	0%	0%	1,79%	30,47%	**40,86%**	26,88%	0%	goes to the public

Source: own study based on the research.

Table 12. Question 9: Do you think that marketing communication in the following sectors are

PUBLIC ADMINISTRATION								
	-3	**-2**	**-1**	**0**	**1**	**2**	**3**	
ineffective	0%	0%	9,10%	22,58%	**34,71%**	33,61%	0%	effective
useless	0%	0%	17,20%	3,94%	**59,86%**	12,51%	6,49%	useful
static	0%	16,85%	19,71%	**31,90%**	15,05%	16,49%	0%	dynamic
incomplete	0%	16,13%	**30,82%**	26,53%	21,50%	5,02%	0%	complete
unbelievable	0%	0%	26,99	**37,28%**	30%	5,73%	0%	credible
does not reach the public	0%	0%	2,15%	**67,96%**	28,47%	1,07%	0,35%	goes to the public

Source: own study based on the research.

Over 40% of respondents believe that marketing firefighters are not very effective. Approximately 45% of the respondents believe the current marketing useful and reliable. Respondents also reported a small dynamics and completeness of the above activities. Approximately 60% of respondents believe that marketing firefighters should do more to reach out to the public (Table 13).

SOLUTIONS AND RECOMMENDATIONS

Taken in the publication of the issue of the application of modern concepts of marketing in your marketing efforts in the public sector in Poland is essential theoretical and practical. The importance of theoretical is allotment of the suitability of the application of the concept of marketing in your marketing efforts in the public sector in Poland. In the literature one can find characteristics of the marketing concept and try to reference it to the public sector in Poland.

The purpose of the publication was to provide a model of interdependence of different marketing concepts in the public sector in Poland. Analysis of the literature allows the conclusion that visible is the penetration of such concepts as social marketing and relationship marketing in the marketing area. The results of the pilot studies showed interest on the part of public institutions which were implement-

Table 13. Question 10: Do you think that marketing communication in the following sectors are

FIRE BRIGADE								
	-3	**-2**	**-1**	**0**	**1**	**2**	**3**	
ineffective	0%	14,34%	20,79%	**41,57%**	22,22%	0,72%	0,36%	effective
useless	0%	0%	0%	26,52%	**49,82%**	23,30%	0,36%	useful
static	0%	0%	2,15%	**60,93%**	19,71%	16,85%	0%	dynamic
incomplete	0%	16,49%	16,13%	**46,59%**	20,79%	0%	0%	complete
unbelievable	0%	0%	0%	29,03%	**44,44%**	26,53%	0%	credible
does not reach the public	0%	0%	1,79%	**36,92%**	35,80%	25,49%	0%	goes to the public

Source: own study based on the research.

ing and enhancing marketing activities. The most developed marketing communications in the public sector in Poland are leading universities. Studies indicate that marketing efforts on the part of the public administration and the police are on the same level. According to research the least active marketing and public institution in Poland is the fire department. However, it should be noted that approximately 56% of respondents do not want to be treated as a customer when it comes to the actions of the fire brigade. In a sense, this phenomenon explains the characteristics of the marketing activities in this area.

Looking at the mechanisms to improve the efficiency of the marketing activities in the various units of the public service should consider the wider cooperation between these entities and the private sector. Very often companies are interested in cooperation with the entities of public administration in the Organization of the action pro-social. The answer to this interest can be conducted together social campaigns.

The aim of the chapter was to develop a framework for the application of marketing in the public sector in Poland (Figure 6). The results of pilot studies allow to create the framework of marketing in the public sector in Poland. Further research in this area will help it development and refinement.

Figure 6. Framework public sector marketing in Poland
Source: own study.

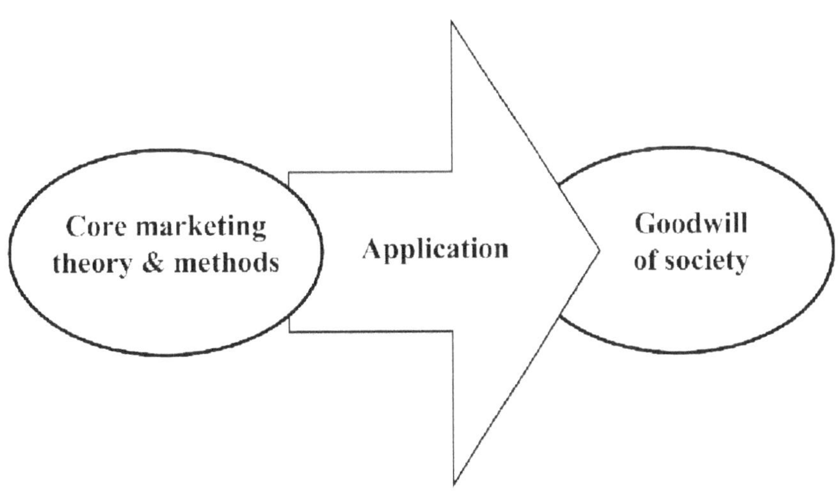

The starting point in the marketing management of the public area is to answer the question: what are the requirements of society in relation to the managers of the public area. Therefore, an important element of the marketing activities conducted by a public entity is continuous monitoring on the expectations and need of society. The society enters into the public sector with certain requirements, which, according to the concept of marketing should be met by "Managing" the public sector. Meeting the requirements can occur in different ways, depending on the same requirements. One way to satisfy social requirements can be administrative decisions (such as lowering taxes). The second way the implementation of the social requirements can become an area of marketing activities in the public sector-public marketing. It should be noted the fact that this term is becoming wider as implemented to newer marketing concepts such as: social marketing, relationship marketing.

Still they interact with the classical tools of marketing. As research shows, often used tools in marketing is a public advertisement or public relations. It should be emphasized that marketing customer orientation plays a big role. In Poland, there is a noticeable change in the perception of the role that society fully in the process of public management. Polish society has passed with the attitude passive to active in higher mentioned process. Meeting the requirements of the society is above all to satisfy its basic needs such as security needs - taking care of health. For this purpose, can be created social campaigns. By this type of event strengthens the relationship between the public and public bodies. Through marketing activities public entities build public confidence. Society increasingly participate in the activities of the public. Such a mechanism could cause it satisfaction and positive feedback about their operation.

FUTURE RESEARCH DIRECTIONS

Dynamics of changes occurring in the public sector, not only in Poland, is very large. The dynamics causes continuous adjustment of processes, improve tools for the management of the sector. Currently, marketing in the public sector has become an area of interest and research because of the great possibilities of implementation concepts and tools in this sector. Conducting research in the future should primarily focus on the alignment of individual tools for specific sub-sectors of public sector. The direction of the development of public marketing should be focused on in-depth analysis of the implementation of various marketing tools in the various sub-sectors due to their diverse nature. It should be examined the effectiveness of various marketing tools. Based on the results of the research will be possible to develop marketing strategies for the various sub-sectors of public.

CONCLUSION

The conclusions of the examination and analysis of the literature allow for verification of the placed in the introduction the following working hypotheses and questions. Hypothesis 1 and 2. Analysis of the results of research and the literature confirms the emergence of a new concept of marketing in the public sector (social marketing, relationship marketing). According to research respondents see the marketing activities of a socially-such as social campaigns. Hypothesis 3. The hypothesis that " The differentiation of marketing activities depending on the type of public organization is increasing " is confirmed by the carried out pilot studies. Respondents said that the most advanced marketing lead Polish universities (over 90%). The subjects saw the marketing activities in the public administration (71,69%), public health units (43,30%) and police (39,78%). Respondents see the least marketing-related activities on the part of the fire department.

Presented work is the result of research on one of the public management processes. Presented the proposals were based on the results of the pilot studies and the analysis of the literature. It seems that the effective implementation of marketing will contribute to the improvement of the activities of public entities in Poland.

Implementation of marketing tools in the public sector in Poland is determined by many variables due to the specificities of this sector. These variables are primarily dependent on the public subsector in which they are deployed. The most important variables include, inter alia, the complexity of the legal subsector, stormy political and legal environment variation, economic barriers. A key step for the implementation of marketing tools in the public sector is to develop a detailed implementation plan that will take into account all the factors changes e. g. economic, social, etc.

Looking at the essence of the introduction of public marketing in public sector special attention should be paid to the social cost and will to change in the society. In Poland, gradually moving away from the bureaucracy in public administration for the relational concept of customer service. Currently Polish society expects from public institutions of new forms and standards in communication. Increasingly, State institutions create social campaigns designed to influence the changing attitudes of society.

In summary, featured can be the following activities in the public sector in Poland:

- Greater openness of public institutions in the direction of their clients;
- Changing the perception of the role of the applicant in the process of management of the country (the transition from passive active society called multipliers);
- Increasing involvement of public institutions in the creation of correct social attitudes such as. creating a healthy lifestyle, road safety etc.;
- Cooperation between public areas to create a common and consistent marketing messages such as. security of citizens;
- Use modern forms of communication, like social media or websites, for example.

REFERENCES

Andersen, A. (1995). *Marketing Social Change: Changing Behavior to Promote Health, Social Development, and the Environment*. San Francisco: Jossey-Bass.

Berry, L. L. (2000). Relationship marketing of Services, Growing Interest, Emerging Perspectives. In J. N. Sheth & A. Parvatiyar (Eds.), *Handbook of Relationship Marketing*. Sage Publictions, Inc. doi:10.4135/9781452231310.n6

Bińczycki, B. (2004). System idealny Nadlera. In M. Romanowska (Ed.), *Leksykon zarządzania*. Warszawa: Difin.

Burke, W., Lake, D. G., & Paine, J. W. (2008). *Organization Change: A Comprehensive Reader*. San Francisco: Jossey-Bass.

Buurma, H. (2001). Public policy marketing: Marketing exchange in the public sector. *European Journal of Marketing*, *35*(11/12), 1287–1302. doi:10.1108/EUM0000000006483

Copulsky, J., & Wolf, M. (1990). Relationship Marketing. Positioning for the Future. *The Journal of Business Strategy*, *11*(4), 16–20. doi:10.1108/eb060069 PMID:10106864

Cyfert, S., Dyduch, W., Latusek-Jurczak, D., Niemczyk, J., & Sopińska, A. (2014). Subdyscypliny w naukach o zarządzaniu – logika wyodrębnienia, identyfikacja modelu koncepcyjnego oraz zawartość tematyczna. *Organizacja i Kierowanie*, *1*(161), 37–49.

Czerw, A. (2010). *Marketing w ochronie zdrowia*. Warszawa: Difin.

Ewing, M. T., & Caruana, A. (1999). An internal marketing approach to public sector management: The marketing and human resources interface. *International Journal of Public Sector Management*, *12*(1), 17–29. doi:10.1108/09513559910262652

Florescu, C., Malcomete, P., & Al Pop, N. (2003). *Marketing. Dictionar explicative*. Bucharest: Editura Economica.

Flynn, N. (2007). Public sector management. *Sage (Atlanta, Ga.)*.

Fonfara, K. (1999). *Marketing partnerski na rynku przedsiębiorstw*. Warszawa: PWE.

French, J., & Blair-Stevens, C. (2005). *Social Marketing Pocket Guide*. London: National Social Marketing Centre of Excellence.

French, J., & Blair-Stevens, C. (2007). *Big Pocket Guide: social marketing*. London: National Social Marketing Centre.

Grönroos, Ch. (1990). *Strategic Management and Marketing, Managing the Moments of Truth in Service Competition*. Lexington: Free Press.

Gummesson, E. (2011). *Total relationship marketing*. Routledge.

Hastings, G. (2010). *Social marketing. Why sholud the Devie have All the Best tunes?* Butterworth-Heinemann.

Hausner, J. (2002). Od idealnej biurokracji do zarządzania publicznego. In M. Maroda (Ed.), *Wymiary życia społecznego. Polska na przełomie XX i XXI wieku*. Warszawa: Wydawnictwo Naukowe Scholar.

Hermel, L., & Romagni, P. (1990). *Le marketing public – Une introduction au marketing des administrations et des organisations publiques*. Paris: Editura Economica.

Kaplan, A. M., & Haenlein, M. (2009). The increasing importance of public marketing: Explanations, applications and limits of marketing within public administration. *European Management Journal*, *27*(3), 197–212. doi:10.1016/j.emj.2008.10.003

Kotler, Ph. (1994). *Marketing. Analiza. Planowanie. Wdrażanie i kontrola*. Warszawa: Gebethner i S-ka.

Kotler, Ph. (2005). *Marketing – Koncepcje marketingu w XXI wieku*. Poznań: REBIS.

Kotler, Ph., & Zaltman, G. (1971). Social Marketing. An Approach to Planned Social Change. *Journal of Marketing*, *35*(3), 3–12. doi:10.2307/1249783 PMID:12276120

Kożuch, B. (2004). *Zarządzanie publiczne. W teorii i praktyce polskich organizacji*. Warszawa: Placet.

Laing, A. (2003). Marketing in the public sector: Towards a typology of public services. *Marketing Theory*, *3*(4), 427–445. doi:10.1177/1470593103042005

Lee, N. R., & Kotler, P. (2011). Social marketing to protect the environment: What works. *Sage (Atlanta, Ga.)*.

Lefebvre, R. C. (2011). An integrative model for docile marketing. *Journal of Social Marketing*, *1*(1), 54–72. doi:10.1108/20426761111104437

Leonard, B. (1983). *Relationship Marketing. Emerging Perspectives on Services Marketing*. Chicago: American Marketing Association.

Lisiecka, K., & Papaj, T. (2009). *Kierunki doskonalenia usług świadczonych przez administrację publiczną*. Katowice: Śląskie Centrum Społeczeństwa Informacyjnego i Akademia Ekonomiczna.

MacMillan, K., Money, K., Money, A., & Downing, S. (2005). Relationship marketing in the not-for-profit sector: An extension and application of the commitment trust theory. *Journal of Business Research*, *58*(6), 806–818. doi:10.1016/j.jbusres.2003.08.008

Mitręga, M., (2005). *Marketing relacji – teoria i praktyka*. Warszawa: CeDeWu.

Morgan, R., & Hunt, S. (1994). Relationship Marketing in the Era of Network Competition. *Marketing Management*, *5*, 18–28.

Morgan, R., & Hunt, S. (1994). The Commitment – Trust Theory of Relationship Marketing. *Journal of Marketing*, *58*(3), 20–38. doi:10.2307/1252308

Mruk, H. (2008). *Marketing – zagadnienia współczesne*. Poznań: Forum Naukowe.

Osborne, S. P., & McLaughlin, K. (2005). The New Public Management in context. In K. McLaughlin, E. Ferlie, & S. Osborne (Eds.), *New Public Management: Current Trends and Future Prospects* (pp. 7–14). London: Routledge.

Osterwalder, A., & Pigneur, Y. (2010). *Business model generation: a handbook for visionaries, game changers, and challengers*. John Wiley & Sons.

Ostrowska, S., Frączkiewicz-Wronka, A., & Bratnicki, M. (2013). Mission Oriented Scorecard: The Experience of Public Hospitals in Poland. In *7th Conference On Performance Measurement And Management Control*. European Institute for Advanced Studies in Management.

Otto, J. (2004). *Marketing relacji*. Warszawa: C.H. Beck.

Palmatier, R. W., Dant, R. P., Grewal, D., & Evans, K. R. (2006). Factors influencing the effectiveness of relationship marketing: A meta-analysis. *Journal of Marketing*, *70*(4), 136–153. doi:10.1509/jmkg.70.4.136

Parvatiyar, A., & Sheth, J. N. (2000). The domain and conceptual foundations of relationship marketing. Handbook of Relationship Marketing, 1, 3-38.

Petrescu, I., & Muscalu, E. (2003). *Tratat de management public. Lucian Blaga*. Sibiu: University.

Proctor, T. (2007). *Public sector marketing*. Pearson Education.

Rudolf, W. (2010). Koncepcja governance i jej zastosowanie – od instytucji międzynarodowych do niższych szczebli władzy. *Acta Universitatis Lodziensis. Folia Oeconomica*, *245*, 73–82.

Rydel, M. & Ronkowski C. (1995). Marketing partnerski. *Marketing i Rynek*, 9.

Sargeant, A. (1999). *Marketing Management for Non profit Organizations*. Oxford, UK: Oxford University Press.

Shani, D., & Chalasani, S. (2013). Exploiting niches using relationship marketing. *Journal of Services Marketing*.

Sheth, J. N., & Parvatiyar, A. (2000). The domain and conceptual foundations of relationship marketing. Handbook of relationship marketing, 3-38.

Sudoł, S. (2014). Podstawowe problemy metodologiczne nauk o zarządzaniu. *Organizacja i Kierowanie*, *1*(161), 11–35.

Szumowski, W. (2014). Zarządzanie publiczne–próba systematyzacji koncepcji. *Nauki o Zarzadzaniu*, *4*(21), 86-98.

Truss, A., Marshall, R., & Blair-Stevens, C. (2010). A history of social marketing. In J. French, C. Blair-Stevens, D. McVey, & R. Merritt (Eds.), Social marketing and public health. Theory and practice. Oxford.

Walsh, K. (1991). Citizens and consumers: Marketing and public sector management. *Public Money & Management*, *11*(2), 9–16. doi:10.1080/09540969109387649

Walsh, K. (1994). Marketing and public sector management. *European Journal of Marketing*, *28*(3), 63–71. doi:10.1108/03090569410057308

Wrzosek, S. (2012). *Administracja publiczna jako przedmiot badań nauki administracji*. Teka Komisji Prawniczej V.

KEY TERMS AND DEFINITIONS

CSR: A concept that builds a network of mutual relations between market participants.

Marketing in Public Administration: It is carrying out activities of public institutions in order to determine the needs of society. The result is products and services beneficial to society.

Public Marketing: These are the integrated tools and marketing activities oriented in the public area, with a view to promoting this sector.

Relationship Marketing: It is carrying out activities in the network interconnections, providing benefits to all of the parties.

Social Marketing: These are actions aimed at social welfare, changing habits and attitudes of society so as to achieve the common good.

Stakeholders in the Public Sector: An individual or a group of people that are interested in the effect of the organization.

TMA: (Total Market Approach) is a system in which all sectors: public and commercial work together to give the public the possibility of enjoying the benefit of health for all social groups.

ENDNOTES

1. Research, 843 people, place test Poland, pilot studies.
2. A study commissioned by the Chancellery of the President of Poland in 2011 on a sample of more than 2,000 people (survey).

Chapter 10
The Productive Leadership Game:
From Theory to Game-Based Learning

Marko Olavi Kesti
University of Lapland, Finland

Jaana Leinonen
University of Lapland, Finland

Terhi Kesti
PlayGain Inc., Finland

ABSTRACT

The aim of the article is to illustrate how new digital leadership simulation game, Productive Leadership Game, can foster the leadership competencies and actions to improve team based and organizational productivity. Game's theoretical roots are based on the human capital production function for public organizations, which is also described in this article. Human capital production function illustrates the significance of the human resource management actions for the organizational performance and the quality of working life connection to team performance. A case study in municipal business is performed using a Finnish mid-sized case organisation. In terms of evidence-based theory of productivity, this article describes methodology of how complex and multitude cause-relationship can be gamified to serious game for learning purposes. The article makes a significant contribution to public sector research and management, as it explains the ontology of the performance and productivity of public organisation and illustrate method to multiply essential learning to practice.

INTRODUCTION

As global economy growth has been slow the public organizations are confronting fiscal distress, and therefore the organization performance improvement has become important issue. There are increased pressures for public sector organizations to improve their performance. Organizations are producing

DOI: 10.4018/978-1-5225-2215-7.ch010

services within the limits of budget constraints, thus it is essential to turn the inputs into high-qualified service products using effective processes, building innovative HRM practices as well as investing in the quality of working life and psychological well-being on the employees. Leadership is one of the major sources of organization's competitive advantage and productivity. Public sector organizations have invested in leadership development training programs which aim to strengthen the leadership cultures, competence, leadership identities and the whole organizational performance. These development programs have usually been implemented as lectures, seminars and workshops comprising theoretical approaches to leadership and testing leadership skills utilizing various questionnaires and self-evaluations. However, the effectiveness of these training programs has remained ambiguous and they are also often disconnected from the distinct management practices. Furthermore, despite of these numerous trainings and management courses aimed for public managers, there are several studies and arguments which yet reveal the leadership problems in public sector organizations. That is, managers lack skills in strategic actions, change management, conflict resolution and in enhancing collaborative and entrepreneurial organizational culture, for example.

Continuous performance improvement in private companies – with the entrepreneur mind-set - has a long history. Due to the productivity and quality pressures, this same kind of entrepreneur mind-set should also be adapted at the public sector and public sector leadership. The public sector entrepreneurship means that organization adapt the culture where continuous improvement is at every employee's mind-set. This does not mean innovative re-engineering projects but rather in everyday orientation to solve problems, act proactively and make innovative improvements at work environment, customer service and, processes. Because culture is deeply rooted at the organization operations, the middle managers' cognitive competence, attitudes and behavior are the most important elements in changing or maintaining the elements of the entrepreneurial behavior and culture in public sector organizations. Middle managers are described as initiators and the main source in advancing public entrepreneurship and entrepreneurial behavior among employees (Borins, 2000; Bernier & Hafsi, 2007). That is, employees act entrepreneurially if they perceive their managers as being supportive of such behavior. The entrepreneurial orientation in management requires active engagement with the stakeholders, systematic human resource management practices, attributes to motivate, delegate, organize and think strategically as well as the ability to develop and manage a flexible, innovatory organization (e.g. Boyett, 1996). That is, managers need new effective methods to really understand their facilitative role, importance and effects on their team performance and on the public sector organization productivity.

However, organizational entrepreneurship as well as leadership is difficult to teach at the classroom. It is also typical, that leadership training programs teach mainly leadership theories, concepts, scientific views and studies. This kind of training doesn't necessarily promote the actual leadership skills, behavior and competences needed in every-day organizational life. Leadership theories, for example, are often quite difficult to implement in organization practices. Also reason-cause relationships are much more complex in real life compared to theoretical views and scientific assumptions. We argue that more practice-based teaching orientation and tools are needed in leadership training in order to give a learner a more pragmatic view into the organizational world. That is, the content of the training should represent the diversity of challenges and address the real-world problems public managers are facing in organizations. As Siewiorek et al. (2013) stress, leadership competence develops only when learner is forced to address the challenge of innovating, inspiring and adapting. According to several studies, various evidence-based practices in leadership development are effective and lead to favorable performance improvement results. Therefore, public sector managers need effective ways to understand and learn the theory of how performance and productivity can be improved with competent leadership behavior which fosters the employee quality of working life and entrepreneurial orientation.

Although the use of simulation and computer games in management training have increased, there is yet a lack of empirical studies that have examined the efficacy and influence of these trainings on human resource management practices, on organizational performance, productivity and organizational cultures. The aim of this chapter is to bring contribution into this area and to illustrate how new digital leadership simulation game, Productive Leadership Game, can foster the leadership competencies and actions to improve team based and organizational productivity, hence to gain entrepreneurial mindset.

The chapter includes data collection, experiences of the public sector middle managers playing the leadership game. Data was gathered during winter and spring 2015-2016, from women leaders from public sector organizations participating Executive MBA course in Lapland's University. They played Productive Leadership simulation game for four months as part of their management course.

Game's theoretical roots are based on the human capital production function (Kesti et al., 2016; Kesti & Syväjärvi, 2015; Kesti, 2013), which can also be applied to public organizations, as described in this chapter. Human capital production function illustrates the significance of the human resource management actions for the organizational performance and the quality of working life connection to team performance. A case study in municipal business is performed using a Finnish mid-sized case organization. In terms of evidence-based theory of productivity, this chapter describes methodology of how complex and multitude cause-relationship can be gamified to serious game for learning purposes. Chapter describes how scientific theory of human capital production function can be utilized at digital simulation learning tool to foster management practice based learning. Therefore the theory is first introduced before presenting the serious game application.

BACKGROUND

Human Capital and Productivity in Public Sector Organizations

There are big differences in how effectively public organizations use their human resources (HR). The differences arise from organizational, cultural and structural factors related to how public services are provided. When increasingly strict conditions of the economy dominate, the public organizations must carefully consider the service task obligations that are needed in relation to capacity and quality. To balance public finance deficits in the long term, municipal organizations have to continuously improve the quality of their service production (Linna et al., 2008; Jääskeläinen & Lönnqvist, 2011), hence also develop entrepreneurial mind-set. In addition, studies have shown that there are significant opportunities for productivity improvement in welfare services (Käpylä et al., 2008). For example, the difference between the most inefficient and efficient health-care center, measured in terms of cost efficiency, was 14% (Kangasharju, 2008).

Recent developments in information and communication technology have been shown to have great potential for further improvements in productivity in both the private and public sectors (e.g. Pohjola, 2014). However, to utilize this technology and support social innovations, new methods of organizing work and new structural solutions are needed to make productivity development possible. The enhancement of public organizations' productivity requires multiple improvement actions, as well as comprehensive knowledge of the cause–effect fundamentals of the municipal organization's productivity system (e.g.

Jääskeläinen, 2010). For public service providers, measurement and analysis have to be emphasized, particularly concerning human capital productivity.

In addition to the development of processes, attention must be paid to the structure of HR, workforce wellbeing and the development of staff know-how – in other words, issues related to HR management. Several studies have indicated that the human resource management with the actions of systematic human resource development practices have positive effects on the innovativeness, motivation, staff turnover, well-being of the employees and to the performance and productivity of the organization (Bae & Lawler, 2000; Guthrie, 2001; Batt, 2002; Hassan et. al., 2006; Baptiste, 2008; Messersmith et. al., 2011). In addition, the experiences of quality of working life seems to support and enhance the organizational productivity (Abdeen, 2002; Warr, 2005; Wright & Bonett, 2007). Also Fabi et. al. (2010) stress that development of HRM capabilities with regard to informing, integrating and rewarding employees enable to improve productivity. Hassan et. al. (2006) observe the manager's significance in managing people in ways that lead to innovation and productivity. That is, manager has an important task in creating operational culture that strengthens human capital productivity and supports learning and development (Laakoli & Peiponen, 2012). It also seems that systematic and consistent training of human resources will help to improve the municipal economy (Virtanen & Stenvall, 2014).

Analyzing Productivity of Public Organization

In the traditional thinking, the development of productivity is a question of how output can be increased with the same input cost or how the required output can be achieved with as little cost as possible. Thus, it is a question of the relationship between the outputs and the inputs. Productivity is created in the real process, in those practical functions in which performance is implemented from different production factors. The productivity concept is clearly determined by this point of view: Productivity is a ratio between the outputs and the inputs, both of which can be measured in monetary units (Meklin, 2008). In reality, however, this definition does not represent the whole picture of municipal organization productivity because the value added is not always so clearly measurable with fiscal metrics.

The contribution outputs (work, capital, energy, etc.) of the product or service production process affect the productivity outcome. Productivity can be examined from the point of view of overall productivity or the productivity ratio. When measuring overall productivity, the total output value of the measured time period is divided by all of the inputs that have been used for production. In contrast, for the productivity ratio, the overall output value is proportioned to one input factor; when estimating, for example, the productivity of the work, the output is divided by the work contribution (the number of people or the labor costs). Input costs are easy to measure when the input contribution is purchased from private sector; in the annual accounting, those costs are included as variable costs. However, the definition of output will be more problematic when the outputs are generated by internal public service providers. In such cases, customers are not charged or the services are offered for nominal compensation; thus, the output cost may not correspond to the real monetary market value of the contribution (Meklin, 2008).

In practice, special number-type indicators have also been used in the evaluation of productivity, for example, the number of doctor visits. Thus, productivity is usually quantitatively defined, in which case it is supposed that the quality will be a constant, or quality is examined as a separate phenomenon from actual productivity (see Grönroos & Ojasalo, 2004). That is, productivity should be understood more deeply and pay further attention especially to the relationship between productivity and quality, productivity measurement and factors that raise the productivity and quality (Linna et. al., 2008.) However, what makes the public organization productivity measurement more complex is the fact that they do not make revenue and operations funding does not necessarily correlate straight forward with operative service production capacity. Thus the strategic productivity analyze can be more complex that what is needed at operative working team and foreman level.

To foster entrepreneurship certain simplification in determining the public organization productivity is needed, however the operative evidence-based reliability must be maintained. At operative level the adequate simplification is achieved when the service capacity is assumed to be in line with the service incomes. This means that from one service hour the team will get constant value incomes. Thus the public organization operative working team level is assumed to operate with "the same phenomenology" as a private enterprise.

GAME BASED LEARNING DEVELOPING PUBLIC MANAGEMENT AND ENTREPRENEURSHIP

Determining Team Level Productivity in Public Organizations

The following example illustrates the principle of team level human capital production function at public organization. Example analyze is from the municipal organization of 20 employees (in Full-Time-Equivalent), and all information used in this example is based on Finnish municipal organization case study. The organization present the hospital emergency services. The fiscal data is presented at US-dollars as the book will be published in the US. Average employee has the theoretical working time, which in this case is 1 950 hours per year. From this theoretical working time the auxiliary working time takes roughly 20% from the theoretical working time. Auxiliary working time (AX) includes vacation, absence, maternity leave, work orientation, staff training, HR-practices (also called leadership practices) and HR-development (HRD).

When auxiliary working time is reduced from theoretical working time we get the time for work. During the time spend at work the intangible human assets have to be considered. Quality of Working Life (QWL) describes the utilizing degree of intangible human capital. If Quality of Working Life is at the level of 60%, then the effective working time share from the time for work is 60% as well. Therefore the calculated effective working time is 80% times 0.6. Thus, 48% from theoretical working time is calculated to be effective.

Other working time can now be calculated and it is 80% - 48% = 32%. This other working time includes work planning, quality assurance, reporting, wasted working time due to errors, and fuss because poor motivation and engagement, and handling customer complaints.

With the effective working time the team produce services which they get funded from the service incomes. There is organization specific coefficient (K), which determines how much incomes the team can make with one effective working hour. This example team makes 2.51 million dollars revenue. According the annual profit and loss calculation the team has variable costs of 0.91 million dollars which is 36.2% from the incomes. Variable costs are materials and purchase services which are needed in making the services. Staff costs are 1.34 million dollars, and other fixed costs, like rents and licenses, are 0.26 million dollars. Finally, when all costs are reduced there will be operating profit which in this case is 0 dollars at the beginning of the productivity analyze.

Team foreman has entrepreneur mind set so he/she wants to improve team's human capital productivity, and therefore invest some time to staff training and HR-practices. For simplicity, in this case we assume that this training and HR-practices time is already included at earlier year auxiliary working time, so the time for work remains the same. When training and HR-practices are done properly, there will be positive effect at the Quality of Working Life (QWL). In practice we have recorded at case studies that 5% improvement at QWL is possible and realistic objective at one year. In this illustrative case it means that the Quality of Working Life improves from 60 to 63%. Because QWL-index determines the Effective Working Time share, there will be equal 5% improvement at the Effective Working Time.

In this simplified operative level productivity analysis the team can utilize the K-factor in making more services or introducing new ones, thus increase incomes. When incomes increases the variable costs are increased as well. Therefore $0.13M more incomes means in this case $0.05M dollars more variable costs. HR-development creates so called virtual talent capacity increase, which means that Human Resource capacity will increase without fixed costs raise. Improving intangible human assets effectively will enable more production capacity without fixed cost increase. Therefore only variable costs increase are reduced from Revenue and the rest $80 000 ($0.08M) will improve the operating profit. As a conclusion, this company makes 4 000 dollars more profit for each employee.

Human capital production function mathematical equation is following:

Incomes = K x HR x TWh x (1-AX) x QWL

where

K = K-coefficient (Human Resource Business Ratio, $/h)
HR = human resource in full time equivalent (FTE)
TWh = theoretical yearly working time (h)
AX = auxiliary working time, this is the same as the time for work (%)
QWL = quality of working life index (%).

Figure 1. Improving company productivity by HR-development

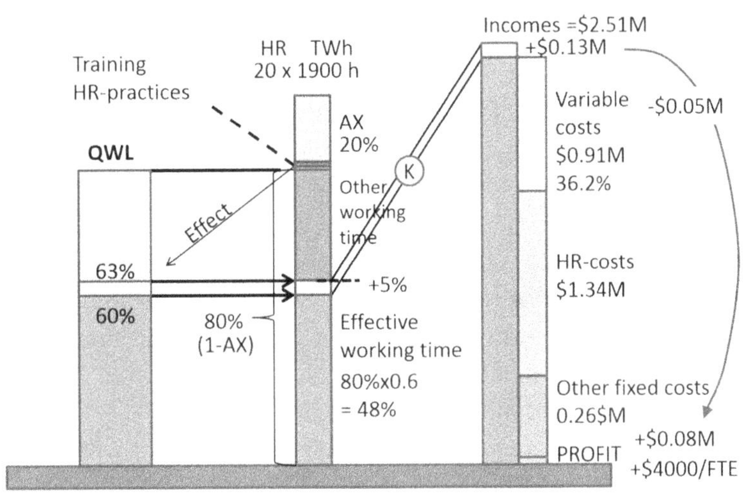

The operating profit, in financial terms Earnings before Interest, Taxes, Depreciation and Amortization, is revenue reduced by all costs.

The Human resource business coefficient, shortly the K-factor, can be calculated from earlier year realization with following equation:

K = Incomes / (HR x TWh x (1 - Ax) x QWL)
= $2 519 000 / (20 x 1900h x (1 – 0.20) * 0.6 = 137.6$/h

When case numbers are inserted at the equation, there will be coefficient of 137.6 dollars per hour. This means that one effective working hour produce $137.6 worth service capacity. Public organization management can influence on K-factor by certain strategic innovations and investing at product development and IC-technology. There is saying that management determines the profitability and leadership determines the productivity. At the Human Capital Production Function this means that management determines the K-factor and leadership the Quality of Working Life and thus the effective working time. Usually when K-factor is improved for example by strategic innovation there will be some reduce at the QWL, causing decline at organization productivity.

Determining the Quality of Working Life

In the human capital production function, there is an index of QWL, which describes the intangible human assets' connection to performance. With moderate reliability, this can be measured through a validated human resource inquiry (also called employee satisfaction survey, ESS). QWL includes the individual's job-related well-being and experiences relating to job satisfaction, benefits, and work environment, the social environment of the organization, organizational performance, employee participation, employee involvement and opportunities to develop human capacity (Greenhous et al., 1987; Sirgy et al., 2001; Totawar & Nambutini, 2014). The QWL seems to be related to operative performance, which leads to better business outcomes (Abdeen, 2002; Wright & Croppanzano, 2004; Warr, 2005; Wright & Bonett, 2007). However, the quantified relationship between QWL and organizational performance is controversial and vague. It has been difficult to quantify QWL results such that the measurement correlates reliably with business performance.

Kesti et al. (2016) argue that there is no single human intangible asset that is generally more important than another, and it is a combination of psychological factors that determines human intangible asset performance. Each psychological factor related to self-esteem has a different effect on working group performance. For example, emotional safety has a different effect on performance than creativity. This can be explained by a simple engineering analogy: Given three length measurements of 1, 2 and 3, it is easy to calculate the average: (1+2+3)/3 = 2. However, if the measurements are carried out using different units (1 ft, 2 in, 3 cm), the average length will be totally different than average result of the measured numbers (Kesti et al., 2016). The same phenomenon seems to emerge in terms of wellbeing inquiry measurement and its connection to performance, although this is somewhat more complicated. The indicator of the QWL gives information about the performance of the human resources, so the average of the results of a traditional HR inquiry is not suitable for this purpose.

To obtain a more reliable index for intangible human capital performance, Kesti et al. (2016) propose a model in which there are three different self-esteem categories, each of which has a unique effect on performance. The self-esteem categories are as follows:

1. Physical and emotional safety (PE);
2. Collaboration and identity (CI); and
3. Objectives and creativity (OC).

According to phenomenography analysis, the chosen categories should be logically related, as well as having logical connections with the phenomenon of interest (Marton, 1986; Uljens, 1996; Marton & Booth, 1997). The QWL self-esteem categories' connection to performance can be formed through the same phenomenon as in Kano's (1984) model, where three different customer opinion categories (must-be, discussed and attractive) affect customer satisfaction. Also Losada's (2004) study shows that positive feelings improve the performance, whereas negative feelings reduce performance.

In Hertzberg's (1958) theoretical model, employee motivation comprises a combination of hygiene factors multiplied by motivation factors. Hygiene factors can be understood as items that cause distress and illness, whereas motivation factors are relate to self-esteem elements that boost performance. Hertzberg's motivation theory is highly respected, but it seems to be incomplete, since it does not consider both self-esteem factors that may cause distress or increase motivation. In the QWL self-esteem model, the missing self-esteem factor is collaboration and identity (CI), which is found on the spectrum between distress and motivation. The self-esteem factors of CI and OC are considered motivators, while self-esteem PE is seen as a hygiene factor. In line with Herzberg's theory, the QWL index is calculated using the following equation (Kesti et al., 2016):

$$QWL = PE(x_1) * ((CI(x_2) + OC(x_3))/2)$$

where

QWL is calculated using the quality of working life index (0 … 1)
$PE(x_1)$ is the value of the function of physical and emotional safety
$CI(x_2)$ is the value of the function of collaboration and identity

Figure 2. Self-esteem categories of QWL functions

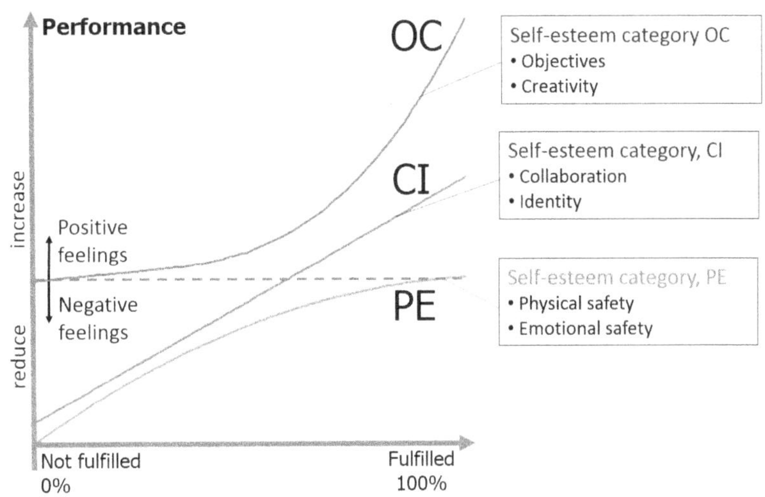

$OC(x_3)$ is the value of the function of objectives and creativity

The functions of the self-esteem categories are adjusted so that the result is always between 0 and 1 (0% … 100%); thus, the QWL index describes the human intangible assets utilization percentage. Kesti et al. (2016) argue that there are no general absolutely right equations to measure the connection between self-esteem and performance; rather, what is at issue, is the best theoretically reliable estimate.

Game-Based Learning and Management Training

Organizations are continually looking for new methods to influence on the behaviors of their employees and to develop their employees also in municipal organizations. During last decades, there has been various management training program experimentations with different teaching and learning methods. There has been, for example, a growing interest in how different forms of play and gamification can be used in organizations in order to develop leadership skills, motivate and engage employees and achieve the organizational goals (e.g. Kark, 2011). Furthermore, the technology development has made it possible to use simulations and digital based learning in management training. Recently, computer game applications have become popular, especially in the medical (e.g. Morrison, 2004) but also in the management training field, involving players to interact within an information technology to achieve specific goals and missions (Wang, 2013). Management simulations are also often used for higher education purposes, in university courses.

In management training and development programs, the purpose of using games as learning tools is to allow learners to sharpen their skills in decision-making, experiment with different goal settings and strategies and overall, gain insights into many areas of organizational operations (Gonen et. al., 2009). Games and simulations are played either individually or in a team learning environments. Team based problem solving entails intensive interaction in decision making between team members. This also implicates that members usually experience a full range of emotions including excitement, frustration, satisfaction, joy, surprise and also disappointments (Lu & Hallinger, 2014). At its best, in team based games a diversity of ideas and solutions can be promoted enhancing the quality of decisions.

Game based learning simulations and games have been associated with several learning theories (e.g. Kebritchi & Hirumi, 2008). Pedagogical foundations are in constructivistic learning theories, which suggest that knowledge is built by the active learner and in practice. That is, learners base their knowledge on personal experiences and on their efforts to give meaning to these experiences. According to this view, the learning environment should enable students to construct their knowledge from trial and error (Gonen et. al., 2009). Another pedagogical foundation, practice-based learning, emphasize that one learns by formulating the doings in real work-life situations (Gherardi & Perrotta, 2014). The underlying pedagogical strategy of game based learning seem to operate within the principles of experiential learning which actually have been common grounds for designers to integrate games with education. Typically, computer game applications are seen as experience-based medium that enables game player to gain experience via game playing (Swartout & Van Lent, 2003). Games are seen as experiential exercises offering concrete experiences to validate and test concepts presented in the gaming environment (Siewiorek et. al., 2013).

Experiential learning is based on the belief that people learn best by doing. That is, one makes the meanings of experiences and learns (Kolb & Kolb, 2010; Guthrie & Jones, 2012). On the other hand, in order to deepen the experience of learning, reflection in forms of discussing, writing and reading are also needed (Guthrie & Jones, 2012; Crookall, 2010; Siewiorek et al., 2012). In experiential learning process knowledge is created through the transformation of experience. That is, knowledge is a result from the combination of grasping and transforming. Experiential learning consists of four stages or events (Kolb, 1984). The cycle starts from the concrete experience which forms the base for the learner's observations and reflections. The reflections transform experiences into abstract concepts, which stimulate new actions. The experiences from the new actions are then tested and reflected creating new concrete experiences. The cycle is then repeated and the learning is a continuous cycle of experiencing and exploring. However, in order for learner to fully engage in learning cycle, it is essential for learner to provide a learning environment which allow learner to be in charge of their own learning, and space for reflections, thinking and actions. As a matter of fact, Kolb & Kolb (2010) stress that play and game environments exemplify the highest forms of the experiential learning process.

According to several studies, simulation and computer games seem to be very effective in education and management training. The proponents of utilizing simulation games argue that simulation based learning enhance complex applied competencies in decision making, management and teamwork and foster skills in higher order thinking and reflection (Lu, Hallinger & Showanasai, 2011). Games offer an environment where learners have a possibility to experience the unpredictability and complexity of organizations. That is, games and simulations provide new opportunities to deal with complex and risky real-life processes in a safe educational context (Siewiorek et. al., 2013). Through the simulated experience, students are able to see how organizational processes develop and change over time. In a certain period learners have the opportunity to gain a wide perspective on the management of organization that would otherwise require years of working experience. Furthermore, Farrell (2005) have compared simulation games with traditional teaching methods for undergraduate students in business management and have found the simulation game as a more effective learning tool. One of the effective key advantages of simulation games is that they provide outcomes and feedback in real time.

Bell et. al. (2008) have observed that well-designed simulations foster learning and applying job-relevant knowledge and skills, and engage learners to solve complex and dynamic management problems. Furthermore, Wang et. al. (2013) stress that game based learning platforms stimulate learning motivation, are efficient and enable learners to achieve improved learning performance. Management simulations and games form a dynamic, constantly changing environment for learners. That is, the initial problem changes its form depending on the solutions and decisions made by the learner. Well-designed computer games and simulations are also cost-effective providing fun learning environments. In game environment learners have a possibility to apply a combination of knowledge to choose from given solutions and to make decisions and to test the outcomes of their actions and decisions. Games help learners to connect theory and practice fostering learners to understand of certain subject area and show greater retention than the traditional classroom (Bee, 2016).

Simulations are a useful pedagogical tool only when they meet certain criteria. To gain the specific learning outcomes, the virtual organizational reality of the computer simulations need to be well-designed. Simulation need to be complex enough and reflect the real-life situations in organizations. Otherwise there is a risk that learner may perceive the simulation to be trivial, boring or unrealistic losing the interest and motivation to play. Simulations should also have good user interfaces. They should be easy enough that the mechanics of their use do not distract users from learning. That is, for learners, it should be easy to input their decisions, read the outputs and what critical areas they need to focus on to improve the results. Simulations should also be realistic in which learner see the direct relationship between the made decisions and results. (Adobor & Daneshfar, 2006.) Solving the problems, learners also need to have an increased interest in the subject matter. They are required to consider both formal and tacit knowledge, and to situate their knowledge in a certain problem context. To proceed and succeed in a game, learners should first understand and define the organizational problem. Secondly, learners should apply acquired knowledge and skills to certain real organizational situation. Playing should also motivate the learner and playing should evoke a positive experience increasing the motivation to learn and using game as a learning method.

There are lots of researches that claim simulations and games are useful and thus the best way to educate leadership (Lopes et.al., 2013; Gentry, 1990; Siewiorek et al., 2012; Siewiorek et. al., 2013). It is also expressed that leadership may only be learned by doing (Lopes et.al., 2013, 525; Siewiorek et. al., 2013). Like we pointed out leadership training is needed for municipal organizations to understand the productivity and to gain entrepreneurial mind set. This is needed for municipal organizations to make it out in public sector service competitions that is going to be the future in most of the welfare countries. The management in public and municipal organizations seems to require more knowledge of productive leadership and entrepreneurial business skills at the double. To do training productively means proper tools because traditional classroom trainings are too slow and time taking for current situation.

SOLUTIONS AND RECOMMENDATIONS

Productive Leadership Game: From Theory to Simulation Tool

Leadership simulation learning game is based on the theory of human capital production function and the quality of working life connection to team performance. For team leader these theories are too complicated to utilize at practical leadership behavior. Digital simulation provide useful tool to teach the practical essentials from these theories. The purpose of the game is to connect productivity, quality of working life and economical result of the company or public organization. In the game the quality of the working life is one production factor. In middle manager's point of view it is important to understand following:

1. Public organization manager's actions or neglects have great effect on entrepreneurial orientation, team performance and productivity;
2. Problems should be solved at the early stage and therefore staff opinions are important;
3. There are several feasible HR-practices for different kind of problems solving and for improving the team performance;
4. Co-operative organization culture bases on effective leadership and heathy leader identity.

Constructing a simulation is challenging since leader activities effect to staff emotions and fiscal business outcomes have to be mathematically formulated. In the simulation the leader is facing different workplace challenges which the staff member comes and discuss with the foreman (Figure 3). Each workplace challenge will reduce one of the self-esteem factor and thus decrease the quality of working life index (QWL). QWL reduce will cause fussing and eat the team performance (effective working time). Leader has several good HR-practices that he/she can utilize at the team. Each HR-practice will improve one of the self-esteem factor, thus improving the QWL-index. Workplace challenges are eating the QWL-index and HR-practices are improving the QWL-index. Furthermore each HR-practice will require certain working time spend from the staff, therefore increasing the Auxiliary working time and

Figure 3. Example of simulation game user-interface

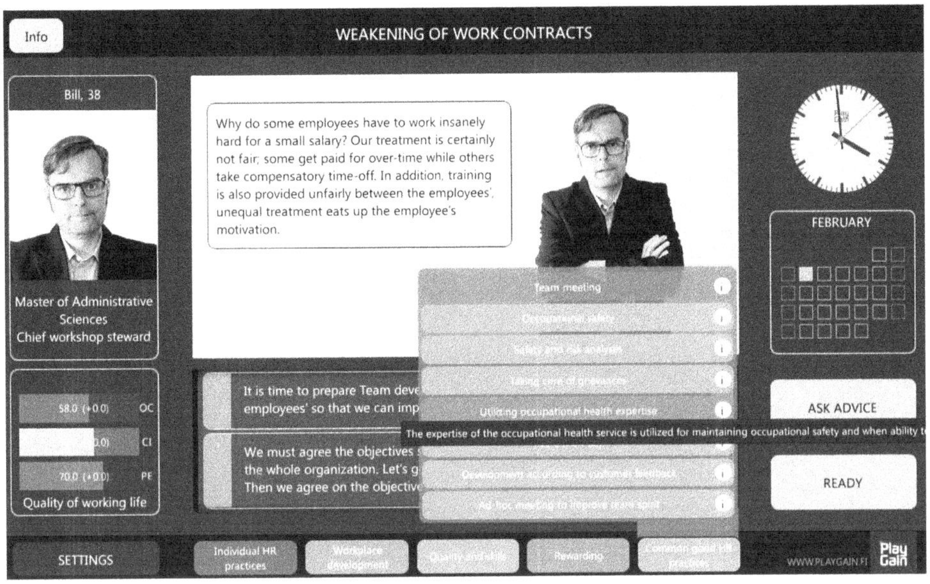

reducing the time for work. Therefore it is possible that chosen HR-practices are reducing the month's profit but increasing the QWL capital to be utilized later. This phenomenon is included at the human capital production function.

When the mathematical formulations are completed for the simulation there has to be made the user-interface. Mathematical equations are hidden at the game motor and only results and effects are shown for the player. Each employee that comes to discuss with the leader has several different facial emotions to describe the human aspect at the simulation. When challenge decrease the QWL-index there will be disappointment with corresponding comment and when HR-practice improve QWL there will be facial emotions showing the degree of satisfaction. After each month player gets the emotional response and also fiscal results from the month played (Figure 4).

Digital simulation learning game has to create high level of engagement and immersion to maintain the motivation to continue and to activate cognitive learning. Therefore there are included several elements of responses, credits, guidance and enjoyment, as well as critical feedback to form creative stress. Feedback comes in form of emotional response and in numbers with graphics (Figure 5). Each player has their individual learning style and leadership style, so the simulation game should adapt to each player's style. Some leaders are more focused on human perspective while others are focused on numbers and performance results. Whatever the leadership style is the excellent results require good HR-practices knowledge, listening of the workers and situation sensitive implementation of different HR-practices.

Formation Creates Professionalism

In simulation game, there are two dimensions of learning. First you have to learn the game as a tool. When the player can operate the game-tool he is ready to deepen his knowledge to subjects of the game. Formation plays a big role in both of these learnings. Formation means trial-error, repeating, reflection, doing again, observing, and while doing this person is formatting his doings and finally getting

Figure 4. Example of positive emotional feedback

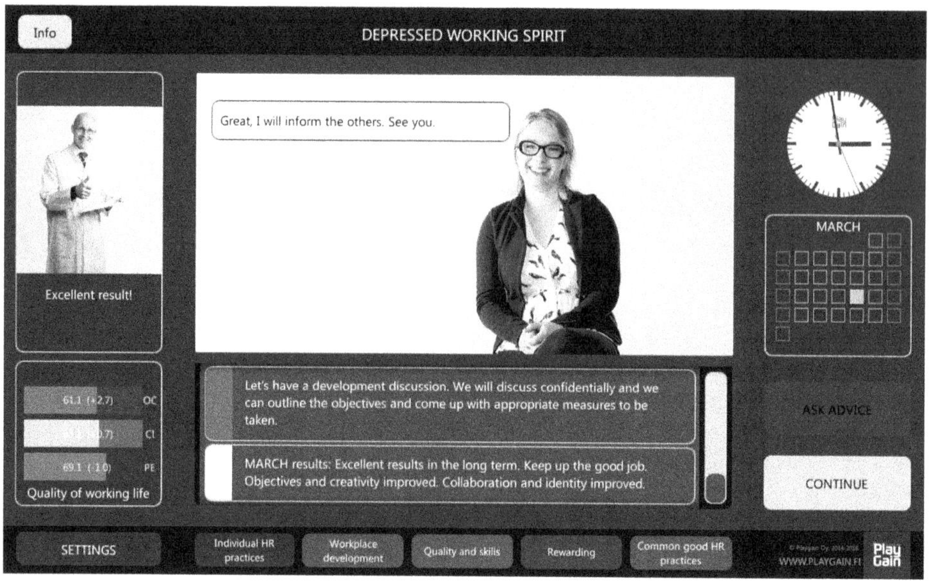

Figure 5. Example of performance numbers and graphs

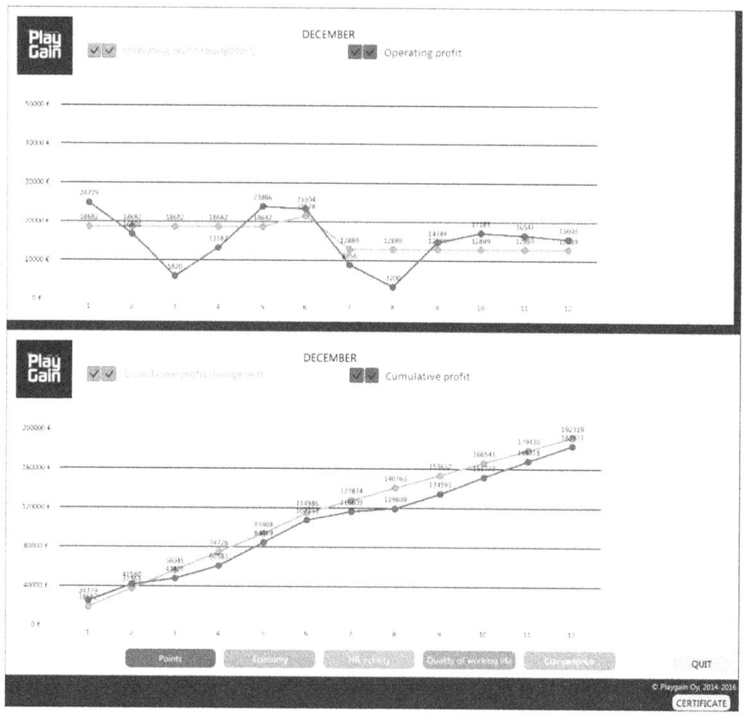

experience and understanding about a new matter. Formation is also a term of practice-based learning. In practice-based approach formation occurs every day again and again when person brings together the human existence (*dasein*), knowledge, tools, work context, and adapts himself, and his knowledge to the present moment of work (Gherardi & Perrotta, 2014). Usually when person learns totally new subjects and doings the situation usually requires formation. Kark (2011) expresses that formation is also one essential phase when developing leader's identity. Leaders' identity is constructed in social relationships among leader and other people. Leader claims actively his or her role while others grant this role. This is an active interplay of claiming and granting whilst working situations where people are taking part.

When learning leadership by simulation person has a possibility to react and operate like in every-day-life of a leader. Player may act, take a role, try, choose, decide, do errors, repeat, observe, think and reflect. Formation is not always enjoyable. Trial-error, repeating and trying presume lots of time and attention. The player may feel frustration if not succeeding right away. But this is part of formation and learning. In formation waste of time and material is inevitable. Like a carpenter he or she might has an idea of an object. It needs several recreations to achieve the one that meets carpenter's professional expectations. These trials take lots of time and produce lots of non-marketable objects that are most of all waste of material. Learning in simulation does not produce waste of material but only waste of time if you want to express it that way. But in the end player has achieved his goals and reached an experience of real-world-like leadership, and is proud about his success and results. These results of knowledge, understanding and experience is then possible to be used in players real working contexts.

The process of formation is like experimentation, playfulness, and attachment to matter and is intrinsically tied to the discovery of materials, techniques, and tools (Gherardi & Perrotta, 2014, 142). So even if the formation is not always easy there is and should be, a bit of play in it. That makes the emotions be involved as well as sensible knowledge, and aesthetic judgement. Kark (2011) has mentioned that when there are rules, structures and goals in play it can often be defined as a game. Kark (2011) also expressed that playfulness frees up imaginations which is one important element of learning. To learn leadership, we need rules and goals but also playfulness, experimentation and attachment. In simulation you are able to encounter all this. The enjoy and freedom offers the environment of play and the rules, and goals of leadership offers the professional dimension. To create a proper simulation game for leadership means balancing between these two aspects. If there is too much bureaucracy and standardization involved the inventive activity is apparently oxymoron (Gherardi & Perrotta, 2014, 145). Leadership simulation is an environment where one may concentrate purely to activity of leadership without aspects that hinder creativity and thus learning.

Practicing in Holding Environment

Leadership simulation is almost as tricky as leadership in everyday life. One may easily fail. But failing in a simulation may be even fun and is part of the play. Failing in every-day life and middle of a tricky situation may be a catastrophe for one's carrier. Learning new things is mostly achieved by trial-error. Trial-error method in every-day life is perceived as a non-desirable issue. It is highly unprofessional and for a leader may even be unforgivable. We need a holding environment to learn leadership. Kolb & Kolb (2010) has pointed out that experiential learning and play could offer safe and supportive spaces where four modes of learning may take place – feeling, reflection, thinking, action – and where learner is respected (Kark, 2011). The environment to learn should be also challenging enough.

Holding environments can provide "safe heavens" (Kark, 2011). In these safe environments player has a possibility to experiment a different kind of provisional leadership images and adopt various leader selves. It is kind of "formation of leader's identity" (Kark, 2011). Player may even develop his or her leadership identity by imagining to be someone different or "adapt a new persona" (Oprescu et. al., 2014). It is one way to claim oneself future leadership role. Simulation game environments may take a huge role in future leadership development process and also further ability to develop as a leader. Productive Leadership -game players found the game scenario representing foreman's responsibilities, and wanted others to play it to find out how tricky issues foreman's work concern.

Game for Several Purposes

Leadership simulation game can be used for several purposes. A game can be played alone or with colleges. When playing together simulation creates a texture of leadership where players may co-formulate leadership practices and develop leadership culture. Like researches express to develop work it has to be described (Gherardi, 2011). It is difficult for a worker to describe his work because there is always some tacit knowledge involved which the worker is not able to express (Gherardi, 2011). When simulation is based on scientific theory it offers a proper texture of the work and may be utilized as base for discussions and work development actions.

Game can also be used to test supervisors and familiarize new talents for supervisors' responsibilities. There are always some newcomers to be in charge. A simulation game is a proper practicing environment where the player gets valuable experience of responsibilities and challenges of a leader and manages to avoid to learn it the hard way. Simulation allows a possibility for individuals to engage in real-world-like scenarios and make decisions and take actions without any real-world consequences (Backus et al., 2010).

The storyline of Productive Leadership game gives an idea of shared leadership and situation sensible leadership where foreman listens to followers' opinions, followers expresses their opinions and shares the ideas, and foreman and followers practice constructive co-operation by sharing and listening. This increases quality of the working life, entrepreneurial mind-set as well as productivity of the whole team.

Features of a Good Simulation

To be a simulation game there has to be certain elements in it. According to Teach (1990) there has to exist scenario, roles, calculating system for results and algorithms to indicate and count the results of cause-effects during the game. In Teach (1990) model the scenario represents the business to be played. In Productive Leadership -game scenario concerns leadership. The language used in the game should be recognized to that scenario. Player sets himself to a role of the game. In Productive Leadership game player's role is to be a leader. Player should recognize the role as well as act according to the role. Calculation system of the game should register game actions and report results during and in end of game. Reports should represent the results of the played scenario. In Productive Leadership -game results follow the form of financial statements. The algorithms should keep count of all actions and their effects during the game history. The effects of earlier done actions reflects during the whole game. There might be even some artificial intelligence in the game. In Productive Leadership -game all actions player selects during the game, player's skill level, and amount of actions all affects in counting.

Games that include large scenarios, have a convincing calculation system of results, may be called as a simulation games. Games that include perhaps convincing game roles but weak scenarios and result calculations, may be called as role playing exercises (Teach, 1990).

Players' Experiences

The players' experiences of playing the leadership game were positive and enthusiastic. When Productive Leadership simulation game was played intensively and experiences reflected and discussed players were able to name several theory-based subjects they adopted from the game. They also named several goals how to develop their leadership skills in the future. These ideas players adopted from game theories and experiences of simulation. It is important that simulation games base on scientific leadership theories (Lopes et. al., 2013). Thus it is possible to evaluate how much player has adopted new ideas and learned. The game produced a master-apprenticeship-like environment (Backus et. al., 2010) by directing and guiding the players and giving them a safe practicing setting where a player can repeat actions as many times as needed without disrupting their real organization.

Course members learned also some new ideas to be used in their work in public and municipal organizations where structure of funding of business and understanding of economical result creates own challengers for leadership and entrepreneurship. Here are two comments from the course members:

Time that has been benefited from the removal of fuss and hassle should be canalized for developing quality of the work. That will improve time and knowledge to produce more and right kind of services and products. When services intensify and more time is saved first idea usually is to produce even more services. In public service companies this idea could be even dangerous if you do not have a possibility to finance these additional services.

It was illustrating to see in Productive Leadership game that whatever leadership actions you chose there were direct effect to the working time. Effective working time increased at the same time when other working time, meaning fuss and hassle, decreased. It was also interesting to follow how these actions affected directly to the economical result by producing surplus or deficit. You could all the time follow up graphs and development of situation by checking budgets. I'm more and more enthusiastic about an idea to clarify human capital productivity situation in my own organization and develop it more focused in the future.

Players named several actions they would like to take in the future to further develop their leadership skills, which also foster development of entrepreneurial mindset. When people played the game together with others they were able to name ideas how to develop their organization and leadership- and entrepreneurship-oriented culture. It would need more research to examine if these development actions really activate in the organization and what kind of results that would bring.

Game-Based Learning: General Reflections from Productive Leadership Game

According to phenomenology in every day practices there are always "co-present activities like work, learning, innovation, communication, negotiation, conflict over goals and their interpretation, and all history" (Nicolini et. al., 2003). There are also material, equipment, tools, and context that human needs for her or his actions. In this human flow of existence there is no distinction between subject and object. Humans are present in this existence and practice their activities among subject, object, context, and knowledge. Like a hammer and a carpenter. It is the tool that makes the professional and the professional that makes the tool, this co-operation makes the practice come alive. And it does it again every time the carpenter takes the hammer in his/her hand. When a professional learns to work with tools he or she does not realize the existence of the tool anymore. They both exist within one setting. Heidegger (1962) used a term *Dasein*, "being there" or "being-in-the-world", where subject and object are indistinguishable. Polynyi (1962) says a professional "knows more than he can tell" and has "tacit and personal knowledge". This is why you may not be a professional to use a hammer just by reading books. You need to practice and experience.

It seems to be assumed that a person becomes a leader just by taking a position of a leader. There is usually lots of professional knowledge person has achieved during his or her career, maybe as a carpenter, but when this carpenter is set to be a leader of other workers it is totally different knowledge and practice that he or she had been doing before. Well how to be a leader then. There are written and published thousands of books about leadership thus there is less educational courses concerning leadership because it is almost impossible to train leadership by traditional learning methods (Jameson, 2013; de Freitas & Routledge, 2013). And to educate leadership it is time and resource consuming because leadership is multidimensional, changing in time, concerning people with emotions, and feelings (Jameson, 2013; de Freitas & Routledge, 2013), and it happens in context of organization bureaucracy, norms, and rules (Kark, 2011). Simulation and game environments and technologies are a natural way to implement this kind of features. In leadership simulation norms, rules and bureaucracy of organization is detached and player may concentrate purely on the content of leadership practices.

FUTURE RESEARCH DIRECTIONS

Like Gherardi (2011) mentioned there is lack of methods to describe work. Structure of the work needs to be described in case it needs to be developed. Productive Leadership simulation game could perhaps be used as a structure of leadership in this case. It would be interesting to find out how simulation game fits for this purpose. It would be also interested to assess Productive Leadership game by ELESS or other existing assessment models. De Freitas and Routledge (2013) has develop an ELESS model to assess use and construct effective design of soft skills and leadership skills in learning game environment. ELESS stands for e-Leadership and Soft Skills Educational Design Model. ELESS may also be used to efficacy of games and to inform effective design of new games that focus to develop leadership and soft skills. E-Leadership stands for advanced models of leadership that can be applied in virtual and physical environment. ELESS consist of three areas: to assess design of the game, to assess the content of learning in the game, and assess implementation of the game.

For further research, it is interesting to study if learnings from the game transfer to player's leadership actions in his or her everyday-life in organization, and how would these results affect working life quality and economical result of player's organization. Most of all it would be interesting to see if municipal organizations may reach their leadership development by simulation tools and would that bring any crucial skills and changes for the organizations to survive in future competitions with private sector companies.

CONCLUSION

This chapter presents a simulation based leadership development tool as a solution for developing municipal organizations' leadership skills to encounter future competitions of services with private sector companies. The predominant understanding of leadership skills in municipal organizations seems to suffer of a knowing-doing cap thus it seems they would need more entrepreneurial mind set and understanding of productivity. The productivity is also dangerous issue if developed with a short-term view. Developing public service efficiency might increase the costs, thus decreasing the fiscal productivity. This is because of public organizations' public funding which defines the organization business limits. Usually, the more services are provided the more they cost, and in public organizations more costs are not acceptable. That is why leaders in municipal organizations needs special kind of understanding of productivity and its' consequences, and therefore entrepreneurial mindset is so important to disseminate. If the productivity is increased by buying less subcontracting and doing more self instead would save money and the productivity is canalized properly. Improving productivity involves improvements in services, clear strategic objectives, systematic human resource management practices and strengthening the competencies and skills of public middle managers. Middle managers, with their many roles as supporters, influencers, facilitators, emotional balancers and initiators, for example, are in demanding task to gain the objectives set for high-performing municipal organization.

Producing productivity and quality of working life is a tricky issue. It is not done by the leader him- or herself but with the whole organization and the leader has a big role in implementing this. Usually one learns to be a professional at universities and institutions but less as a leader. Still many of graduated students find a way to a leading position. To these moments a simulation is a good solution to maintain one's skills to encounter the challenges of leadership and entrepreneurial spirit. In simulation player learns by practice, rehearsing in an every-day-like environment. Simulation game may provide an individual learning environment where any kind of learners find a suitable master-apprenticeship-like environment just for him.

In Productive Leadership Simulation Game example, we illustrated how a theoretical and mathematical model and theory may be developed to a digital simulation tool. It may be one tool to close the knowing-doing cap of leadership skills some municipal organizations still have, and to gain more entrepreneurial mindset for public managers. The chapter explains the performance and productivity of public organization and illustrates method to multiply essential learning to practice. A simulation tool might be even a faster way to foster skills there is still lack of. Anyway games are a playful way to experience it.

REFERENCES

Abdeen, T. (2002). Company performance: Does quality of work life really matter? *Management Research News*, *25*, 8–10.

Adobor, H., & Daneshfar, A. (2006). Management simulations: Determining their effectiveness. *Journal of Management Development*, *25*(2), 151–168. doi:10.1108/02621710610645135

Backus, C., Keegan, K., Gluck, C., & Gulick, L. M. V. (2010). Accelerating leadership development via immersive learning and cognitive apprenticeship. *International Journal of Training and Development*, *14*(2), 144–148. doi:10.1111/j.1468-2419.2010.00347.x

Bae, J., & Lawler, J. J. (2000). Organizational and HRM strategies in Korea: Impact on firm performance in an emerging economy. *Academy of Management Journal*, *13*(3), 502–517. doi:10.2307/1556407

Baptiste, N. R. (2008). Tightening the link between employee wellbeing at work and performance: A new dimension for HRM. *Management Decision*, *46*(2), 284–309. doi:10.1108/00251740810854168

Batt, R. (2002). Managing customer services: Human resource practices, quit rates and sales growth. *Academy of Management Journal*, *45*(3), 587–597. doi:10.2307/3069383

Bee, L., Benson, O., & Teck-Heng, L. (2016). Game-based learning in construction management courses: A case of bidding game. *Engineering, Construction, and Architectural Management*, *23*(1), 4–19. doi:10.1108/ECAM-02-2015-0029

Bell, B. S., Kanar, A. M., & Kozlowski, S. W. (2008). Current issues and future directions in simulation-based training in North America. *International Journal of Human Resource Management*, *19*(8), 1416–1434. doi:10.1080/09585190802200173

Bernier, L., & Hafsi, T. (2007). The changing nature of public entrepreneurship. *Public Administration Review*, *67*(3), 488–503. doi:10.1111/j.1540-6210.2007.00731.x

Borins, S. (2000). Loose cannons and rule breakers, or enterprising leaders? Some evidence about innovative public managers. *Public Administration Review*, *60*(6), 498–507. doi:10.1111/0033-3352.00113

Boyett, I. (1996). The public sector entrepreneur - a definition. *International Journal of Public Sector Management*, *9*(2), 36–51. doi:10.1108/09513559610119546

Crookall, D. (2010). Serious games, debriefing, and simulation/gaming as a discipline. *Simulation & Gaming*, *41*(6), 898–920. doi:10.1177/1046878110390784

de Freitas, S., & Routledge, H. (2013). Designing leadership and soft skills in educational games: The e-leadership and soft skills educational games design model (ELESS). *British Journal of Educational Technology*, *44*(6), 951–968. doi:10.1111/bjet.12034

Fabi, B., Lacoursière, R., Raymond, L., & St-Pierre, J. (2010). HRM capabilities as a determinant and enabler of productivity for manufacturing SMEs. *Human Systems Management*, *29*(3), 115–125.

Farrell, C. (2005). Perceived effectiveness of simulations in international business pedagogy: An exploratory analysis. *Journal of Teaching in International Business*, *16*(3), 71–88. doi:10.1300/J066v16n03_05

Gentry, J. W. (1990). *Guide to Business Gaming and Experiential Learning*. London: GP Publishing, East Brunswick.

Gherardi, S. (2011). Organizational learning: The sociology of practice. In M. Easterby-Smith & M. Lyles (Eds.), *Handbook of Organizational Learning and Knowledge Management* (pp. 43–66). West Sussex, UK: John Wiley & Sons.

Gherardi, S., & Perrotta, M. (2014). Between the hand and the head. How things get done, and how in doing the ways of doing are discovered. *Qualitative Research in Organizations and Management*, *9*(2), 134–150. doi:10.1108/QROM-06-2012-1079

Gonen, A., Brill, E., & Frank, M. (2009). Learning through business games – an analysis of successes and failures. *On the Horizon*, *17*(4), 356–367. doi:10.1108/10748120910998434

Greenhaus, J., Bedian, A., & Mossholder, K. (1987). Work experiences, job performances, and feelings of personal and family well-being. *Journal of Vocational Behavior*, *31*(2), 200–215. doi:10.1016/0001-8791(87)90057-1

Grönroos, C., & Ojasalo, K. (2004). Service productivity: Towards a conceptualization of the transformation of inputs into economic results in services. *Journal of Business Research*, *57*(4), 414–423. doi:10.1016/S0148-2963(02)00275-8

Guthrie, J. P. (2001). High-involvement work practices, turnover and productivity: Evidence from New Zealand. *Academy of Management Journal*, *44*(1), 180–190. doi:10.2307/3069345

Guthrie, K. L., & Jones, T. B. (2012, Winter). Teaching and learning: Using experiential learning and reflection for leadership education. *New Directions for Student Services*, *2012*(140), 53–63. doi:10.1002/ss.20031

Hassan, M., Hagen, A., & Daigs, I. (2006). Strategic human resources as a strategic weapon for enhancing labour productivity: Empirical evidence. *Academy of Strategic Management Journal*, *5*, 75–96.

Herzberg, F., Mausner, B., & Snyderman, B. B. (1959). *The Motivation to Work* (2nd ed.). New York: John Wiley & Sons.

Jääskeläinen, A. (2010). *Productivity measurement and management in large public service organizations*. Publication 927. Tampere: Tampereen teknillinen yliopisto.

Jääskeläinen, A., & Lönnqvist, A. (2011). Public service productivity: How to capture outputs? *International Journal of Public Sector Management*, *24*(4), 289–302. doi:10.1108/09513551111133461

Jameson, J. (2013). Special issue on e-Leadership[Editorial]. *British Journal of Educational Technology*, *44*(6), 883–888. doi:10.1111/bjet.12106

Kangasharju, A. (2008). *Tuottavuus osana tuloksellisuutta*. Helsinki: The Association of Finnish Local and Regional Authorities.

Kano, N., Seraku, N., Takahashi, F., & Tsuji, S. (1984). Attractive quality and must-be quality. *Hinshitsu: The Journal of the Japanese Society for Quality Control*, *14*(2), 39–48.

Käpylä, J., Jääskeläinen, A., Seppänen, S. K., Vuolle, M., & Lönnqvist, A. (2008). *Tuottavuuden kehittäminen Suomessa. Haasteet ja tutkimustarpeet. The Finnish Work Environment Fund. Työsuojelurahaston selvityksiä 1. Tampere University of Technology. Department of Information Management and Logistics*. Helsinki: Edita.

Kark, R. (2011). Games management play: Play as a form of leadership development. *Academy of Management Learning & Education*, *10*(3), 507–527. doi:10.5465/amle.2010.0048

Kebritchi, M., & Hurumi, A. (2008). Examining the pedagogical foundations of modern educational computer games. *Computers & Education, 51*(4), 1729–1743. doi:10.1016/j.compedu.2008.05.004

Kesti, M. (2013). Human capital production function. *GSTF Journal on Business Review, 3*(1), 22–32. doi:10.7603/s40706-013-0001-7

Kesti, M., Leinonen, J., & Syväjärvi, A. (2016). A multidisciplinary critical approach to measure and analyze human capital productivity. In M. Russ (Ed.), *Quantitative Multidisciplinary Approaches in Human Capital and Asset Management* (pp. 1–22). Hershey, PA: IGI Global. doi:10.4018/978-1-4666-9652-5.ch001

Kesti, M., & Syväjärvi, A. (2015). Human capital production function in strategic management. *Technology and Investment, 6*(01), 12–21. doi:10.4236/ti.2015.61002

Kolb, A. Y., & Kolb, D. A. (2010). Learning to play, playing to learn: Case study of a ludic learning space. *Journal of Organizational Change Management, 23*(1), 26–50. doi:10.1108/09534811011017199

Laakoli, K., & Peiponen, A. (2012). Tuottavuusmatriisi johtamisen välineenä vanhusten ympärivuorokautisissa palveluissa. *Kunnallistieteellinen Aikakauskirja, 40*(1), 9–23.

Linna, P., Pekkola, S., Ukko, J., & Melkas, H. (2008). Defining and measuring productivity in the public sector: Managerial perceptions. *International Journal of Public Sector Management, 23*(5), 479–499. doi:10.1108/09513551011058493

Lopes, M. C., Fialho, F. A. P., Cunha, C. J. C. A., & Niveiros, S. I. (2013). Business games for leadership development: A systematic review. *Simulation & Gaming, 44*(4), 523–543. doi:10.1177/1046878112471509

Losada, M., & Heaphy, E. (2004). The role of positivity and connectivity in the performance of business teams: A nonlinear dynamics model. *The American Behavioral Scientist, 2*(1), 71–87.

Lu, J., Hallinger, P., & Showanasai, P. (2014). Simulation-based learning in management education. A longitudinal quasi-experimental evaluation of instructional effectiveness. *Journal of Management Development, 33*(3), 218–244. doi:10.1108/JMD-11-2011-0115

Marton, F. (1986). Phenomenography – A research approach investigating different understandings of reality. *Journal of Thought, 21*(2), 28–49.

Marton, F., & Booth, S. (1997). *Learning and awareness*. Mahwah, NJ: Lawrence Erlbaum Associates.

Meklin, P. (2008). Tuottavuuden mittaaminen verorahoitteisessa toiminnassa. *Kunnallistieteellinen Aikakauskirja, 36*(4), 386–395.

Messersmith, J. G., Patel, P. C., Lepak, D. P., & Gould-Williams, J. S. (2011). Unlocking the black box: Exploring the link between high-performance work systems and performance. *The Journal of Applied Psychology, 96*(6), 1105–1118. doi:10.1037/a0024710 PMID:21787040

Morrison, J. (2004). Where now for problem based learning? *Lancet, 363*(9403), 174. doi:10.1016/S0140-6736(03)15298-1 PMID:14726191

Nicolini, D., Gherardi, S., & Yanow, D. (2003). *Knowing in Organization. A Practice-based approach.* Armonk, NY: E Sharpe.

Oprescu, F., Jones, C., & Katsikitis, M. (2014). I play at work – ten principles for transforming work processes through gamification. *Frontiers in Psychology*, *5*(14). doi:10.3389/fpsyg.2014.00014 PMID:24523704

Pohjola, M. (2014). *Suomi uuteen nousuun. ICT ja digitalisaatio tuottavuuden ja talouskasvun lähteinä.* Vantaa: Julkaisumonistamo Eteläranta Oy.

Siewiorek, A., Gegenfurtner, A., Lainema, T., Saarinen, E., & Lehtinen, E. (2013). The effects of computer-simulation game training on participants opinions on leadership styles. *British Journal of Educational Technology*, *44*(6), 1012–1035. doi:10.1111/bjet.12084

Siewiorek, A., Saarinen, E., Lainema, T., & Lehtinen, E. (2012). Learning leadership skills in a simulated business environment. *Computers & Education*, *58*(1), 121–135. doi:10.1016/j.compedu.2011.08.016

Sirgy, M. J., Efraty, D., Siegel, P., & Dong-Jin, L. (2001). A new measure of quality of work life (QWL) based on need satisfaction and spillover theories. *Social Indicators Research*, *55*(3), 241–302. doi:10.1023/A:1010986923468

Swartout, W., & Van Lent, M. (2003). Making game of system design. *Communications of the ACM*, *46*(7), 32–39. doi:10.1145/792704.792727

Teach, R. D (1990). Designing Business Simulations. In Guide to Business Gaming and Experiential Learning. Association for Business Simulation and Experiential Learning (ABSEL).

The evolution of public services from co-production to co-creation and beyond: New Public Management's unfinished trajectory? (2014). *International Journal of Leadership in Public Services, 10*(2), 91–107.

Totawar, A. K., & Nambudiri, R. (2014). Can fairness explain satisfaction? Mediation of quality of work life (QWL) in the influence of organizational justice on job satisfaction. *South Asian Journal of Management*, *21*(2), 101–122.

Uljens, M. (1996). On the philosophical foundation of phenomenography. In G. Dall'Alba & B. Hasselgren (Eds.), *Reflections on Phenomenography* (pp. 105–130). Goteborg: Acta Universitatis Gothenburgensis.

Wang, C.-S., Li, U.-C., & Tzeng, Y.-R. (2015). How to replicate the cognitive process in computer game-based learning units. *Information Technology & People*, *28*(2), 327–343. doi:10.1108/ITP-03-2013-0053

Warr, P. (2005). Work, well-being, and mental health. In J. Barling, E. K. Kelloway, & M. R. Frone (Eds.), *Handbook of Work Stress* (pp. 547–574). Thousand Oaks, CA: Sage. doi:10.4135/9781412975995.n23

Wright, T. A., & Bonett, D. G. (2007). Job satisfaction and psychological well-being as non-additive predictors of workplace turnover. *Journal of Management*, *2*(33), 141–160. doi:10.1177/0149206306297582

Wright, T. A., & Cropanzano, R. (2004). The role of psychological well-being in job performance: A fresh look at an age-old quest. *Organizational Dynamics*, *33*(4), 338–351. doi:10.1016/j.orgdyn.2004.09.002

KEY TERMS AND DEFINITIONS

Human Resource: Employees of an organization (in this chapter if used at the equation it is staff number in full time equivalent).

Incomes: Monetary value of public organization produced services. In this chapter the incomes means the same as revenue.

IC-Technology: Information and communication technology.

K-Factor: Human capital business ratio that describe how much incomes does one effective working hour produce.

Middle Managers: Managers that are responsible for operative level performance.

Profit: The same as operating profit which is incomes reduced by costs.

Quality of Working Life Index (QWL): Describes the human intangible assets utilization performance.

Revenue: Monetary value of production which in this case is the same as incomes.

Compilation of References

Abdeen, T. (2002). Company performance: Does quality of work life really matter? *Management Research News*, *25*, 8–10.

Abdelkafi, N., & Täuscher, K. (2016). Business Models for Sustainability From a System Dynamics Perspective. *Organization & Environment*, *29*(1), 74–96. doi:10.1177/1086026615592930

Achtenhagen, L., Melin, L., & Naldi, L. (2013). Dynamics of business models – Strategizing, critical capabilities and activities for sustained value creation. *Long Range Planning*, *46*(6), 427–442. doi:10.1016/j.lrp.2013.04.002

Act of Law on Associations. (1989). *Journal of Laws.*

Act of Law on Public Benefit and Volunteer Work. (2003). *Journal of Laws.*

Adizes, I. (2004). *Managing Corporate Lifecycles*. Santa Barbara, CA: Adizes Institute.

Adobor, H., & Daneshfar, A. (2006). Management simulations: Determining their effectiveness. *Journal of Management Development*, *25*(2), 151–168. doi:10.1108/02621710610645135

Afuah, A., & Tucci, C. (2003). *Biznes internetowy strategie i modele*. Kraków: Oficyna Ekonomiczna.

Afuah, A., & Tucci, C. L. (2000). *Internet Business Models and Strategies: Text and Cases* (1st ed.). Mcgraw-Hill College.

Afuah, A., & Tucci, C. L. (2001). *Internet business models*. New York: McGraw-Hill/ Irwin.

Agarwal, R., & Selen, W. (2009). Dynamic capability building in service value networks for achieving service innovation. *Decision Sciences*, *40*(3), 431–475. doi:10.1111/j.1540-5915.2009.00236.x

Agarwal, R., Selen, W., Sajib, S., & Scerri, M. (2014). Dynamic capability building in service networks: An exploratory case study. *Journal of New Business Ideas and Trends*, *12*(1), 27–41.

Agencia Vasca de la Innovacion. (2016). *Social Innovation in Basque Country*. Retrieved July 1, 2016, from: https://webgate.ec.europa.eu/socialinnovationeurope/sites/default/files/sites/default/files/social_innovation_in_the_basque_country.pdf

Albach, H. (1962). *Investition und Finanzierung*. Wiesbaden: Gabler.

Alberti, A., & Bertucci, G. (2006). Replicating Innovations in Governance: An Overview. In *Innovations in Governance and Public Administration: Replicating what works* (pp. 1–24). New York: United Nations.

Al-debei, M. M., El-Haddadeh, R., & Avison, D. (2008). Defining the Business Model in the New World of Digital Business. In *Proceedings of the Fourteenth Americas Conference on Information Systems* (pp. 1–11). Retrieved from http://bura.brunel.ac.uk/bitstream/2438/2887/1/AMCIS2008.pdf

Alford, J. (1993). *Towards a New Pubuc Management Model: Beyond "Managerialism" and Its Critics. Australian Journal of Public Administration*. doi:10.1111/j.1467-8500.1993.tb00263.x

Alford, J. (1998). A Public Management Road Less Travelled : Clients as Co-producers of Public Services. *Australian Journal of Public Administration*, *57*(December), 128–137. doi:10.1111/j.1467-8500.1998.tb01568.x

Alford, J. (2002). Defining the Client in the Public Sector: A Social-Exchange Perspective. *Public Administration Review*, *62*(3), 337–346. doi:10.1111/1540-6210.00183

Alt, R., & Zimmerman, H. D. (2001). Introduction to special section on business models. *Electronic Markets*, *11*(1), 3–9.

Amit, R., & Zott, C. (2001). Value creation in e-business. *Strategic Management Journal*, *22*(2), 493–520. doi:10.1002/smj.187

Amit, R., & Zott, C. (2012). Creating Value Through Business Model Innovation. *MIT Sloan Management Review*, *53*(53310), 41–49. doi:10.2139/ssrn.1701660

Andel, N. (1998). *Finanzwissenschaft*. Tübingen: Mohr.

Andersen, A. (1995). *Marketing Social Change: Changing Behavior to Promote Health, Social Development, and the Environment*. San Francisco: Jossey-Bass.

Andersen, S. K., & Mailand, M. (2002). *Multipartite Social Partnerships. A new Role for Employers and Trade Unions*. Copenhagen: The Copenhagen Centre for Partnership Studies.

Anderson, J., Narus, J., & Narayandas, D. (2009). *Business market management: Understanding, creating, and delivering value* (3rd ed.). Upper Saddle River, NJ: Prentice Hall.

Ansoff, H. I. (1985). *Zarządzanie strategiczne*. Warszawa: PWE.

Appadurai, A. (2013). *The Future as Cultural Fact*. London: Verso.

Applegate, L. M. (2000). E-business Models: Making sense of the internet business landscape. In G. Dickson & G. Desanctis (Eds.), Information Technology and the Future Enterprise, New Models for Managers. Prentice-Hall.

Applegate, L. M. (2001). Emerging e-business models. *Harvard Business Review*, *79*(1), 79–87. PMID:11189465

Applegate, L. M. (2001). *Emerging e-business models: Lessons from the field*. HBS No.9-801-172. Harvard Business School.

Aranha, E. A., Abudd, N., Garcia, P., & Corrêa, G. (2015). Open Innovation and Business Model : A Brazilian Company Case Study. *Journal of Technology Management & Innovation*, *10*(4), 91–98. doi:10.4067/S0718-27242015000400010

Armstrong, J., & Ford, R. (2001). Public Sector Innovations and Public Interest Issues. *The Innovation Journal: The Public Sector Innovation Journal*, *6*(1), 1–21.

Asmild, M., Paradi, J. C., Reese, D. N., & Tam, F. (2007). Measuring Overall Efficiency and Effectiveness using DEA. *European Journal of Operational Research*, *178*(1), 305–321. doi:10.1016/j.ejor.2006.01.014

Austin, J., Stevenson, H., & Wei-Skillern, J. (2006). Social and commercial entrepreneurship: Same, different or both? *Entrepreneurship Theory and Practice*, *1*(1), 1–22. doi:10.1111/j.1540-6520.2006.00107.x

Austin, J., Stevenson, H., & Wei-Skillern, J. (2006). Social Entrepreneurship and Commercial Entrepreneurship: Same, Different, or Both? *Entrepreneurship Theory into Practice*, *30*(1), 1–22.

Backus, C., Keegan, K., Gluck, C., & Gulick, L. M. V. (2010). Accelerating leadership development via immersive learning and cognitive apprenticeship. *International Journal of Training and Development*, *14*(2), 144–148. doi:10.1111/j.1468-2419.2010.00347.x

Baden-Württemberg (2016). *Gesetz über kommunale Zusammenarbeit (GKZ) in der Fassung vom 16*. September 1974 GBl. 1974, 408, ber. 1975 S. 460, ber. 1976 S. 408 Gliederungs-Nr: 2805-1.

Baeck, P., & Collins, L. (2015). *Crowdfunding public services - tapping into the crowd to finance public projects*. Retrieved from http://www.nesta.org.uk/blog/crowdfunding-public-services-tapping-crowd-finance-public-projects

Bae, J., & Lawler, J. J. (2000). Organizational and HRM strategies in Korea: Impact on firm performance in an emerging economy. *Academy of Management Journal*, *13*(3), 502–517. doi:10.2307/1556407

Bankvall, L., Dubois, A., & Lind, F. (2014). Towards a network-based business model concept and the case of a truck OEM considering its business model in India. *IMP Journal*, *8*(3), 44–50.

Baptiste, N. R. (2008). Tightening the link between employee wellbeing at work and performance: A new dimension for HRM. *Management Decision*, *46*(2), 284–309. doi:10.1108/00251740810854168

Bardach, E. (1998). *Getting Agencies to Work Together. The Practice and Theory of Managerial Craftsmanship*. Washington, DC: Brookings Institution Press.

Baregheh, A., Rowley, J., & Sambrook, S. (2009). Towards a multidisciplinary definition of innovation. *Management Decision*, *47*(8), 1323–1339. doi:10.1108/00251740910984578

Bartholomae, F. W., & Popescu, A. M. (2008). Technologischer Fortschritt als Entscheidungskriterium für Kooperation: Europäische Regionale Zusammenarbeit mittels FOCJ. In P. Friedrich, & F. Bartholomae (Eds.), FOCJ als Mittel regionaler Kooperation (pp. 62-72). Universität der Bundeswehr München, Fakultät für Wirtschafts- und Organisationswissenschaften Lehrstuhl für Finanzwissenschaft, Discussion paper No. 50. Neubiberg.

Barzelay, M., & Thompson, F. (2010). Back to the Future: Making Public Administration a Design Science. *Public Administration Review*, 70.

Battilana, J., & Lee, M. (2014). Advancing Research on Hybrid Organizing: Insights form the Study of Social Enterprises. *The Academy of Management Annals*, *8*(1), 397–441. doi:10.1080/19416520.2014.893615

Batt, R. (2002). Managing customer services: Human resource practices, quit rates and sales growth. *Academy of Management Journal*, *45*(3), 587–597. doi:10.2307/3069383

Bauby, P. (2012). Local services of general economic interest in Europe. Water services: What are the challenges? *Annals of Public and Cooperative Economics*, *83*(4), 561–583. doi:10.1111/j.1467-8292.2012.00477.x

Bee, L., Benson, O., & Teck-Heng, L. (2016). Game-based learning in construction management courses: A case of bidding game. *Engineering, Construction, and Architectural Management*, *23*(1), 4–19. doi:10.1108/ECAM-02-2015-0029

Beinare, D., & McCarthy, M. (2012). Civil society organisations, social innovation and health research in Europe. *European Journal of Public Health*, *22*(6), 889–893. doi:10.1093/eurpub/ckr152 PMID:22117053

Bekkers, V. J., Tummers, J. M., Stuijfzand, L. G., & Voorberg, W. (2013). *Social Innovation in the Public Sector: An integrative framework*. LIPSE Working papers (no. 1). Rotterdam: Erasmus University Rotterdam.

Bekkers, V., Edwards, A., Moody, R., & Beunders, H. (2011). Cauht by surprise? Micro-mobilization, new media and management of strategic surprises. *Public Management Review*, *13*(7), 1003–1022. doi:10.1080/14719037.2011.589615

Bell, B. S., Kanar, A. M., & Kozlowski, S. W. (2008). Current issues and future directions in simulation-based training in North America. *International Journal of Human Resource Management*, *19*(8), 1416–1434. doi:10.1080/09585190802200173

Bernier, L., & Hafsi, T. (2007). The changing nature of public entrepreneurship. *Public Administration Review*, *67*(3), 488–503. doi:10.1111/j.1540-6210.2007.00731.x

Berry, L. L. (2000). Relationship marketing of Services, Growing Interest, Emerging Perspectives. In J. N. Sheth & A. Parvatiyar (Eds.), *Handbook of Relationship Marketing*. Sage Publictions, Inc. doi:10.4135/9781452231310.n6

Betz, F. (2002). Strategic business models. *Engineering Management Journal, 14*(1), 21–27. doi:10.1080/10429247.2002.11415145

Beyer, H., & Holtzblatt, K. (1998). *Contextual Design: Defining Customer-Centered Systems* (1st ed.). San Francisco, CA: Morgan Kaufmann.

Bińczycki, B. (2004). System idealny Nadlera. In M. Romanowska (Ed.), *Leksykon zarządzania*. Warszawa: Difin.

Binek, J. et al., (2011). *Obce a regiony pro podnikatele; Podnikatelé pro obce a regiony*. Brno: GaREP.

Björkdahl, J. (2009). Technology cross-fertilization and the business model: The case of integrating ICTs in mechanical engineering products. *Research Policy, 38*(9), 1468–1477. doi:10.1016/j.respol.2009.07.006

Blankart, Ch., & Borck, R. (2004). Local Public Finance. In J. G. Backhaus, & R. E. Wagner (Eds.), Handbook of Public Finance (pp. 441-476). Boston: Kluwer Academic Publisher.

Blankart, Ch. B. (2002). Steuern als Preise: Eine finanzwissenschaftliche Untersuchung mit einer Anwendung auf die EU-Zinsbesteuerung. *Schweizerische Zeitschrift fur Volkswirtschaft und Statistik, 138*(1), 19–38.

Blöchliger, H., & Charbit, C. (2008). Fiscal Equalisation. *OECD Economic Studies,* 44. Retrieved June, 2016, from https://www.oecd.org/eco/42506135.pdf

Blumentrath, U. (1969). *Investitions- und Finanzplanung mit Hilfe von Endwertmodellen*. Wiesbaden: Gabler. doi:10.1007/978-3-663-02599-3

Bocken, N. M. P., Short, S. W., Rana, P., & Evans, S. (2014). A literature and practice review to develop sustainable business model archetypes. *Journal of Cleaner Production, 65*, 42–56. doi:10.1016/j.jclepro.2013.11.039

Bodenhoff, H. (2002). *Einführung in die Finanzierungs- und Investitionsrechnung*. Verlag München.

Bognetti, G., & Obermann, G. (2008). Liberalization and privatization of public utilities: Origins of the debate, current issues and challenges for the future. *Annals of Public and Cooperative Economics, 79*(3/4), 461–485. doi:10.1111/j.1467-8292.2008.00367.x

Boiteux, M. (1951). Le revenue distribuable et les pertes économique. *Econometrica, 19*(2), 112–133. doi:10.2307/1905729

Boons, F., & Lüdeke-Freund, F. (2013). Business models for sustainable innovation: State-of-the-art and steps towards a research agenda. *Journal of Cleaner Production, 45*, 9–19. doi:10.1016/j.jclepro.2012.07.007

Booz, A. (n.d.). *Hamilton*. Retrieved from http://www.strategyand.pwc.com/media/file/The_Customer_Centric_Organization.pdf

Borins, S. (2000). Loose cannons and rule breakers, or enterprising leaders? Some evidence about innovative public managers. *Public Administration Review, 60*(6), 498–507. doi:10.1111/0033-3352.00113

Borins, S. (2002). Leadership and innovation in the public sector. *Leadership and Organization Development Journal, 23*(8), 467–476. doi:10.1108/01437730210449357

Bös, D. (1986). Public Enterprise Economics: Theory and Application. Amsterdam: Oxford.

Boschee, J. (1998). *What does it take to be a Social Entrepreneur?* National Centre for Social Entrepreneurs. Retrieved May 10, 2016, from www.socialentrepreneurs.org/whatdoes.html

Bosehee, J., & Mc Clurg, J. (2003). *Toward a better understanding of social entrepreneurship: Some important distinctions*. Retrieved May 10, 2016, from http://www.se-alliance.org/better _understanding.pdf

Bossidy, L., & Charan, R. (2002). *Execution: The Discipline of Getting Things Done*. Crown Business.

Bossidy, L., & Charan, R. (2010). *Szósty zmysł w zarządzaniu firmą. Tworzenie wykonalnych planów i modeli biznesowych*. Warszawa: MT Biznes.

Boudreau, K. J., & Lakhani, K. R. (2009). How to manage outside innovation. *MIT Sloan Management Review, 50*.

Bovaird, T. (2007). Beyond engagement and participation: User and community coproduction of public services. *Public Administration Review*, *67*(5), 846–860. doi:10.1111/j.1540-6210.2007.00773.x

Bovaird, T., & Loffler, E. (Eds.). (2009). *Public management and governance*. London: Routledge.

Bowen, S., McSeveny, K., Lockley, E., Wolstenholme, D., Cobb, M., & Dearden, A. (2013). How was it for you? Experiences of participatory design in the UK health service. *Codesign-International Journal of Cocreation in Design and the Arts*, *9*(4), 230–246. doi:10.1080/15710882.2013.846384

Boyd, S. (2002). *Partnership Working: European Social Partnership Models*. Retrieved June 10, 2016, from: http://www.gov.scot/resource/doc/25954/0028681.pdf

Boyer, E. (1996). The scholarship of engagement. *Journal of Public Service & Outreach*, *1*(1), 11–20.

Boyett, I. (1996). The public sector entrepreneur - a definition. *International Journal of Public Sector Management*, *9*(2), 36–51. doi:10.1108/09513559610119546

Boyne, G. A., Gould-Williams, J. S., Law, J., & Walker, R. M. (2005). Explaining the adoption of innovation: An empirical analysis of public management reform. *Environment and Planning. C, Government & Policy*, *23*(3), 419–435. doi:10.1068/c40m

Bozeman B. (2002), Public-Value Failure: When Efficient Markets May Not Do. *Public Administration Review*. DOI: 10.1111/0033-3352.00165

Brandenburger, A., & Harborne Stuart, S. (1996). Value-based business strategy. *Journal of Economics & Management Strategy*, *5*(1), 5–25. doi:10.1111/j.1430-9134.1996.00005.x

Brandsen, T., & Karré, P. M. (2011). Hybrid organizations: No cause for concern? *International Journal of Public Administration*, *34*(13), 827–836. doi:10.1080/01900692.2011.605090

Brhlíková, P. (2004). *The Nonprofit Sector in the Czech Republic*. CERGE-EI Discussion Paper, no. 128.

Brown, S. J., & Sibley, D. S. (1986). *The Theory of Public Utility Pricing*. Cambridge, UK: Cambridge University Press. doi:10.1017/CBO9780511666773

Brown, W. (2015). *Undoing the Demos. Neoliberalism's Stealth Revolution*. Brooklyn, NY: Zone Books.

Brümmerhoff, D., & Büttner, T. (2015). *Finanzwissenchaft*. München: De Gruyter, Oldenbourg.

Bryson, J. M. (2004). *Strategic planning for public and nonprofit organizations*. San Francisco: John Wiley & Sons.

Brzóska, J., Durczok, M., Jerzok, I., Zarębski, A., & Żbik, W. (Eds.). (2015). *Formaty biznesowe*. Katowice: Stowarzyszenie Współpracy Regionalnej.

Brzóska, J., & Jerzok, I. (2014). *Koncepcje i aplikacje modeli biznesu w organizacjach ekonomii społecznej*. Gliwice: Wydawnictwo Politechniki Śląskiej w Gliwicach.

Budget Code of the Russian Federation. (2016a). Retrieved May, 2016, from http://www.consultant.ru/document/cons_doc_LAW_19702/

Burke, W., Lake, D. G., & Paine, J. W. (2008). *Organization Change: A Comprehensive Reader*. San Francisco: Jossey-Bass.

Buurma, H. (2001). Public policy marketing: Marketing exchange in the public sector. *European Journal of Marketing*, *35*(11/12), 1287–1302. doi:10.1108/EUM0000000006483

Cambini, C., Filippini, M., Piacenza, M., & Vannoni, D. V. (2011). Corporatization and firm performance: Evidence from publicly-provided local utilities. *Review of Law & Economics*, *7*(1), 195–217. doi:10.2202/1555-5879.1428

Canvanizer. (2016). *Business Model Canvas vs Lean Model*. Retrieved October 10, 2016, from https://canvanizer.com/how-to-use/business-model-canvas-vs-lean-canvas

Cao, G., Wiengarten, F., & Humphreys, P. (2011). Towards a contingency resource-based view of IT business value. *Systemic Practice and Action Research*, *24*(1), 85–106. doi:10.1007/s11213-010-9178-0

Cappelli, P., & Sherer, P. D. (1991). The missing role of context in OB. The need for meso-level approach. In L.L. Cummings & B.M. Staw (Eds.), Research in Organizational Behavior. Greenwich, CT: JAI Press.

Casadesus-Masanell, R., & Ricart, J. E. (2010). From Strategy to Business Models and onto Tactics. *Long Range Planning*, *43*(2-3), 195–215. doi:10.1016/j.lrp.2010.01.004

Casadesus-Masanell, R., & Zhu, F. (2013). Business Model Innovation and Competitive Imitation: The Case of Sponsor-Based Business Models. *Strategic Management Journal*, *34*(4), 464–482. doi:10.1002/smj.2022

Caulier-Grice, D. J. A., Patrick, R., & Norman, W. (2012) Social Innovation Overview: A deliverable of the project: "The theoretical, empirical and policy foundations for building social innovation in Europe" (TEPSIE), European Commission – 7th Framework Programme. Brussels: European Commission, DG Research. The Young Foundation.

Caulier-Grice, J., Davies, A., Patrick, R., & Norman, W. (2013). *Defining social innovation. A deliverable project "The theoretical, empirical and policy foundation for building social innovation in Europe". (TEPSIE), European Commission – 7th Framework Programme*. Brussels: European Commission, DG Research.

Chabanet, D., & Faniel, J. (Eds.). (2012). *The Mobilization of the Unemployed in Europe: from Acquiescence to Protest?* Palgrave Macmillan. doi:10.1057/9781137011862

Chell, E. (2007). Social enterprise and entrepreneurship: Towards a convergent theory of the entrepreneurial process. *International Small Business Journal*, *25*(1), 5–26. doi:10.1177/0266242607071779

Chesbrough, H. (2007). Why companies should have open business models. *MIT Sloan Management Review*, *48*(2), 22–29. Retrieved from http://doi.org/Article

Chesbrough, H. (2013). *Open Innovation. The New Imperative for Creating and Profiting from Technology*. Boston: Harvard Business School Press. http://doi.org/<ALIGNMENT.qj></ALIGNMENT>10.1017/CBO9781107415324.004

Chesbrough, H.W., & Appleyard, M.M. (2007). *Open Innovation and Strategy*. Portland State University PDXScholar.

Chesbrough, H. (2007). Business model innovation: Its not just about technology anymore. *Strategy and Leadership*, *35*(6), 12–17. doi:10.1108/10878570710833714

Chesbrough, H. (2010). Business model innovation: Opportunities and barriers. *Long Range Planning*, *43*(2–3), 354–363. doi:10.1016/j.lrp.2009.07.010

Chesbrough, H. W. (2007). Why Companies Should Have Open Business Models. *MIT Sloan Management Review*, *48*(2), 22–28.

Chesbrough, H., & Rosenbloom, R. (2002). The Role of Business Model in Capturing Value from Innovations: Evidence from Xerox Corporations Technology Spin-off Companies. *Industrial and Corporate Change*, *11*(3), 529–555. doi:10.1093/icc/11.3.529

Chodyński, A. (2011). *Odpowiedzialność ekologiczna w proaktywnym rozwoju przedsiębiorstw. Krakowska Akademia im.* Kraków: Jana Frycza Modrzejewskiego.

Ciepielewska-Kowalik, A., Pieliński, B., Starnawska, M., & Szymańska, A. (2015). *Looking for Social Enterprise Models in Poland: Institutional and Historical Context*. ICSEM Working Papers, No. 2015-11. Retrieved May 19, 2016, from http://www.iap-socent.be/sites/default/files/Poland%20-%20Ciepielewska-Kowalik%20et%20al.pdf

Civil Code of the Russian Federation. (2016b). Retrieved June, 2016, from http://www.consultant.ru/document/cons_doc_LAW_5142/

Clarkson, M. (1994). A risk Based Model of Stakeholder Theory. In *Proceedings of the Second Toronto Conference of Stakeholder Theory*. Centre for Corporate Social Performance and Ethics, University of Toronto.

Clifford, A., & List, M. & Theobald, C. (2010). *Evaluating Economic*. London: Alliance of Sector Skills Councils.

Coblence, E., Normandin, F., & Poisson-de Haro, S. (2014). Sustaining Growth through Business Model Evolution: The Industrialization of the Montreal Museum of Fine Arts (1986–2012). *The Journal of Arts Management, Law, and Society*, *44*(3), 126–144. doi:10.1080/10632921.2014.936077

Colvin, R. A. (2006). Understanding Policy Adoption and Gay Rights: The role of the media and other factors. *The Innovation Journal: The Public Sector Innovation Journal, 11*(2).

Conteh, Ch. (2013). Strategic inter-organizational cooperation in complex environments. *Public Management Review*, *15*(4), 501–521. doi:10.1080/14719037.2012.674424

Cooney, K., & Lynch-Cerullo, K. (2014). Measuring the Social Returns of Nonprofits and Social Enterprises: The Promise and Perils of the SROI. *Nonprofit Policy Forum*, *5*(2), 367–393.

Cooney, K. (2011). An exploratory study of social purpose business models in the United States. *Nonprofit and Voluntary Sector Quarterly*, *40*(1), 185–196. doi:10.1177/0899764009351591

Cooper, J. R. (1998). A multidimensional approach to the adoption of innovation. *Management Decision*, *36*(8), 493–502. doi:10.1108/00251749810232565

Copulsky, J., & Wolf, M. (1990). Relationship Marketing. Positioning for the Future. *The Journal of Business Strategy*, *11*(4), 16–20. doi:10.1108/eb060069 PMID:10106864

Cornes, R., & Hartley, R. (2001). *Joint Production and Share Functions*. Working Paper, Department of Economics. Keele University.

Cox, H. (2008). Public enterprises and service providers in institutional competition and undergoing structural change. New challenges to the theory of public economics and public services in Germany. *Annals of Public and Cooperative Economics*, *79*(3/4), 527–547. doi:10.1111/j.1467-8292.2008.00369.x

Crilly, D., Hansen, M., & Zollo, M. (2016). The grammar of decoupling: A cognitive-linguistic perspective on firms sustainability claims and stakeholders. *Academy of Management Journal*, *59*(2), 705–729. doi:10.5465/amj.2015.0171

Crookall, D. (2010). Serious games, debriefing, and simulation/gaming as a discipline. *Simulation & Gaming*, *41*(6), 898–920. doi:10.1177/1046878110390784

Crossan, M. M., & Apaydin, M. (2010). A multi-dimensional framework of organizational innovation: A systematic review of the literature. *Journal of Management Studies*, *47*(6), 1154–1191. doi:10.1111/j.1467-6486.2009.00880.x

Crucke, S., Moray, N., & Vallet, N. (2015). Internal representation and factorial faultiness as antecedents for board performance in social enterprises. *Annals of Public and Cooperative Economics*, *86*(2), 385–400. doi:10.1111/apce.12076

CSO. (2015). *Veřejná database*. Retrieved May 03, 2016, from: https://vdb.czso.cz/vdbvo2/faces/cs/index.jsf?page=statistiky&filtr=G~F_M~F_Z~F_R~F_P~_S~_null_null_

Cyfert, S., & Krzakiewicz, K. (2011). Wykorzystanie koncepcji modeli biznesu w zasobowej teorii firmy. In Rozwój szkoły zasobowej zarządzania strategicznego. Prace Naukowe Wałbrzyskiej Wyższej Szkoły Zarządzania i Przedsiębiorczości.

Cyfert, S., Dyduch, W., Latusek-Jurczak, D., Niemczyk, J., & Sopińska, A. (2014). Subdyscypliny w naukach o zarządzaniu – logika wyodrębnienia, identyfikacja modelu koncepcyjnego oraz zawartość tematyczna. *Organizacja i Kierowanie*, *1*(161), 37–49.

Cyr, H. (2014). *Autonomy, subsidiarity, solidarity: foundations of cooperative federalism.* Constitutional Forum 20. doi:10.2139/ssrn.2512429

Czakon, W. (2005). Więzi rekurencyjne a więzi sieciowe przedsiębiorstwa. In J. Pyka (Ed.), Nowoczesność przemysłu i usług. Współczesne koncepcje i metody zarządzania przedsiębiorstwami. Katowice: TNOiK.

Czemiel-Grzybowska, W. (2010). *Zarządzanie przedsiębiorstwem społecznym – raport z badań*. Białystok: Regionalny Instytut Społeczno-Ekonomiczny.

Czerw, A. (2010). *Marketing w ochronie zdrowia*. Warszawa: Difin.

Dacin, M. T., Dacin, P. S., & Tracey, P. (2011). Social entrepreneurship: A critique and future directions. *Organization Science*, *22*(5), 1203–1213. doi:10.1287/orsc.1100.0620

Dafflon, B. (2006). The assignment of function to decentralized government: from theory to practice. In E. Ahmad & G. Brosio (Eds.), *Handbook of Fiscal Federalism* (pp. 271–301). Cheltenham, UK: Edward Elgar. doi:10.4337/9781847201515.00020

Daglio, M., Gerson, D., & Kitchen, H. (2015). *Building Organisational Capacity for Public Sector Innovation.* Paper presented at OECD Conference "Innovating the Public Sector: from Ideas to Impact", Paris, France.

Damanpour, F. (1996). Organizational complexity and innovation: Developing and testing multiple contingency models. *Management Science*, *42*(5), 693–716. doi:10.1287/mnsc.42.5.693

Dart, R. (2004). Legitimacy of Social Enterprise. *Nonprofit Management & Leadership*, *14*(4), 411–424. doi:10.1002/nml.43

Daud Ahmed, M., & Sundaram, D. (2005). Sustainability modeling and reporting: From roadmap to implementation. *Decision Support Systems*, *53*, 2012.

de Freitas, S., & Routledge, H. (2013). Designing leadership and soft skills in educational games: The e-leadership and soft skills educational games design model (ELESS). *British Journal of Educational Technology*, *44*(6), 951–968. doi:10.1111/bjet.12034

de Lancer Julnes, P. (2016). The Study of Innovation. State of Art and Framework for Analysis. In Innovation in the Public and Nonprofit Sector (pp. 12–32). New York: Routledge.

de Witt, B., & Meyer, R. (2007). *Synteza strategii. Tworzenie strategii konkurencyjnej przez tworzenie paradoksów*. Warszawa: PWE.

Dees, J. G., Elias, J., & Bowie, N. E. (1998). The challenges of combining social and commercial enterprise. *Business Ethics Quarterly*, *8*(1), 165–178. doi:10.2307/3857527

Defourny, J., & Nyssens, M. (2010). Conceptions of Social Enterprise and Social Entrepreneurship in Europe and the United States: Convergences and Divergences. *Journal of Social Entrepreneurship*, *1*(1), 32–53. doi:10.1080/19420670903442053

Defourny, J., & Nyssens, M. (2013). Social Coops: When Social Enterprises meet the Cooperative tradition. *Journal of Entrepreneurial and Organizational Diversity*, *2*(2), 11–33.

Deloitte. (2014). *Dual Purpose Corporate Structure Legislation: Stakeholder Engagement Report*. Ontario: Author.

DEMA. (2016). *Centro Especial de Empleo*. LISMI Law. Retrieved July 2, 2016, from: http://www.demacee.com/en/cumplo-la-ley-lismi/

Demil, B., & Lecocq, X. (2010). Business Model Evolution: In Search of Dynamic Consistency. *Long Range Planning*, *43*(2-3), 227–246. doi:10.1016/j.lrp.2010.02.004

Dempsey, S. E., & Barge, J. K. (2013). Engaged Scholarship and Democracy. In D. K. Mumby & L. L. Putnam (Eds.), *Handbook of Organizational Communication: Advances in Theory, Research, and Methods* (pp. 665–688). Thousand Oaks, CA: Sage.

Denhardt, J. V., & Denhardt, R. B. (2003). *New Public Service. Serving, not steering*. Armonk: M.E. Sharpe.

Denhardt, R. B. (2011). *Theories of Public Organization*. Boston: Wadsworth Cengage Learning.

Deodhar, S.J., Saxena, K., Gupta, R.K., Ruohonen, M. (2012). Strategies for software-based hybrid business models. *Journal of Strategic Information Systems*, (4), 274–294.

Design Commission. (2013). *Restarting Britain 2: Design and Public Services*. Retrieved from files/101/DC_Restarting_Britain_2_report - interactive.pdf

DESIS Network. (2013). *Public and Collaborative: Exploring the intersection of Design, Social Innovation and Public Policy* (E. Manzini & E. Staszowski, Eds.). DESIS Network.

Detig, St., Feng, X., & Friedrich, P. (2002). FOCJ als Grundlage für Förderungs-institutionen. Aufbau-Ost und Bevölkerung-Ost. In P. Friedrich (Ed.), Öffentliche Unternehmen im Standortwettbewerb für den Aufbau Ost (pp. 82-115). Diskussionspapier Nr. 41, Lehrstuhl für Finanzwissenschaft, Universität der Bundeswehr München, Neubiberg.

Detig, S. (2004). *Die kommunale Anstalt des öffentlichen Rechts als Wirtschaftsförderungsinstitution*. Baden-Baden: Nomos.

Dibben, P., James, P., Roper, I., & Wood, G. (Eds.). (2007). Modernising Work in Public Services. Redefining Roles and Relationships in Britain's Changing Workshops. Palgrave Macmillan.

Dijk, P. V., & Weiler, B. (2009). An assessment of the outcomes of a Chinese-language interpretive tour experience at a heritage tourism attraction. *Tourism Analysis*, *14*(1), 49–63. doi:10.3727/108354209788970144

Ditoiu, M., Stancioiu, A., Teodorescu, N., Onisor, L., & Radu, A. (2014). Sensory experience - between the tourist and the marketer. *Theoretical & Applied Economics*, *21*(12), 37–50.

Dobele, L. (2015). Factors Which Influence the Development of Social Innovation in Latvia. *Proceedings of the 2015 International Conference "Economic Science for Rural Development"*. Jelgava: LLU ESAF.

Doherty, B., Haugh, H., & Lyon, F. (2014). Social Enterprises as Hybrid Organizations: A Review and Research Agenda. *International Journal of Management Reviews*, *16*(4), 417–436. doi:10.1111/ijmr.12028

Dohnalová, M., Guri, D., Hrabětová, J., Legnerová, K., & Šlechtová, V. (2015). *Social Enterprise in the Czech Republic.* ICSEM Working Papers, No. 24. Liege: The International Comparative Social Enterprise Models (ICSEM) Project.

Donetto, S., Pierri, P., Tsianakas, V., & Robert, G. (2015). Experience-based Co-design and Healthcare Improvement: Realizing Participatory Design in the Public Sector. *Design Journal, 18*(2), 227–248. http://doi.org/<ALIGNMENT.qj></ALIGNMENT>10.1111/j.1747-4949.2010.00571.x

Downs, A. (1964). *Inside Bureaucracy.* Rand Corporation Research Study. http://doi.org/<ALIGNMENT.qj></ALIGNMENT>10.2307/447012

Drucker, P. (1954). *The Practice of Management.* Harper and Row Publishers.

Drucker, P. (1977). *An Introductory View of Management.* New York, NY: Harper College Press.

Drucker, P. (1985). *Innovation and Entrepreneurship: Practice and Principles.* London: Butterworth-Heinemann.

DTI. (2002). *Social Enterprise: A Strategy for Success.* Retrieved May 10, 2016, from www.seeewiki.co.uk/~wiki/images/5/5a/SE_Strategy_for_success.pdf

Dubosson-Torbay, M., Osterwalder, A., & Pigneur, Y. (2001). E-business model design, classification and measurements. *Thunderbird International Business Review*, *44*(1), 5–23. doi:10.1002/tie.1036

Duca, I. (2013). Efficiency, Effectiveness and Profitability – Concepts Used in Assessing Public Expenditures in Education. *Journal of Knowledge Management Economics and Information Technology*, *3*(6), 63–75.

Duchemin, C., & Weber, T. (2013). Social Partners Involvement in pension reform in the EU. In *Eurofund; EurWork: European Observatory Of Working Life.* Retrieved June 20, 2016, from: http://www.eurofound.europa.eu/observatories/eurwork/comparative-information/social-partners-involvement-in-pension-reform-in-the-eu

Dyllick, T., & Hockerts, K. (2002). Beyond the business case for corporate sustainability. *Business Strategy and the Environment*, *11*(2), 2. doi:10.1002/bse.323

Eckardt, M., & Gritsch, M. (2016). Governance of cross-border regional cooperation by the EGTC – An explorative analysis with a focus on Hungarian EGTCs. *Public Governance Journal for Public and Non-Profit Services*, 39(1-2), 200-215.

Eichhorn, P. (1974). *Liquiditätsplanung und Gelddisposition in öffentlichen Haushalten.* Baden-Baden: Nomos.

Eichhorn, P., & Friedrich, P. (1976). *Verwaltungsökonomie.* Baden-Baden: Nomos.

Eisenhardt, K. M. (1989). Building theories from case study research. *Academy of Management Review*, *14*(4), 532–550.

Eisenhardt, K. M. (1989). Building Theories from Case Study Research. *Academy of Management Review*, *14*(4), 532–550.

Ellen MacArthur Foundation. (2013a). Towards the Circular Economy: Opportunities for the consumer goods sector (vol. 2). Author.

Ellen MacArthur Foundation. (2013b). Towards the Circular Economy: Economic and business rationale for an accelerated transition (Vol. 1). Author.

Ellen MacArthur Foundation. (2015a). *Delivering the circular economy a toolkit for policymakers.* Author.

Ellen MacArthur Foundation. (2015b). Growth within: A circular economy vision for a competitive. Author.

Enkel, E., & Mezger, F. (2013). Imitation processes and their application for business model innovation: An explorative study. *International Journal of Innovation Management*, *17*(1), 1340005. doi:10.1142/S1363919613400057

Epstein, M. J., & Rejc Buhovac, A. (2010). Solving the sustainability implementation challenge. *Organizational Dynamics*, 39.

Eriksson, M., Einarsson, T., & Wijkström, F. (2014). *Report on the European social innovation policy framework in light of third sector and civil society actors. Deliverable 1.2 of the project: 'Impact of the Third Sector as Social Innovation' (ITSSOIN), European Commission – 7th Framework Programme*. Brussels: European Commission, DG Research.

Euchner, J. (2014). Business Model Innovation in Practice. *Research Technology Management*, (November-December), 33–39. http://doi.org/<ALIGNMENT.qj></ALIGNMENT>10.5437/08956308X5706013

European Commission. (2010). *A European strategy for smart, sustainable and inclusive growth*. Brussels: Author.

European Commission. (2013). *Guide to Social Innovation*. Author.

European Commission. (2013a). *Powering European Public Sector Innovation, Towards New Architecture, Report of the Expert Group on Public Sector Innovation*. Luxembourg: Publications Office of the European Union. Retrieved June 13, 2016, from https://ec.europa.eu/research/innovation-union/pdf/psi_eg.pdf://

European Commission. (2013b). *Executive Summary. Powering Public Sector Innovation, Towards New Architecture*. Luxembourg: Publications Office of the European Union. Retrieved June 13, 2016, from http://psi_eg.pdf

European Commission. (2013c). *Successful partnerships in delivering Public Employment Services*. Retrieved June 13, 2016, from http://ec.europa.eu/social/BlobServlet?docId=11376&langId=en

European Commission. (2014). *A map of social enterprises and their eco-systems in Europe. Country Reports: Poland*. Brussels: European Commission Directorate General for Employment, Social Affairs and Inclusion. Retrieved May 30, 2016, from http://ec.europa.eu/social/main.jsp?langId=en&catId=89&newsId=2149

European Union. (2012). *Social Dialog. Social Europe Guide, 2*. Luxemburg: Publication Office of the EU. Retrieved June 13, 2016, from: http://ec.europa.eu/social/main.jsp?catId=738&langId=en&pubId=6352&type=2&furtherPubs=no

EUSTAT. (2015). *Population of the Basque Country by province, according to place of birth*. Retrieved June 20, 2016, from: http://en.eustat.eus/elementos/ele0011400/ti_Population_of_the_Basque_Country_by_province_according_to_place_of_birth/tbl0011425_i.html#axzz4EOZjzwzM

Ewing, M. T., & Caruana, A. (1999). An internal marketing approach to public sector management: The marketing and human resources interface. *International Journal of Public Sector Management*, *12*(1), 17–29. doi:10.1108/09513559910262652

Fabi, B., Lacoursière, R., Raymond, L., & St-Pierre, J. (2010). HRM capabilities as a determinant and enabler of productivity for manufacturing SMEs. *Human Systems Management*, *29*(3), 115–125.

Farrell, C. (2005). Perceived effectiveness of simulations in international business pedagogy: An exploratory analysis. *Journal of Teaching in International Business*, *16*(3), 71–88. doi:10.1300/J066v16n03_05

Ferrari, P. A., & Manzi, G. (2014). Citizens evaluate public services: A critical overview of statistical methods for analysing user satisfaction. *Journal of Economic Policy Reform*, *17*(3), 236–252. doi:10.1080/17487870.2014.909313

Fiet, J.O., & Patel, P.C. (2008). *Forgiving Business Models for New Ventures*. Doi: 10.1111/j.1540-6520.2008.00252.x

Filep, S., & Deery, M. (2010). Towards a picture of tourists happiness: An insight from psychology. *Tourism Analysis*, *15*(4), 399–410. doi:10.3727/108354210X12864727453061

Filion, L. (2011). Defining the entrepreneur. In L.-P. Dana (Ed.), *World Encyclopedia of Entrepreneurship* (pp. 41–52). Cheltenham, UK: Edward Elgar. doi:10.4337/9781849808453.00011

Fink, A. (2012). The Hanseatic League and the Concept of Functional Overlapping Competing Jurisdictions. *Kyklos, 65*(2), 194–217. doi:10.1111/j.1467-6435.2012.00534.x

Fjeldstad, Q. D., Snow, C. C., Miles, R. E., & Letti, C. (2012). The Architecture of Collaboration. *Strategic Management Journal, 33*(6), 734–750. doi:10.1002/smj.1968

Fladung, L., & Friedrich, P. (2008). Finanzierung eines Wirtschaftsförderungs-FOCJ. In P. Friedrich, & F. Bartholomae (Eds.), FOCJ als Mittel regionaler Kooperation (pp. 21-46). Universität der Bundeswehr München, Lehrstuhl für Finanzwissenschaft, Discussion paper No. 50, Neubiberg.

Florescu, C., Malcomete, P., & Al Pop, N. (2003). *Marketing. Dictionar explicative*. Bucharest: Editura Economica.

Florida, R. (2005). *Cities and the Creative Class*. New York: Routledge.

Florio, M. (2014). Contemporary public enterprises: innovation, accountability, governance. *Journal of Economic Policy Reform*, 201-208.

Florio, M. (2013). Rethinking on public enterprise: Editorial introduction and some personal remarks on the research agenda. *International Review of Applied Economics, 27*(2), 135–149. doi:10.1080/02692171.2013.785664

Flynn, N. (2007). Public sector management. *Sage (Atlanta, Ga.)*.

Fonfara, K. (1999). *Marketing partnerski na rynku przedsiębiorstw*. Warszawa: PWE.

Frączkiewicz-Wronka, A. (2011). Efektywność sieci jako czynnik krytyczny rozwoju organizacji realizujących usługi społeczne. In J. Skalik (Ed.), Zmiana warunkiem sukcesu. Przełamywanie barier rozwoju i wzrostu przedsiębiorstw. Prace Naukowe Uniwersytetu Ekonomicznego we Wrocławiu.

Frączkiewicz-Wronka, A. (2012). Analiza interesariuszy jako metoda wspierająca formułowanie strategicznych zamierzeń w organizacji. *Metody badań problemów zarządzania strategicznego, Prace Naukowe Wałbrzyskiej Wyższej Szkoły Zarządzania i Przedsiębiorczości*, (17), 269-292.

Frączkiewicz-Wronka, A. (2014). *Zarządzanie usługami społecznymi Studium partnerstw publiczno-społecznych*. Warsaw: Difin.

Frączkiewicz-Wronka, A., Szołtysek, J., & Kotas, M. (2012). Key success factors of social services organizations in the public sector. *Management, 16*(2), 231–255. doi:10.2478/v10286-012-0068-7

Franck, G. (1998). *Oekonomie der Aufmerksamheit; ein Entwurf*. Munich: Carl Hanser Verlag.

Franck, G. (2005). *Mentaler Kapitalismus*. Munich: Carl Hanser Verlag.

Frankenberger, K., Weiblen, T., Csik, M., & Gassmann, O. (2013). The 4I-framework of business model innovation: A structured view on process phases and challenges. *International Journal of Product Development, 18*(3/4), 249–273. doi:10.1504/IJPD.2013.055012

Freeman, E., & Reed, D. (1998). Stockholders and Stakeholders: A New Perspective on Corporate Governance. In B. De Wit & R. Meyer (Eds.), Strategy. Process, Conten. Context.

Freeman, R. E. (1984). *Strategic Management: A Stakeholder Approach*. Boston: Pitman.

French, J., & Blair-Stevens, C. (2005). *Social Marketing Pocket Guide*. London: National Social Marketing Centre of Excellence.

French, J., & Blair-Stevens, C. (2007). *Big Pocket Guide: social marketing*. London: National Social Marketing Centre.

Frenkel, M. (1986). *Federal theory. The Australian National University*. Canberra: Centre for Research on Federal Financial Relations.

Frey, B. (2000). *A Utopia? Government without Territorial Monopoly*. Institute for Empirical Research in Economics of the University of Zurich, Working paper No. 47. University of Zurich.

Frey, B. (1999). *The New Democratic Federalism for Europe*. Cheltenham, UK: Edward Elgar.

Frey, B. (2005). Functional, Overlapping and Competing Jurisdictions: Redrawing of geographic boarders of administration. *European Journal of Law Reform*, *5*(3/4), 543–555.

Frey, B. S. (1997). *Ein neuer Föderalismus für Europa: Die Idee der FOCJ*. Tübingen: Mohr.

Frey, B., & Eichenberger, R. (1995). Competition among Jurisdictions: The Idea of FOCJ. In L. Gerken (Ed.), *Competition among Institutions* (pp. 209–229). London: Macmillan. doi:10.1007/9781349242627_8

Frey, B., & Eichenberger, R. (1996). FOCJ: Competitive Governments for Europe. *International Review of Law and Economics*, *16*(3), 315–327. doi:10.1016/0144-8188(96)00019-1

Frey, B., & Eichenberger, R. (2002). Democratic Governance for a Globalized World. *Kyklos*, *55*(2), 265–287. doi:10.1111/1467-6435.00186

Frey, B., & Eichenberger, R. (2006). Functional, Overlapping and Competing Jurisdictions (FOCJ): A Complement and Alternative to Today's Federalism. In E. Ahmad & G. Brosio (Eds.), *Handbook of Fiscal Federalism* (pp. 154–181). Cheltenham, UK: Edward Elgar.

Fric, P. (2005). The third sector and the policy process in the Czech Republic. *Third Sector European Policy Working Paper*, *6*(1), 1-55.

Friedrich, P. (2011). *Legal European Company Forms to Realize FOCJ – Functional Overlapping Competitive Jurisdictions*. Retrieved May, 2016, from https://www.researchgate.net/publication/254457385_Legal_European_Company_Forms_to_Realize_Cross_Border_FOCJ_asA_Functional_Overlapping_Competing_Jurisdictions

Friedrich, P., & Reiljan, J. (2011). *FOCJ as model for financing the Estonian general Education*. Retrieved June, 2016, from http://papers.ssrn.com/sol3/papers.cfm?abstract_id=1998055

Friedrich, P., & Ukrainski, K. (2012). *Measuring Public Entrepreneurship of Quangos*. The University of Tartu Faculty of Economics and Business Administration Working Paper No. 89-2012. Retrieved from https://ssrn.com/abstract=2172801

Friedrich, P., & Ukrainski, K. (2013). Quangos as a tool of public management to promote public sector entrepreneurship. *Zeitschrift für öffentliche und gemeinwirtschaftliche Unternehmen*, *43*, 37–51.

Friedrich, P., Kaltschütz, A., & Nam, C. W. (2004). *Significance and determination of fees for municipal finance*. CESinfo Working Paper Nr. 1357. München: Info-institute.

Friedrich, P., Nam, C. W., & Reiljan, J. (2009). *Local fiscal equalization in Estonia: is a reform necessary?* CES Ifo Working Paper No. 2800. Munich: Ifo-institute.

Friedrich, P., Ukrainski, K., & Timpmann, K. (2014). *Teaching Public Management: Between Academic Disciplines*. Chair of Public Economics and Policy, University of Tartu. Retrieved from https://ec.europa.eu/epale/sites/epale/files/tu_avaliku_sektori_pm_book_friedrich_ukrainski_timpmann_isbn.pdf

Friedrich, P. (2006). *FOCJ: Ein neuer Körperschaftstyp im Gesundheitswesen? Liberalisierung im Gesundheitswesen. In Schriftenreihe der Gesellschaft für öffentliche Wirtschaft* (Vol. 53, pp. 133–154). Baden-Baden: Nomos.

Friedrich, P., & Eckardt, M. (2014). *European grouping of territorial cooperation (EGTC) as a legal form to realize cross border FOCJ. In Public Sector in Transition* (pp. 106–141). University of Tartu.

Friedrich, P., & Popescu, A. M. (2006). FOCJ as institution for population policy. In T. Markowski & M. Turala (Eds.), *New members – new challenges for the European regional development policy* (pp. 228–242). Warsaw: Polish Academy of Sciences.

Fyrberg Yngfalk, A. (2013). Its not us, its them! – Rethinking value co-creation among multiple actors. *Journal of Marketing Management*, *29*(9/10), 1163–1181. doi:10.1080/0267257X.2013.796318

Gabbe, J. (2008). *Legal status of cross-border co-operation structures – past, present and prospects. Gronau: Associations of European border regions*. AEBR.

Gambardella, A., & McGahan, A. M. (2010). Business-Model Innovation: General Purpose Technologies and their Implications for Industry Structure. *Long Range Planning*, *43*(2-3), 262–271. doi:10.1016/j.lrp.2009.07.009

Gartner. (2003). Available at: http://www3.gartner.com

Gauthier, C., & Gilomen, B. (2015). Business Models for Sustainability: Energy Efficiency in Urban Districts. *Organization & Environment*. doi:10.1177/1086026615592931

Gentry, J. W. (1990). *Guide to Business Gaming and Experiential Learning*. London: GP Publishing, East Brunswick.

Gerometta, J., Haussermann, H., & Longo, G. (2005). Social Innovation and Civil Society in Urban Governance: Strategies for an Inclusive City. *Urban Studies (Edinburgh, Scotland)*, *42*(11), 2007–2021. doi:10.1080/00420980500279851

Gherardi, S. (2011). Organizational learning: The sociology of practice. In M. Easterby-Smith & M. Lyles (Eds.), *Handbook of Organizational Learning and Knowledge Management* (pp. 43–66). West Sussex, UK: John Wiley & Sons.

Gherardi, S., & Perrotta, M. (2014). Between the hand and the head. How things get done, and how in doing the ways of doing are discovered. *Qualitative Research in Organizations and Management*, *9*(2), 134–150. doi:10.1108/QROM-06-2012-1079

Gherghina, R., Văduva, F., & Postole, M. (2009). The Performance Management in Public institutions of Higher Education and the Economic Crisis. *Annales Universitatis Apulensis Series Oeconomica*, *11*(2), 639–645.

Gibson, E., & Jetter, A. (2014). *Towards a Dynamic Process for Business Model Innovation: A Review of the State-of-the-Art. Proceeedings of PICMET '14: Infrastructure and Service Integration* (pp. 1230–1238). IEEE.

Giesen, E., Berman, S. J., Bell, R., & Blitz, A. (2007). Three ways to successfully innovate your business model. *Strategy and Leadership*, *35*(6), 27–33. doi:10.1108/10878570710833732

Gilmore, J. H., & Pine, B. J. II. (1998). Welcome to the experience economy. *Harvard Business Review*, *76*(4), 97–105. PMID:10181589

Głód, G. (2015). Measurement of Public Entrepreneurship in the Polish Health Sector. In *Proceedings of the Management, Knowledge and Learning International Conference, MakeLearn* (pp. 665–673). Academic Press.

Glor, E. D. (Ed.). (2002). Is Innovation a Question of Will or Circumstance? An Exploration of the Innovation Process Through the Lens of the Blakeney Government in Saskatchewan. *Innovation*.

Głowacki, J., & Jelonek, M. (2013). *Badanie społecznej wartości dodanej podmiotów ekonomii społecznej w Polsce za pomocą narzędzia ESometr*. Kraków: MSAP.

Gnan, L., Hinna, A., Monteduro, F., & Scarozza, D. (2013). Corporate governance and management practices: Stakeholder involvement, quality and sustainability tools adoption. *Journal of Management & Governance*, *17*(4), 907–937. doi:10.1007/s10997-011-9201-6

GoGreen Portland. (n.d.). *The Social Enterprise Business Model Canvas*. Retrieved May 27, 2016, from http://portland.gogreenconference.net/downloads/GGPDX_Guided_Learning_Entrepreneurship(FRONT).pdf

Gołębiowski, T., Dudzik, T. M., Lewandowska, M., & Witek-Hajduk, M. (2008). *Modele biznesu polskich przedsiębiorstwa*. Warszawa: Szkoła Główna Handlowa.

Gonen, A., Brill, E., & Frank, M. (2009). Learning through business games – an analysis of successes and failures. *On the Horizon*, *17*(4), 356–367. doi:10.1108/10748120910998434

Gordon, R. J. (2016). *The Rise and Fall of American Growth. The U.S. Standard of living Since the Civil War*. Princeton, NJ: Princeton University Press. doi:10.1515/9781400873302

Greenhaus, J., Bedian, A., & Mossholder, K. (1987). Work experiences, job performances, and feelings of personal and family well-being. *Journal of Vocational Behavior*, *31*(2), 200–215. doi:10.1016/0001-8791(87)90057-1

Greiling, D., & Grüb, B. (2014). Sustainability reporting in Austrian and German local public enterprises. *Journal of Economic Policy Reform*, *17*(3), 209–223. doi:10.1080/17487870.2014.909315

Greiner, R., & Ang, S. (2012). Biotechnology collaborations: Does business model matter? *Journal of Management & Governance*, *16*(3), 377–392. doi:10.1007/s10997-010-9156-z

Grewiński, M. (2014). Social Services Mix State jako nowatorski i przyszłościowy model polityki społecznej w Polsce. In E. Trafiałek (Ed.), *Innowacje w polityce społecznej XXI wieku. Przedsiębiorczość, zmiana, rozwój, polityka socjalna* (pp. 90–111). Kielce: Wydawnictwo Uniwersytetu Jana Kochanowskiego w Kielcach.

Grindle, M. S., & Hilderbrand, M. E. (1995). Building sustainable capacity in the public sector: What can be done? *Public Administration and Development*, *15*(5), 441–463. doi:10.1002/pad.4230150502

Grönroos, C., & Ojasalo, K. (2004). Service productivity: Towards a conceptualization of the transformation of inputs into economic results in services. *Journal of Business Research*, *57*(4), 414 423. doi:10.1016/S0148-2963(02)00275-8

Grönroos, Ch. (1990). *Strategic Management and Marketing, Managing the Moments of Truth in Service Competition*. Lexington: Free Press.

Gummesson, E. (2011). *Total relationship marketing*. Routledge.

Gupta, S. (2014). Modeling Regional Disparities for a Balanced Quality of life and Apportioning Public Funding – a Graph Theoretical Approach. *Applied Research in Quality of Life*, *10*(3), 473–493. doi:10.1007/s11482-014-9327-1

Gureak. (2016). *Gureak website*. Retrieved July 10, 2016, from: http://www.gureak.com/es/quienessomos

Guthrie, J. P. (2001). High-involvement work practices, turnover and productivity: Evidence from New Zealand. *Academy of Management Journal*, *44*(1), 180–190. doi:10.2307/3069345

Guthrie, K. L., & Jones, T. B. (2012, Winter). Teaching and learning: Using experiential learning and reflection for leadership education. *New Directions for Student Services*, *2012*(140), 53–63. doi:10.1002/ss.20031

Gutu, G. (2014). General administration vs. right to manage of local community public property. *Juridical Current, 17*(2), 54–59.

Guy, P.B. (1998). Governance Without Government? Rethinking Public Administration. *Journal of Public Administration Research and Theory, 2*, 223-243.

Haigh, N., & Hoffman, A. J. (2014). The New Heretics: Hybrid Organizations and the Challenges They Present to Corporate Sustainability. *Organization & Environment, 16*(4), 417–436.

Hämäläinen, M. (2014). *Customer-centric and value-based business model design Impacts of the additive manufacturing technology on firm's business model.* Jyväskylän yliopisto Tietojenkäsittelytieteiden laitos.

Hamel, G. (2001). *Leading the revolution.* Boston, MA: Harvard Business School Press.

Harrison, J. G. (1994). *Strategic Management of Organisation and Stakeholder.* St. Paul, MN: West.

Hassan, M., Hagen, A., & Daigs, I. (2006). Strategic human resources as a strategic weapon for enhancing labour productivity: Empirical evidence. *Academy of Strategic Management Journal, 5*, 75–96.

Hastings, G. (2010). *Social marketing. Why sholud the Devie have All the Best tunes?* Butterworth-Heinemann.

Haugh, H. (2006). Social enterprise: Beyond economic outcomes and individual returns. In J. Mair, J. Robinson, & K. Hockerts (Eds.), *Social entrepreneurship* (pp. 180–205). Basingstoke, UK: Palgrave Macmillan. doi:10.1057/9780230625655_12

Hausner, J. (2002). Od idealnej biurokracji do zarządzania publicznego. In M. Maroda (Ed.), *Wymiary życia społecznego. Polska na przełomie XX i XXI wieku.* Warszawa: Wydawnictwo Naukowe Scholar.

Hausner, J. (2007). Ekonomia społeczna jako kategoria rozwoju. In J. Hausner (Ed.), *Ekonomia Społeczna a rozwój* (pp. 11–28). Kraków: Małopolska Szkoła Administracji Publicznej.

Hausner, J., Laurisz, N., & Mazur, S. (2007). Przedsiębiorstwo społeczne – konceptualizacja. In J. Hausner (Ed.), *Zarządzanie podmiotami ekonomii społecznej* (pp. 9–22). Kraków: Małopolska Szkoła Administracji Publicznej.

Hax, H. (1964). Investitions- und Finanzplanung mit Hilfe der linearen Programmierung. Zeitschrift für betriebswirtschaftliche. *Forschung, 16*, 430–446.

Hedman, J., & Kalling, T. (2003). The business model concept: Theoretical underpinnings and empirical illustrations. *European Journal of Information Systems, 12*(1), 49–59. doi:10.1057/palgrave.ejis.3000446

Heinig, K. (1951). *Das Budget.* Tübingen: Mohr.

Herman, R. D., & Renz, D. O. (2008). Advancing nonprofit organizational effectiveness research and theory: Nine theses. *Nonprofit Management & Leadership, 18*(4), 399–415. doi:10.1002/nml.195

Hermel, L., & Romagni, P. (1990). *Le marketing public – Une introduction au marketing des administrations et des organisations publiques.* Paris: Editura Economica.

Herzberg, F., Mausner, B., & Snyderman, B. B. (1959). *The Motivation to Work* (2nd ed.). New York: John Wiley & Sons.

Hidalgo, C. (2015). *Why Information Grows. The Evolution of Order from Atoms to Economies.* New York: Basic Books.

Hildreth, W. B., & Hildreth, R. P. (1989). The Business of Public Management. *Public Productivity Review, 12*(3), 303–321. doi:10.2307/3380120

Hilgers, D., & Ihl, C. (2010). Citizensourcing: Applying the Concept of Open Innovation to the Public Sector. *The International Journal of Public Participation, 4*(1), 67–88.

Hirschman, A. O. (1970). *Exit, voice, and loyalty: responses to decline in firms, organisations, and states*. Cambridge, MA: Harvard University Press.

Hjorth, D. (2013). Public entrepreneurship: desiring social change, creating sociality. *Entrepreneurship & Regional Development, 25*, 34–51.

Hockerts, K. (2005). Bootstrapping Social Change, Towards an Evolutionary Theory of Social Entrepreneurship.*1st International Social Entrepreneurship Research Conference (ISERC)*. Barcelona: IESE.

Hollanders, H., Arundel, A., Buligescu, B., Peter, V., Roman, L., Simmonds, P., & Es-Sadki, N. (2013). European Public Sector Innovation Scoreboard 2013. A pilot exercise. Belgium: Academic Press.

Hood, C. (1991). A New Public Management for all seasons? *Public Administration, 69*(1), 3–19. doi:10.1111/j.1467-9299.1991.tb00779.x

Hoque, F. (2002). *The Alignment Effect: How to Get Real Business Value Out of Technology*. Financial Times Prentice Hall.

Horowitz, A. S. (1996). The real value of VARS: Resellers lead a movement to a new service and support. *Mark Comput, 16*(4), 31–36.

Horton, S., & Farnham, D. (Eds.). (2002). Public Management in Britain. Palgrave.

Howaldt, J., & Schwarz, M. (2010). *Social innovation: concepts, research fields and international trends*. Dortmund: Sozialforschungsstelle Dortmund.

Hsieh, A.-T., Chou, C.-H., & Chen, C.-M. (2002). Job standardization and service quality: A closer look at the application of total quality management to the public sector. *Total Quality Management, 13*(7), 899–912. doi:10.1080/0954412022000017012

Hughes, A., Moore, K., & Kataria, N. (2011). *Innovation in Public Sector Organisations: A pilot survey for measuring innovation across the public sector*. London: NESTA. Retrieved from http://scholar.google.com/scholar?hl=en&btnG=Search&q=intitle:Innovation+in+Public+Sector+Organisations+A+pilot+survey+for+measuring+innovation+across+the+public+sector#0

Hughes, A., Moore, K., & Kataria, N. (2011). *Innovation in Public Sector Organisations: A pilot survey for measuring innovation across the public sector*. London: NESTA. Retrieved from https://www.nesta.org.uk/sites/default/files/innovation_in_public_sector_orgs.pdf

Hyman, D. N. (1993). *Public Finance: A Contemporary Application of Theory to Policy*. Fort Worth, TX: The Dryden Press.

ILO. (2016). *About ILO*. Retrieved July 2, 2016, from: http://www.ilo.org/global/about-the-ilo/lang--en/index.htm

Innerarity, D. (2013). *The Democracy of Knowledge*. New York: Bloomsbury.

Innobarometer. (2010). *Analytical Report Innovation in Public Administration*. Author.

International Monetary Fund (IMF). (2001). *Guidelines for public debt management*. Retrieved May 2016, from http://www.imf.org/external/np/mae/pdebt/2000/eng/

International Monetary Fund. (2010). *Toolkit manual. Inter-municipal cooperation*. Retrieved June, 2016, from http://www.municipal-ooperation.org/images/4/4c/IMC_Toolkit_Manual.pdf

ITC. (2012). *Social Dialogue. A Manual for Trade Union Education*. International Trade Centre of International Labour Organization. International Training Centre of the ILO, Retrieved June 13, 2016, from http://www.lpsk.lt/lpsk-web/wp-content/uploads/2014/06/Social-dialogue-manual.pdf

Jääskeläinen, A. (2010). *Productivity measurement and management in large public service organizations*. Publication 927. Tampere: Tampereen teknillinen yliopisto.

Jääskeläinen, A., & Lönnqvist, A. (2011). Public service productivity: How to capture outputs? *International Journal of Public Sector Management*, *24*(4), 289–302. doi:10.1108/09513551111133461

Jabłoński, A., & Jabłoński, M. (2013). Projektowanie sieciowych modeli biznesu. *Ekonomika i Organizacja Przedsiębiorstwa*, *12*, 30.

Jabłoński, A. (2008). *Modele biznesu w sektorach pojawiających się i schyłkowych. Tworzenie przewagi konkurencyjnej przedsiębiorstwa opartej na jakości i kryteriach ekologicznych*. Wydawnictwo Wyższej Szkoły Biznesu w Dąbrowie Górniczej, Dąbrowa Górnicza.

Jabłoński, A. (2013). *Modele zrównoważonego biznesu w budowie długoterminowej wartości przedsiębiorstw z uwzględnieniem ich społecznej odpowiedzialności*. Warszawa: Difin.

Jabłoński, A. (2015). Design and Operationalization of Technological Business Models. *Acta Universitatis Agriculturae et Silviculturae Mendelianae Brunensis*, *63*(3), 927–935. doi:10.11118/actaun201563030927

Jabłoński, A. (2015). *Spójność hybrydy strategicznej w środowisku sieciowym*. Warszawa: Difin.

Jałocha, B. (2012). Projectification of the European Union and its implications for public labour market organisations in Poland. *Journal of Project Program & Portfolio Management*, *3*(2), 1–16.

Jameson, J. (2013). Special issue on e-Leadership[Editorial]. *British Journal of Educational Technology*, *44*(6), 883–888. doi:10.1111/bjet.12106

Janson, C. (1981). Some Problems of Longitudinal Research in the Social Sciences. In F. Schulsinger, S. Mednick, & J. Knop (Eds.), *Longitudinal Research: Methods and Uses in Behavioral Science*. Boston: Martinus Nijhoff Publishing. doi:10.1007/978-94-009-8147-8_2

Jepson, D., & Sharpley, R. (2015). More than sense of place? Exploring the emotional dimension of rural tourism experiences. *Journal of Sustainable Tourism*, *23*(8/9), 1157–1178. doi:10.1080/09669582.2014.953543

Johannes Sang Un, C., & Hedman, J. (2015). Business models for NFC based mobile payments. *Journal of Business Models*, *3*(1), 29–48.

Johns, G. (2006). The essential impact of context on organizational behavior. *Academy of Management Review*, (31): 384–388.

Johnson, M.W., Christensen, C.M., & Kagermann, H. (2009). Jak fundamentalnie zmienić model biznesowy firmy. *Harvard Business Review Polska*, 160–161.

Johnson, M. W. (2010). *Seizing the white space - Business model innovation for growth and renewal*. Harvard Business Press.

Johnson, M. W., Christensen, C. M., & Kagermann, H. (2008). Reinventing your business model. *Harvard Business Review*, *86*(12). doi:10.1111/j.0955-6419.2005.00347.x

Jones-Evans, D. (1998). *Universities, technology transfer and spin-off activities – academic entrepreneurship in different European regions*. Targeted Socio-Economic Research Project No 1042. Retrieved from http://cordis.europa.eu/docs/publications/7864/78645511-6_en.pdf

Jong, E. de, Engelaer, F., & Mendoza, M. (2015). *Realizing opportunities of a circular business model*. Academic Press.

Josselin, J.-M., & Marciano, A. (2004). Federalism and Conflicts over Principalship. Some Insights into the American Constitutional History. *Constitutional Political Economy*, *15*(13), 281–304. doi:10.1023/B:COPE.0000040433.37364.3e

Jourdan, Z., Rainer, K., & Marshall, T. (2008). Business Intelligence: An Analysis of the Literature. *Information Systems Management*, 25.

Joustra, D. J., de Jong, E., & Engelaer, F. (2013). *Guided Choices towards a Circular Business Model*. Academic Press.

Julnes, P. de L., Gibson, E., & Park, S. (2016). Refining Our Understanding of the Process of Innovation in Public and Nonprofit Organizations. In P. de L. Julnes & E. Gibson (Eds.), *Innovation in the Public and Nonprofit Sector* (pp. 245–260). New York: Routledge.

Jun, J. S. (1997). Dialectic between the Private Realm and the Public Realm: Renewing Debate on the Public Good. *Administrative Theory & Praxis*, *19*(2), 238–245. doi:10.1097/EDE.ObO13e31812e5535

Ju-Yeon, L., Shrihari, S., Henderson, M. C., Palmatier, R.W. (2014). Effect of Customer-Centric Structure on Long- Term Financial Performance. *Marketing Science*. .10.1287/mksc.2014.0878

Kadeřábková, M., & Moghadam-Saman, S. (2013). *Evaluation of social innovations –their characteristics and impacts, cross-country comparison and implications for policy support. Social Frontiers: The next edge of social innovation research Evaluation*. Prague: Centre for Innovation Studies.

Kallio, T. J., & Kuoppakangas, P. (2013). Bandwagoning municipal enterprises: Institutional isomorphism and the search for the third way. *Policy Studies*, *34*(1), 19–35. doi:10.1080/01442872.2012.731842

Kangasharju, A. (2008). *Tuottavuus osana tuloksellisuutta*. Helsinki: The Association of Finnish Local and Regional Authorities.

Kankainen, A., Vaajakallio, K., Kantola, V., & Mattelmäki, T. (2012). Storytelling Group – a co-design method for service design. *Behaviour & Information Technology*, *31*(3), 221–230. doi:10.1080/0144929X.2011.563794

Kano, N., Seraku, N., Takahashi, F., & Tsuji, S. (1984). Attractive quality and must-be quality. *Hinshitsu: The Journal of the Japanese Society for Quality Control*, *14*(2), 39–48.

Kaplan, A. M., & Haenlein, M. (2009). The increasing importance of public marketing: Explanations, applications and limits of marketing within public administration. *European Management Journal*, *27*(3), 197–212. doi:10.1016/j.emj.2008.10.003

Kaplan, S. (2011). Business Models Aren't Just For Business. *Harvard Business Review*, 11.

Kapstein, E., & Kim, R. (2011). *The Socio-Economic Impact of Newmont Ghana Gold Limited*. Retrieved May 03, 2016, from http://www.newmont.com/files/doc_downloads/africa/ahafo/environmental/Socio_Economic_Impact_of_Newmont_Ghana_Gold_July_2011_0_0.pdf

Käpylä, J., Jääskeläinen, A., Seppänen, S. K., Vuolle, M., & Lönnqvist, A. (2008). *Tuottavuuden kehittäminen Suomessa. Haasteet ja tutkimustarpeet. The Finnish Work Environment Fund. Työsuojelurahaston selvityksiä 1. Tampere University of Technology. Department of Information Management and Logistics*. Helsinki: Edita.

Kark, R. (2011). Games management play: Play as a form of leadership development. *Academy of Management Learning & Education*, *10*(3), 507–527. doi:10.5465/amle.2010.0048

Kastberg, G. (2016). Trust and control in network relations: A study of a public sector setting. *Financial Accountability & Management*, *32*(1), 33–56. doi:10.1111/faam.12079

Kearney, C., Hisrich, R., & Roche, F. (2007). Facilitating Public Sector Corporate Entrepreneurship Process: A Conceptual Model. *Journal of Enterprising Culture*, *15*(3), 275–299. doi:10.1142/S0218495807000150

Kearney, J. D. Sr. (2012). Strategic vs. financial buyer: Choosing the best option. *Financial Executive*, *28*(6), 30–33.

Kebritchi, M., & Hurumi, A. (2008). Examining the pedagogical foundations of modern educational computer games. *Computers & Education*, *51*(4), 1729–1743. doi:10.1016/j.compedu.2008.05.004

Kelleher, C., & Peppard, J. (2011). Consumer experience of value creation: A phenomenological perspective. *Advances In Consumer Research - European Conference Proceedings*.

Kesti, M. (2013). Human capital production function. *GSTF Journal on Business Review*, *3*(1), 22–32. doi:10.7603/s40706-013-0001-7

Kesti, M., Leinonen, J., & Syväjärvi, A. (2016). A multidisciplinary critical approach to measure and analyze human capital productivity. In M. Russ (Ed.), *Quantitative Multidisciplinary Approaches in Human Capital and Asset Management* (pp. 1–22). Hershey, PA: IGI Global. doi:10.4018/978-1-4666-9652-5.ch001

Kesti, M., & Syväjärvi, A. (2015). Human capital production function in strategic management. *Technology and Investment*, *6*(01), 12–21. doi:10.4236/ti.2015.61002

Kettl, D. F. (2012). *The Politics of the Administrative Process*. Washington, DC: CQ Press.

Khan, M. A. (2008). Reinventing public enterprises in public enterprises. In *Public enterprises: Unresolved Challenges and New Opportunities*. New York: United Nations.

Kickert Walter, J.M. (2002). *Public Governance in the Netherlands: An Alternative to Anglo-American 'Managerialism'*. doi: 10.1111/1467-9299.00084

Kimberly, J. R. (1976). Issues in the Design of Longitudinal Organizational Research. *Sociological Methods & Research*, 4.

Kim, J., & Fesenmaier, D. R. (2015). Measuring emotions in real time: Implications for tourism experience design. *Journal of Travel Research*, *54*(4), 419–429. doi:10.1177/0047287514550100

Kindström, D. (2010). Towards a service-based business model – Key aspects for future competitive advantage. *European Management Journal*, *28*(6), 479–490. doi:10.1016/j.emj.2010.07.002

Klaus, P., & Maklan, S. (2011). Bridging the gap for destination extreme sports: A model of sports tourism customer experience. *Journal of Marketing Management*, *27*(13/14), 1341–1365. doi:10.1080/0267257X.2011.624534

Klein, P. G., Mahoney, J. T., Mcgahan, A. M., & Pitelis, C. N. (2010). Toward a theory of public entrepreneurship. *European Management Review*, *7*(1), 1–15. doi:10.1057/emr.2010.1

Klijn, E. H., & Teisman, G. R. (2008). Complexity theory and public management. *Public Management Review*, *10*(3), 287–297. doi:10.1080/14719030802002675

Klofsten, M., & Jones-evans, D. (2000). Academic Comparing in Europe - Entrepreneurship The Case of Sweden and Ireland. *Small Business Economics*, *14*(4), 299–309. doi:10.1023/A:1008184601282

Koch, P. M., & Hauknes, J. (2005). *On innovation in the public sector – today and beyond.* Publin Report No. D20. Retrieved from http://citeseerx.ist.psu.edu/viewdoc/download?doi=10.1.1.104.3988&rep=rep1&type=pdf

Kolb, A. Y., & Kolb, D. A. (2010). Learning to play, playing to learn: Case study of a ludic learning space. *Journal of Organizational Change Management*, *23*(1), 26–50. doi:10.1108/09534811011017199

Köllner, W. (2001). *Der interne Finanzausgleich in den niedersächsischen Samtgemeinden* (Dissertation). München: Universität der Bundeswehr München.

Kotler, P., Kartajaya, H., & Setiawan, I. (2010). Marketing 3.0, Dobry produkt? Zadowolony klient? Spełniony Człowiek! MT Biznes Sp. zo.o.

Kotler, Ph. (1994). *Marketing. Analiza. Planowanie. Wdrażanie i kontrola*. Warszawa: Gebethner i S-ka.

Kotler, Ph. (2005). *Marketing – Koncepcje marketingu w XXI wieku*. Poznań: REBIS.

Kotler, Ph., & Zaltman, G. (1971). Social Marketing. An Approach to Planned Social Change. *Journal of Marketing*, *35*(3), 3–12. doi:10.2307/1249783 PMID:12276120

Koźmiński, A. K., & Jemielniak, D. (2008). *Zarządzanie od podstaw. Podręcznik akademicki*. Warszawa: Wydawnictwa Akademickie i Profesjonalne.

Koźmiński, A., & Jemielniak, D. (2008). *Zarządzanie od podstaw. In Podręcznik akademicki*. Warsaw: Wydawnictwa Akademickie i Profesjonalne.

Kożuch, B., Kożuch, A. (2014). Responsywność w zarządzaniu usługami publicznymi. *Prace Naukowe Uniwersytetu Ekonomicznego we Wrocławiu*, *355*.

Kożuch, B., Kożuch, A. (2016). Partycypacja publiczna a skuteczność zarządzania publicznego. *Przedsiębiorczość i Zarządzanie*, *16*(12).

Kożuch, B., Kożuch, A., & Sienkiewicz-Małyjurek, K. (2016). Contemporary faces of Public Management. *Przegląd Organizacj*, 3.

Kożuch, B. (2004). *Zarządzanie publiczne. W teorii i praktyce polskich organizacji*. Warszawa: Placet.

Kożuch, B. (2009). Innowacyjność w sektorze publicznym - bariery i możliwości rozwoju. In B. Kryk & K. Piech (Eds.), *Innowacyjność w skali makro i mikro*. Warszawa: Instytut Wiedzy i Innowacji.

Kożuch, B. (2011). *Skuteczne współdziałanie organizacji publicznych i pozarządowych. In Monografie i Studia Instytutu Spraw Publicznych*. Kraków: Uniwersytet Jagielloński.

Kożuch, B. (2015). Zarządzanie publiczne jako teoria i praktyka zarządzania. In B. Kożuch & Ł. Sułkowski (Eds.), *Instrumentarium zarządzania publicznego*. Warszawa: Difin.

Kożuch, B., & Dobrowolski, Z. (2014). *Creating Public Trust. An Organisational Perspective*. Frankfurt am Main: Peter Lang. doi:10.3726/978-3-653-02970-3

Kożuch, B., & Dobrowolski, Z. (2014). *Public Trust. Organisational Perspective*. Frankfurt am Main: Peter Lang.

Kożuch, B., & Sienkiewicz-Małyjurek, K. (2013). Mierzenie skuteczności współpracy międzyorganizacyjnej w zarządzaniu bezpieczeństwem publicznym. *Studia Ekonomiczne Uniwersytetu Ekonomicznego w Katowicach*, *168*, 82–95.

KPRES. (2014). *The National Programme for the Development of Social Economy, Ministry of Labour and Social Policy, Warsaw*. Retrieved May 30, 2016, from http://www.ekonomiaspoleczna.pl/files/wiadomosci.ngo.pl/public/korespondenci/portal_ekonomiaspoleczna/KPRES_23.06.2014.pdf

Kraan, D., Bergvall, D., Hawkesworth, I., Kostyleva, V., & Witt, M. (2008). Budgeting in Russia. *OECD Journal on Budgeting*, *8*(2). Retrieved June, 2016, from https://www.oecd.org/russia/42007227.pdf

Kuoppakangas, P. (2014). Adopting a municipal enterprise form: Expectations and outcomes – Three cases and dilemmas. *Journal of Change Management*, *14*(1), 126–145. doi:10.1080/14697017.2013.778895

Laakoli, K., & Peiponen, A. (2012). Tuottavuusmatriisi johtamisen välineenä vanhusten ympärivuorokautisissa palveluissa. *Kunnallistieteellinen Aikakauskirja*, *40*(1), 9–23.

Lache, C., & Alexa, E. L. (2011). Particular aspects of the tourist products consumers' loyalty. *Agronomy Series Of Scientific Research*, *54*(1), 234–238.

Laing, A. (2003). Marketing in the public sector: Towards a typology of public services. *Marketing Theory*, *3*(4), 427–445. doi:10.1177/1470593103042005

Lamb, C. W. Jr. (1987). Public Sector Marketing is Different. *Business Horizons*, *30*(4), 56–60. doi:10.1016/0007-6813(87)90066-8

Lane, J. E. (2009). *State Management. An enquiry into models of public administration and management*. London: Routledge Taylor and Francis Group.

Lang, F. (1947). Insurance research. *Journal of Marketing*, *12*(1), 66–71. doi:10.2307/1246301

Langtry, B., (1994). Stakeholders and Moral Responsibilities of Business. *Business Ethics Quarterly, 4*.

Lauritzen, J. K. R. (2012). *Social Innovation in Local Government – experiences from Denmark*. Retrieved October 23, 2015, from: http://www.euclidnetwork.eu/news-and-events/sector-news/990-social-innovation-in-local-government-experiences-from-denmark.html

Leadbeater, C. (1998). *The rise of Social Entrepreneur*. London: Demos.

Lee, I. (2015). A Social Enterprise Business Model for Social Entrepreneurs: Theoretical Foundations and Model Development. *Social Entrepreneurship and Innovation*, *3*(4), 269–301. doi:10.1504/IJSEI.2015.069351

Lee, I. W., Feiock, R. C., & Lee, Y. (2012). Competitors and Cooperators: A Micro-Level Analysis of Regional Economic Development Collaboration Networks. *Public Administration Review*, *72*(2), 253–262. doi:10.1111/j.1540-6210.2011.02501.x

Lee, N. R., & Kotler, P. (2011). Social marketing to protect the environment: What works. *Sage (Atlanta, Ga.)*.

Lee, N. R., & Kotler, P. T. (2007). *Marketing in the Public Sector: A Roadmap for Improved Performance*. Pearson Education.

Lee, S. M., Hwang, T., & Choi, D. (2012). Open innovation in the public sector of leading countries. *Management Decision*, *50*(1), 147–162. doi:10.1108/00251741211194921

Lefebvre, R. C. (2011). An integrative model for docile marketing. *Journal of Social Marketing*, *1*(1), 54–72. doi:10.1108/20426761111104437

Leitenstorfer, A., & Török, A. (2008). *Geeignete Aufgaben und Rechtsformen für FOCJ*. FOCJ als Mittel regionaler Kooperation. Lehrstuhl für Finanzwissenschaft, Universität der Bundeswehr München Discussion Paper, 50, 5-22.

Leixnering, S., & Bramböck, S. (2013). Public-Corporate-Governance-Kodizes: Die Kupenickiade der Beteiligungsverwaltung. *Zeitschrift für öffentliche und gemeinwirtschaftliche Unternehmen, 36*(2), 170-190.

León, R. L., Simmonds, P., & Roman, L. (2012). *Trends and Challenges in Public Sector Innovation in Europe*. Thematic Report 2012 under Specific Contract for the Integration of INNO Policy Trend Chart with ERAWATCH (2011–2012). Inno Policy Trend Chart. European Commission. Retrieved June 13, 2016, from: http://www.technopolis-group.com/wp-content/uploads/2014/06/1447_INNOERATREND-2011-12.pdf

Leonard, B. (1983). *Relationship Marketing. Emerging Perspectives on Services Marketing*. Chicago: American Marketing Association.

Leś, E. (2007). Rola trzeciego sektora w polityce społecznej. In G. Firlit-Fesnak & M. Szylko-Skoczny (Eds.), *Polityka społeczna* (pp. 361–372). Warszawa: PWN.

Lewandowski, M. (2013). Introduction to Academic Entrepreneurship. In A. Szopa, W. Karwowski, & P. Ordóñez de Pablos (Eds.), Academic Entrepreneurship and Technological Innovation (pp. 1–28). IGI Global. http://doi.org/doi:10.4018/978-1-4666-2116-9.ch001

Lewandowski, M. (2013b). *Zarządzanie strategiczne w instytucjach kultury*. Katowice: Wydawnictwo Con Arte.

Lewandowski, M. (2015). Types of Innovations in Cultural Organizations. *International Journal of Contemporary Management*, *14*(1), 67–78.

Lewandowski, M. (2016). Designing the Business Models for Circular Economy — Towards the Conceptual Framework. *Sustainability*, *8*(1), 43. doi:10.3390/su8010043

Leydesdorff, L., & Etzkowitz, H. (1998). The Triple Helix as a model for innovation studies. *Science and Public Policy Beech Tree Publishing*, *25*(3), 195–203. doi:10.1016/j.vaccine.2010.03.035

Liebetrau, A. M. (1983). *Measures of association*. Newbury Park, CA: Sage Publications. doi:10.4135/9781412984942

Linder, J. C., & Cantrell, S. (2001). *Changing Business Models: Surveying the Landscape*. Working Paper. Institute for Strategic Change, Accenture.

Linder, J., & Cantrell, S. (2000). *Changing Business Models: Surveying the Landscape* (Vol. 34). Retrieved from http://scholar.google.com/scholar?hl=en&btnG=Search&q=intitle:Changing+Business+Models:+Surveying+the+Landscape#0

Linder, J. C., & Cantrell, S. (2002). It's All in the Mind(set). *Across the Board*, *39*(3).

Linna, P., Pekkola, S., Ukko, J., & Melkas, H. (2008). Defining and measuring productivity in the public sector: Managerial perceptions. *International Journal of Public Sector Management*, *23*(5), 479–499. doi:10.1108/09513551011058493

Lisein, O., Pichault, F., & Desmecht, J. (2009). Les business models des sociétés de service actives dans le secteur Open Source. Systémes d'Information et Management, 14(2).

Lisiecka, K., & Papaj, T. (2009). *Kierunki doskonalenia usług świadczonych przez administrację publiczną*. Katowice: Śląskie Centrum Społeczeństwa Informacyjnego i Akademia Ekonomiczna.

Loogma, K., Tafel-Viia, K., & Ümarik, M. (2012). Conceptualising educational changes: A social innovation approach. *Journal of Educational Change*, 1–19.

Lopes, M. C., Fialho, F. A. P., Cunha, C. J. C. A., & Niveiros, S. I. (2013). Business games for leadership development: A systematic review. *Simulation & Gaming*, *44*(4), 523–543. doi:10.1177/1046878112471509

Losada, M., & Heaphy, E. (2004). The role of positivity and connectivity in the performance of business teams: A nonlinear dynamics model. *The American Behavioral Scientist*, *2*(1), 71–87.

Louis, K. S., Blumenthal, D., Gluck, M. E., & Stoto, M. A. (1989). Entrepreneurs in Academe : An Exploration of Behaviors among Life Scientists. *Administrative Science Quarterly*, *34*(1), 110–131. doi:10.2307/2392988

Love, J. H., Roper, S., & Vahter, P. (2014). Learning from openness: The dynamics of breadth in external innovation linkages. *Strategic Management Journal*, *35*(11), 1703–1716. doi:10.1002/smj.2170

Low, J. (2000). The value creation index. *Journal of Intellectual Capital*, *1*(3), 252–262. doi:10.1108/14691930010377919

Lüdeke-Freund, F. (2010). *Towards a conceptual framework of business models for sustainability*. Lüneburg: Centre for Sustainability Management.

Luhmann, N. (1994). *Teoria polityczna państwa bezpieczeństwa socjalnego*. Warszawa: Wydawnictwo Naukowe PWN.

Lu, J., Hallinger, P., & Showanasai, P. (2014). Simulation-based learning in management education. A longitudinal quasi-experimental evaluation of instructional effectiveness. *Journal of Management Development*, *33*(3), 218–244. doi:10.1108/JMD-11-2011-0115

Luke, B., Verreynne, M. L., & Kearins, K. (2010). Innovative and entrepreneurial activity in the public sector: The changing face of public sector institutions. *Innovation: Management. Policy & Practice*, *12*(2), 138–153. doi:10.5172/impp.12.2.138

Mack, W. R., Green, D., & Vedlitz, A. (2008). Innovation and implementation in the public sector: An examination of public entrepreneurship. *Review of Policy Research*, *25*(3), 233–252. doi:10.1111/j.1541-1338.2008.00325.x

MacMillan, K., Money, K., Money, A., & Downing, S. (2005). Relationship marketing in the not-for-profit sector: An extension and application of the commitment trust theory. *Journal of Business Research*, *58*(6), 806–818. doi:10.1016/j.jbusres.2003.08.008

Macpherson, A., Jones, O., & Zhang, M. (2004). Evolution or revolution? Dynamic capabilities in a knowledge-dependent firm. *R & D Management*, *34*(2), 161–177. doi:10.1111/j.1467-9310.2004.00331.x

Magala, S. (2009). *The Management of Meaning in Organizations*. Basingstoke, UK: Palgrave. doi:10.1057/9780230236691

Magala, S. (2016). From Linnaeus to Darwin. From Cultural grid to "Going Native". In M. Rozkwitalska, L. Sulkowski, & S. Magala (Eds.), *Intercultural Interactions in the Multicultural Workplace. Traditional and Positive Organizational Scholarship*. Springer International Publishing AG.

Magretta, J. (2002). Why Business Models Matter. *Harvard Business Review*, *80*(5), 86–92. PMID:12024761

Magretta, J. (2002). Why business models matter. *Harvard Business Review*, *80*, 86–92. PMID:12024761

Mahadevan, B. (2000). Business Models for Internet based E-Commerce, An Anatomy. California Management Review, 42(4).

Mahadevan, B. (2000). Business Models for Internet-Based E-Commerce: An Anatomy. *California Management Review*, *42*(4), 55–69. doi:10.2307/41166053

Mair, J., & Noboa, E. (2003). *Working Paper. The Emergence of Social Enterprises and Their Place in the New Organizational Landscape*. Navarra: University of Navarra, IESE Business School.

Mair, J., & Martí, I. (2009). Social entrepreneurship in and around institutional voids. *Journal of Business Venturing*, *24*(5), 419–435. doi:10.1016/j.jbusvent.2008.04.006

Maklan, S., & Klaus, P. (2011). Customer experience. *International Journal of Market Research*, *53*(6), 771–792. doi:10.2501/IJMR-53-6-771-792

Malhotra, Y. (2000). Knowledge management and new organization forms: A framework for business model innovation. *Information Resources Management Journal*, *13*(1), 5–14. doi:10.4018/irmj.2000010101

Malik, N. J., & Mahmood, R. B. (2012). Facilitating corporate entrepreneurship in public sector higher education institutions: A conceptual model. *Issues In Social & Environmental Accounting*, *6*(1/2), 26–49.

Margetts, H. (2009). The internet and public policy. *Policy and Internet*, *1*(1), 1–21. doi:10.2202/1944-2866.1029

Markides, C. (2006). Disruptive innovation: In need of better theory. *Journal of Product Innovation Management*, *23*(1), 19–25. doi:10.1111/j.1540-5885.2005.00177.x

Markides, C., & Charitou, C. D. (2004). Competing with dual business models: A contingency approach. *The Academy of Management Executive*, *18*(3), 22–36. doi:10.5465/AME.2004.14776164

Marton, F. (1986). Phenomenography – A research approach investigating different understandings of reality. *Journal of Thought*, *21*(2), 28–49.

Marton, F., & Booth, S. (1997). *Learning and awareness*. Mahwah, NJ: Lawrence Erlbaum Associates.

Maruyama, Y., Nishikido, M., & Iida, T. (2007). The rise of community wind power in Japan: Enhanced acceptance through social innovation. *Energy Policy*, *35*(5), 2761–2769. doi:10.1016/j.enpol.2006.12.010

Massa, L., & Tucci, L. C. (2013). Business Model Innovation. In M. Dodgson, D. M. Gann, & N. Phillips (Eds.), *The Oxford Handbook of Innovation Management* (pp. 420–441). Oxford, UK: Oxford University Press.

Massa, L., & Tucci, L. C. (2013). *Business Model Innovation. The Oxford Handbook of Innovation Management*. doi:10.1002/9781118466421.ch4

Massaro, M., Dumay, J., Guthrie, J. (2015). On the Shoulders of Giants: Undertaking a Structured Literature Review in Accounting, Accounting. *Audit. Account. J.*, 267 – 284.

Mc Donough, W., & Braugnart M. (1998, October). The next industrial revolution. *The Atlantic Monthly*.

McCurdy, H. E. (1991). Organizational Decline: NASA and the Life Cycle of Bureaus. *Public Administration Review*, *51*(4), 308–315. doi:10.2307/976746

McDonald, D. A., & Ruiters, G. (2012). *Alternatives to privatization. Public options for essential services in the global south*. New York: Routledge.

McGuire, L. (2012). Slippery concepts in context. Relationship marketing and public services. *Public Management Review*, *14*(4), 541–555. doi:10.1080/14719037.2011.649975

McLaughlin, K., Osbourne, S., & Chew, C. (2009). Developing the marketing function in UK public service organizations: The contribution of theory practice. *Public Money and Management*, *29*(1), 35–42. doi:10.1080/09540960802617343

Meklin, P. (2008). Tuottavuuden mittaaminen verorahoitteisessa toiminnassa. *Kunnallistieteellinen Aikakauskirja*, *36*(4), 386–395.

Melecký, L. & Staníčková, M. (2011). Evaluation of the Czech Republic regions' competitiveness in the context of the Lisbon strategy. *Ekonomická revue - Central European Review of Economic Issues*, *14*(3), 183-200.

Memmi, A. (2006). *Decolonization and the Decolonized*. Minneapolis, MN: University of Minnesota Press.

Mentink, B. (2014). *Circular business model innovation: A process framework and a tool for business model innovation in a circular economy* (MSc Thesis). Delft University of Technology & Leiden University.

Mertens, D. (2006). *Portfolio-Optimierung nach Markowitz*. Frankfurt am Main: Frankfurt School Verlag.

Messersmith, J. G., Patel, P. C., Lepak, D. P., & Gould-Williams, J. S. (2011). Unlocking the black box: Exploring the link between high-performance work systems and performance. *The Journal of Applied Psychology*, *96*(6), 1105–1118. doi:10.1037/a0024710 PMID:21787040

Metis. (2014). *Monitoring report 2013, Towards the New Cohesion policy*. European Union. Retrieved June, 2016 from a.http://cor.europa.eu/en/documentation/studies/Documents/EGTC-monitoring-report-2013.pdf

Meynhardt, T., & Diefenbach, F. E. (2012). What drives entrepreneurial orientation in the public sector? Evidence from Germanys Federal Labor Agency. *Journal of Public Administration: Research and Theory*, *22*(4), 761–792. doi:10.1093/jopart/mus013

Michelini, L., & Fiorentino, D. (2012). New business models for creating shared value. *Social Responsibility Journal*, *8*(4), 561–577. doi:10.1108/17471111211272129

Micheli, P., Schoeman, M., Baxter, D., & Goffin, K. (2012). *New Business Models for Public-Sector Innovation Successful Technological Innovation for Government*. Research-Technology Management. doi:10.1007/978-3-642-32150-4

Mihaiu, D. M., Opreana, A., & Cristescu Pompiliu, M. (2010). Efficiency, Effectiveness and Performance of the Public Sector. *Journal for Economic Forecasting Institute for Economic Forecasting*, *13*(4), 132–147.

Minařík, B., Borůvková, J., & Vystrčil, M. (2013). *Analýzy v regionálním rozvoji*. Praha: Professional Publishing.

Mitchell, D. W., & Coles, C. B. (2004). Business model innovation breakthrough moves. *The Journal of Business Strategy*, *25*(1), 16–26. doi:10.1108/02756660410515976

Mitchell, D., & Coles, C. (2003). The ultimate competitive advantage of continuing business model innovation. *The Journal of Business Strategy*, *24*(5), 15–21. doi:10.1108/02756660310504924

Mitchell, R. K. (1997). Toward A Theory of Stakeholder. Identification and Salience: Defining the Principle of Who and What Really Counts. *Academy of Management Review*.

Mitręga, M., (2005). *Marketing relacji – teoria i praktyka*. Warszawa: CeDeWu.

MMR. (2013). *Strategie regionálního rozvoje ČR na období 2014–2020*. Praha: MMR.

Mohanty, J. (2013). Are government owned organizations deliberately demonized? The organizational evolution or revolution? Citizenship behavior indicators. *Advances in Management & Applied Economics*, *3*(4), 177–191.

Moingeon, B., & Lehmann-Ortega, L. (2010). Creation and implementation of a new business model: A disarming case study. *Management*, *13*(4), 266–297.

Moizer, J., & Tracey, P. (2010). Strategy making in social enterprise: The role of resource allocation and its effects on organisational sustainability. *Systems Research and Behavioral Science*, *27*(3), 252–266. doi:10.1002/sres.1006

Molyneaux, D. (2004). Accountability and volunteers at social businesses: A role for ethical checklists. *Business Ethics (Oxford, England)*, *13*(1), 14–25. doi:10.1111/j.1467-8608.2004.00345.x

Monteduro, F. (2014). Public-private versus public ownership and economic performance: Evidence from Italian local utilities. *Journal of Management & Governance*, *18*(1), 29–49. doi:10.1007/s10997-012-9235-4

Moody, M., Littlepage, L., & Paydar, N. (2015). Measuring Social Return on Investment. *Nonprofit Management & Leadership*, *26*(1), 19–37. doi:10.1002/nml.21145

Moon, M. J. (2002). The Evolution of E-Government among Municipalities: Rhetoric or Reality? *Public Administration Review*, *62*(4), 424–433. doi:10.1111/0033-3352.00196

Moore, M., & Hartley, J. (2008). Innovations in Governance. *Public Management Review*, *10*(1), 3-20.

Moore, M., & Khagram, S. (2004). *On Creating Public Value: What Business Might Learn from Government About Strategic Management*. Working paper of the Corporate Social Responsibility Initiative, Kennedy School of Government. Cambridge, MA: Harvard University Press.

Moore, M., & Khagram, S. (2004). *On Creating Public Value: What Business Might Learn from Government About Strategic Management*. Working paper of the Corporate Social Responsibility Initiative. Kennedy School of Government, Harvard University, Cambridge.

Moore, M. (1995). *Creating Public Value: Strategic Management in Government*. Cambridge: Harvard University Press.

Moore, M. H. (1995). *Creating public value: Strategic management in government*. Cambridge, MA: Harvard University Press.

Moore, M. H. (2000). *Creating Public Value. Strategic Management i Government*. Cambridge, MA: Harvard University Press.

Morgan, R., & Hunt, S. (1994). Relationship Marketing in the Era of Network Competition. *Marketing Management*, *5*, 18–28.

Morgan, R., & Hunt, S. (1994). The Commitment – Trust Theory of Relationship Marketing. *Journal of Marketing*, *58*(3), 20–38. doi:10.2307/1252308

Morris, M. H., & Jones, F. F. (1999). Entrepreneurship in established organizations: The case of the public sector. *Entrepreneurship Theory and Practice*, *24*(1), 73–93.

Morris, M., Schindehutte, M., & Allen, J. (2005). The entrepreneurs business model: Toward a unified perspective. *Journal of Business Research*, *58*(6), 726–735. doi:10.1016/j.jbusres.2003.11.001

Morrison, J. (2004). Where now for problem based learning? *Lancet*, *363*(9403), 174. doi:10.1016/S0140-6736(03)15298-1 PMID:14726191

Mort, G. S., Weerawardena, J., & Carnegie, K. (2003). Social entrepreneurship: Towards Conceptualisation. *International Journal of Nonprofit and Voluntary Sector Marketing*, *8*(1), 76–88. doi:10.1002/nvsm.202

Moulaert, F., Martinelli, F., Swyngedouw, E., & González, S. (2005). Towards alternative model(s) of local innovation. *Urban Studies (Edinburgh, Scotland)*, *42*(11), 1969–1990. doi:10.1080/00420980500279893

Mouzas, S. (2006). Efficiency Versus Effectiveness in Business Networks. *Journal of Business Research*, *59*(10/11), 1124–1132. doi:10.1016/j.jbusres.2006.09.018

Mruk, H. (2008). *Marketing – zagadnienia współczesne*. Poznań: Forum Naukowe.

Muegge, S. (2012). *Business Model Discovery by Technology Entrepreneurs*. Technology Innovation Management Review.

Mueller, D. C. (2003). *Public Choice*. Austria: Universitat Wien. doi:10.1017/CBO9780511813771

Mulgan, G., Tucker, S., Ali, R., & Sanders, B. (2007). *Social Innovation: what it is, why it matters, how it can be accelerated*. Retrieved July 10, 2016 from: http://youngfoundation.org/publications/social-innovation-what-it-is-why-it-matters-how-it-can-be-accelerated

Mulgan, J. (2006). *Social innovation: what it is, why it matters and how it can be accelerated*. Oxford, UK: Oxford Said Business School.

Murray, R., Caulier-Grice, J., & Mulgan, G. (2010). *The open book of social innovation*. National Endowment for Science, Technology and the Art.

Musgrave, R. A., & Musgrave, P. (1989). *Public Finance in Theory and Practice.* McGraw-Hill.

Nag, R., Hambrick, D. C., & Chen, M. J. (2007). What is strategic management, really? Inductive derivation of a consensus definition of the field. *Strategic Management Journal*, *28*(9), 935–955. doi:10.1002/smj.615

Nam, T., & Pardo, T. A. (2011). Conceptualizing Smart City with Dimensions of Technology, People, and Institutions. In *Proceedings of the 12th Annual International Conference on Digital Government Research* (pp. 282–291). http://doi.org/ doi:10.1145/2037556.2037602

Napp, H. G. (1994). *Kommunale Finanzautonomie und ihre Bedeutung für eine effiziente lokale Finanzwissenschaft.* Frankfurt am Main: Lang.

Neely, A., & Delbridge, R. (2007). *Effective Business Models: What Do They Mean for Whitehall.* Retrieved from www.nationalschool.gov.uk/sunningdaleiinstitute

Nelson, J. R. (1964). *Marginal Cost Pricing in Practice.* Englewood Cliffs, NJ: Prentice-Hall.

Neumark, F. (1970). *Grundsätze gerechter und ökonomisch rationaler Steuerpolitik.* Tübingen: Mohr.

Newth, F. (2013). *Business Models and Strategic Management: A New Integration.* Business Expert Press.

Nicholls, A. (2006). Playing the Field: A New Approach to the Meaning of Social Entrepreneurship. *Social Enterprise Journal*, *2*(1), 1–5.

Nicholls, A., & Murdock, A. (2012). *Social Innovation: Blurring Boundaries to Reconfigure Markets.* Palgrave Macmillan. doi:10.1057/9780230367098

Nicolini, D., Gherardi, S., & Yanow, D. (2003). *Knowing in Organization. A Practice-based approach.* Armonk, NY: E Sharpe.

Niemczyk, J. (2010). *Modele biznesowe.* Warszawa: C.H. Beck.

Nogalski, B. (2009). Modele biznesu jako narzędzie reorientacji strategicznej przedsiębiorstw. *Master of Business Administrati*on, *2*.

Nogalski, B. (2009, March/April). Modele biznesu jako narzędzie reorientacji strategicznej przedsiębiorstw. *Master of Business Administration, 2*(97), 3-16.

Norman-Major, K. (2011). Balancing the Four Es; or Can We Achieve Equity for Social Equity in Public Administration?. *Journal of Public Affairs Education*, *17*(2), 233–252.

Nowak, M. J. (2011). Municipality companies in Poland. Formula objectives and scope of activities. *Samorząd Terytorialny*, *1*(2), 17–26.

Nowotny, E. (1999). *Der öffentliche Sektor.* Berlin: Spinger. doi:10.1007/978-3-662-06538-9

Obłój, K. (2010). *Pasja i dyscyplina strategii, Jak z marzeń i decyzji zbudować sukces firmy.* Poltext.

Obłój, K. (2002). *Tworzywo skutecznych strategii.* Warszawa: PWE.

Oborn, E., Barrett, M., & Exworthy, M. (2011). Policy entrepreneurship in the development of public sector strategy: The case of London health reform. *Public Administration*, *89*(2), 325–344. doi:10.1111/j.1467-9299.2010.01889.x PMID:22069793

OECD. (2011). *Corporate Governance Corporate Governance in Estonia*. Retrieved June, 2016, from https://books.google.ru/books?id=199U7QplPzIC&pg=PA31&lpg=PA31&dq=corporate+governance+code+in+estonia&source=bl&ots=XFz_du0lsl&sig=85GbWAaa8z64kn3h2fKuK_Z0Wmo&hl=en&sa=X&ved=0ahUKEwiezbGbv8rNAhXFDiwKHTJ3A7gQ6AEIRjAG#v=onepage&q=corporate%20governance%20code%20in%20estonia&f=false

OECD. (2014). *Spain From Administrative Reform to Continuous Improvement. OECD Public Governance Reviews*. OECD Publishing. doi:10.1787/9789264210592-en

OECD. (2016). *Fiscal federalism 2016: making decentralization work*. Paris: OECD Publishing. Retrieved May, 2016, from https://books.google.ru/books?id=FiJ3DAAAQBAJ&pg=PA115&lpg=PA115&dq=the+solidarity+principle+fiscal+federalism&source=bl&ots=890qZl2pPM&sig=5Xd9cekGJ32kQePiL9hTG

OECD. (2016a). *Country Profile: Spain*. Observatory of Public Social Innovation (OPSI). Retrieved June 15, 2016, from: https://www.oecd.org/governance/observatory-public-sector-innovation/countryprofiles/spain/

OECD. (2016b) *About OECD*. Retrieved July 3, 2016, from: http://www.oecd.org/about/

OECD/Eurostat. (2005). *Oslo manual: Guidelines for collecting and interpreting innovation data*. <ALIGNMENT.qj></ALIGNMENT>10.1787/9789264013100-en

Office of the Regional Council South-East. (2014). *Operational Programmes Allocations*. Retrieved October 23, 2015, from: http://www.jihovychod.cz/en/eu-funds/operational-programmes-allocations

OGP Action Plan. (2012). *Open Government Partnership. Action Plan: Spain*. Retrieved June 13, 2016 from: http://www.opengovpartnership.org/country/spain

OGP Spain. (2016). *Open Government Partnership. Spain*. Retrieved June 13, 2016 from: http://www.opengovpartnership.org/country/spain

OGP. (2016). *About OGP*. Retrieved July 2, 2016, from: http://www.opengovpartnership.org/

Olson, M. (1969). The principle of fiscal equivalence. The division of responsibilities among different levels of government. *The American Economic Review*, *59*(2), 479–487.

Omachonu, V. K., & Einspruch, N. G. (2010). Innovation in Healthcare Delivery Systems : A Conceptual Framework. *The Innovation Journal: The Public Sector Innovation Journal*, *15*(1), 1–20.

Oprescu, F., Jones, C., & Katsikitis, M. (2014). I play at work – ten principles for transforming work processes through gamification. *Frontiers in Psychology*, *5*(14). doi:10.3389/fpsyg.2014.00014 PMID:24523704

Osborne, S.P., & Strokosch, K. (2013). It takes two to tango? Understanding the co-production of public services by integrating the services management and public administration perspectives. *British Journal of Management, 24*(SI), S31-S47.

Osborne, D., & Gaebler, T. (1992). *Reinventing Government: How the Entrepreneurial Spirit Is Transforming the Public Sector*. Addison-Wesley Publishing, Company, Inc.

Osborne, S. (2010). *The New Public Governance? Emerging Perspectives on the Theory and Practice of Public Governance*. London: Routledge.

Osborne, S. P., & McLaughlin, K. (2005). The New Public Management in context. In K. McLaughlin, E. Ferlie, & S. Osborne (Eds.), *New Public Management: Current Trends and Future Prospects* (pp. 7–14). London: Routledge.

Osborne, S. P., Radnor, Z., Vidal, I., & Kinder, T. (2014). A sustainable business model for public service organizations? *Public Management Review*, *16*(2), 165–172. doi:10.1080/14719037.2013.872435

Ospina, S., Diaz, W., & OSullivan, J. F. (2002). Negotiating accountability: Managerial lessons from identity-based nonprofit organizations. *Nonprofit and Voluntary Sector Quarterly*, *31*(1), 5–31. doi:10.1177/0899764002311001

Osterwalder, A. (2004). The business model ontology: a proposition in a design science approach. InBusiness Model Generation, A Handbook for Visionaries, Game Changers, and Challengers. John Wiley & Sons, Inc.

Osterwalder, A., & Pigneur, Y. (2002). *An e-Business Model Ontology for Modeling e-Business*. 15th Bled Electronic Commerce Conference e-Reality: Constructing the e-Economy, Bled, Slovenia.

Osterwalder, A., & Pigneur, Y. (2004). An ontology for e-Business Models. In W. Currie (Ed.), *Value Creation from E-Business Models* (pp. 65–97). Oxford, UK: Elsevier Butterworth- Heinemann. doi:10.1016/B978-075066140-9/50006-0

Osterwalder, A., & Pigneur, Y. (2010). *Business model generation –A handbook for visionaries, game changers, and challengers*. John Wiley & Sons, Inc.

Osterwalder, A., & Pigneur, Y. (2010). *Business Model Generation*. John Wiley & Sons, Inc.

Osterwalder, A., & Pigneur, Y. (2010). *Business Model Generation: A Handbook for Visionaries, Game Changers, and Challengers*. Hoboken, NJ: John Wiley and Sons.

Osterwalder, A., & Pigneur, Y. (2010). *Business model generation: a handbook for visionaries, game changers, and challengers*. John Wiley & Sons.

Osterwalder, A., Pigneur, Y., Bernarda, G., & Smith, A. (2014). *Value Proposition Design: How to Create Products and Services Customers Want*. Hoboken, NJ: John Wiley and Sons.

Osterwalder, A., Pigneur, Y., & Tucci, C. L. (2005). Clarifying Business Models: Origins, Present, and Future of the Concept. *Communications of the Association for Information Systems*, *16*, 1–25. http://doi.org/10.1.1.83.7452

Ostrom, E. (1965). *Public Entrepreneurship: A Case Study in Ground Water Basin Management* (Doctoral dissertation). University of California. Retrieved from http://dlc.dlib.indiana.edu/dlc/handle/10535/3581

Ostrowska, S., Frączkiewicz-Wronka, A., & Bratnicki, M. (2013). Mission Oriented Scorecard: The Experience of Public Hospitals in Poland. In *7th Conference On Performance Measurement And Management Control*. European Institute for Advanced Studies in Management.

Otrusinova, M., & Pastuszkova, E. (2012). Concept of 3 E's and Public Administration Performance. International Journal of Systems Applications. *Engineering & Development*, *6*(2), 171–178.

Otto, J. (2004). *Marketing relacji*. Warszawa: C.H. Beck.

Özcan, S., & Reichstein, T. (2009). Transition to entrepreneurship from the public sector: Predispositional and contextual effects. *Management Science*, *55*(4), 604–618. doi:10.1287/mnsc.1080.0954

Özel, Ç. H., & Kozak, N. (2012). Motive based segmentation of the cultural tourism market: A study of Turkish domestic tourists. *Journal of Quality Assurance in Hospitality & Tourism*, *13*(3), 165–186. doi:10.1080/1528008X.2012.645199

Pădurean, M. A., Nica, A. M., & Nistoreanu, P. (2015). Entrepreneurship in Tourism and Financing through the Regional Operational Programme. *Amfiteatru Economic*, *17*(38), 180–194.

Palmatier, R. W., Dant, R. P., Grewal, D., & Evans, K. R. (2006). Factors influencing the effectiveness of relationship marketing: A meta-analysis. *Journal of Marketing*, *70*(4), 136–153. doi:10.1509/jmkg.70.4.136

Palo, T., & Tahtinen, J. (2011). A network perspective on business models for emerging technology based services. *Journal of Business and Industrial Marketing*, *26*(5), 377–388. doi:10.1108/08858621111144433

Papakiriakopoulos, D. A., Poylumenakou, A. K., & Doukidis, G. J. (2001). Building e-Business Models: An Analytical Framework and Development Guidelines. In *14th Bled Electronic Commerce Conference* (pp. 446–464). Retrieved from http://domino.fov.uni-mb.si/proceedings.nsf/0/5ba2fc04f2174ec8c1256e9f0030b818/$FILE/28_Papakiri.pdf

Papenfuß, U. (2013). Ein Public Corporate Governance Kodex ist aus theoretischer wie praktischer Perspektive nützlich– Eine Replik zu einer Köpenickade. *Zeitschrift für öffentliche und gemeinwirtschaftliche Unternehmen/Journal of Public and Nonprofit Services, 4*, 302-322.

Parvatiyar, A., & Sheth, J. N. (2000). The domain and conceptual foundations of relationship marketing. Handbook of Relationship Marketing, 1, 3-38.

Pateli, A. G., & Giaglis, G. M. (2004). A research framework for analysing eBusiness models. *European Journal of Information Systems*, *13*(4), 302–314. doi:10.1057/palgrave.ejis.3000513

PDR. (2013). *Designing Effective Public Services: A practical guide for public service managers*. PDR.

Pearce, J. (2003). *Social Enterprise in Anytown*. London: Calouste Gulbenkian Foundation.

Peda, P., Argento, D., & Grossi, G. (2013). Governance and performance of a mixed public-private enterprise: An assessment of a company in the Estonian water sector public. *Organization Review*, *13*(2), 185–196. doi:10.1007/s11115-013-0233-z

Perechuda, K. (2013). Dyfuzja wiedzy w przedsiębiorstwie sieciowym, wizualizacja i kompozycja. Wydawnictwo Uniwersytetu Ekonomicznego we Wrocławiu.

Perridon, L., & Steiner, M. (2009). *Finanzwirtschaft der Unternehmung*. München: Vahlen.

Perrini, F., & Vurro, C. (2006). Social entrepreneurship: Innovation and social change across theory and practice. In J. Mair, J. Robinson, & K. Hockerts (Eds.), *Social entrepreneurship* (pp. 57–85). Basingstoke, UK: Palgrave Macmillan. doi:10.1057/9780230625655_5

Perrone Pinheiro, F. A., Ferreira Savoia, J. R., & de Angelo, C. F. (2016). A comparative analysis of the public and private water supply and sanitation service providers' work in Brazil. *Brazilian. Business Review (Federal Reserve Bank of Philadelphia)*, *13*(1), 115–136.

Peters, B. G. (1998). Managing horizontal government: The politics of co-ordination. *Public Administration*, *76*(3), 295–316. doi:10.1111/1467-9299.00102

Petrescu, I., & Muscalu, E. (2003). *Tratat de management public. Lucian Blaga*. Sibiu: University.

Petrovic, O., Kittl, C., & Teksten, R. D. (2001). Developing business models for e-business.*International Electronic Commerce Conference*.

Philips, S. D., Laforest, R., & Graham, A. (2010). From shopping to social innovation: Getting public financing right in Canada. *Policy and Society*, *29*(3), 189–199. doi:10.1016/j.polsoc.2010.06.001

Phills, J., Deiglmeier, K., & Miller, T. (2008). Rediscovering Social Innovation. *Stanford Social Innovation Review*, *6*(4), 34–43.

Pine, B. J. II, & Gilmore, J. (2016). Integrating experiences into your business model: Five approaches. *Strategy and Leadership*, *44*(1), 3–10. doi:10.1108/SL-11-2015-0080

Pohjola, M. (2014). *Suomi uuteen nousuun. ICT ja digitalisaatio tuottavuuden ja talouskasvun lähteinä*. Vantaa: Julkaisumonistamo Eteläranta Oy.

Pollitt, C. (1990). *Managerialism and the Public Services: The Anglo-American Experience*. Basil Blackwell.

Pollitt, C., & Bouckaert, G. (2000). *Public management reform*. Oxford, UK: Oxford University Press.

Pollitt, C., & Bouckaert, G. (2004). *Public Management Reform. A Comparative Analysis*. Oxford, UK: Oxford University Press.

Pollock, K. (2008). The Four Pillars of Innovation: An Elementary School Perspective. *The Innovation Journal: The Public Sector Innovation Journal, 13*(2), 1–20. Retrieved from http://www.innovation.cc/peer-reviewed/pollack_innovative2.pdf

Postlep, R. (1993). *Gesamtwirtschaftliche Analyse kommunaler Finanzpolitik*. Baden-Baden: Nomos.

Potluka, O., & Liddle, J. (2014). Managing European Union Structural Funds: Using a Multilevel Governance Framework to Examine the Application of the Partnership Principle at the Project Level. *Regional Studies*, *48*(8), 1434–1447. doi:10.1080/00343404.2014.898837

Poulsson, H. G., & Kale, H. S. (2004). The experience economy and commercial experience. *The Marketing Review*, *4*(3), 267–277. doi:10.1362/1469347042223445

Prebensen, N. K., & Rosengren, S. (2016). Experience value as a function of hedonic and utilitarian dominant services. *International Journal of Contemporary Hospitality Management*, *28*(1), 113–135. doi:10.1108/IJCHM-02-2014-0073

PricewaterhouseCoopers. (n.d.). *Redefining local government*. Author.

Priem, R. L., Butler, J. E., & Li, S. (2013). Toward reimagining strategy research: Retrospection and prospection on the 2011 Amr Decade Award Article. *Academy of Management Review*, *38*(4), 471–489. doi:10.5465/amr.2013.0097

Proctor, T. (2007). *Public sector marketing*. Pearson Education.

Psacharopoulos, G. (1995). *The Profitability of Investment in Education: Concepts and Methods*. New York: World Bank. Retrieved June 13, 2015, from https://www.commdev.org/userfiles/files/1863_file_Profitability_Investment_Edu.pdf

Public Administration Select Committee. (2008). *User Involvement in Public Services*. Sixth Report of Session 2007–08.

Radine, L. B. (1987). Organization Theory in Administrative Law: A Proposal for a Design Science. *The American Sociologist*, *18*(3), 278–283. doi:10.1007/BF02691772

Rageh, A., Melewar, T., & Woodside, A. (2013). Using an ethnography research method to reveal the underlying dimensions of the customer/tourist experience. *Qualitative Market Research: An International Journal*, *16*(2), 126–149. doi:10.1108/13522751311317558

Ramamurti, R. (1986). Public Entrepreneurs: Who They Are and How They Operate. *California Management Review*, *28*(3), 142–158. doi:10.2307/41165207

Rappa, M. (2002). *Managing the digital enterprise: Business models on the web*. Retrieved from http://digitalenterprise.org/models/models.html

Ravneberg, B. (2009). Identity politics by design: Users, markets and the public service provision for assistive technology in Norway. *Scandinavian Journal of Disability Research*, *11*(2), 101–115. doi:10.1080/15017410902753904

Rayport, J. F., & Jaworski, B. J. (2001). *E-commerce*. New York: McGraw-Hill/Irwin.

Redman, T., Mathews, B., Wilkinson, A., & Snape, E. (1995). Quality management in services: Is the public sector keeping pace? *International Journal of Public Sector Management*, *8*(7), 21–34. doi:10.1108/09513559510103166

Rees, R. (Ed.). (2006). *The economics of public utilities*. Cheltenham, UK: Edward Elgar.

Rehm, H. (2001). *Management der öffentlichen Schuld – Befund, Probleme, Perspektiven. Wissenschaft für die Praxis Abt. 3 Analysen* (Vol. 14, pp. 24–60). Stuttgart: Deutscher Sparkassen Verlag.

Rehm, H., & Matern-Rehm, S. (2003). *Kommunale Finanzwirtschaft*. Frankfurt am Main: Lang.

Reichard, C. (2002). Marketisation of Public Services in Germany. *International PublicManagement Review*, *3*(2), 63–80.

Reingewertz, Y. (2014). *Fiscal Decentralization - a Survey of the Empirical Literature*. University of Haifa. Retrieved June, 2016, from https://mpra.ub.uni-muenchen.de/59889/1/MPRA_paper_59889.pdf

Reis, M., & Augusto, G. (2013). The Terminal Value (TV) Performing in Firm Valuation: The Gap of Literature and Research Agenda. *Journal of Modern Accounting and Auditing*, *9*(12), 1622–1636.

Richmond, B. J., Mook, L., & Quarter, J. (2003). Social accounting for nonprofits: Two models. *Nonprofit Management & Leadership*, *13*(4), 308–324. doi:10.1002/nml.2

Ritchie, J. B., & Hudson, S. (2009). Understanding and meeting the challenges of consumer/tourist experience research. *International Journal of Tourism Research*, *11*(2), 111–126. doi:10.1002/jtr.721

Rivette, K. G., Nothhaft, H. R., & Kline, D. (2000, January-February). *Discovering New Value in Intellectual Property. Harvard Business Review*.

Roome, N., & Louche, C. (2015). Journeying Toward Business Models for Sustainability: A Conceptual Model Found Inside the Black Box of Organisational Transformation. *Organization & Environment*, 1–24. doi:10.1177/1086026615595084

Rorty, R. (1982). *Consequences of pragmatism*. University of Minnesota Press.

Rosenman, M. (2007). Charity Fundraiser-in-Chief: The government's competition with nonprofits. *Stanford Sociar Innovation Review*. Retrieved from http://ssir.org/articles/entry/charity_fundraiser_in_chief#

Rosenman, M. (2015). *Why Crowdfunding Government Is a Bad Idea*. Retrieved June 20, 2016, from http://philanthropynewsdigest.org/commentary-and-opinion/why-crowdfunding-government-is-a-bad-idea

Ross, V. E., Kleingeld, A. W., & Lorenzen, L. (2004). a Topographical Map of the Innovation Landscape. *The Innovation Journal: The Public Sector Innovation Journal*, *9*(2), 1–19. Retrieved from http://linkinghub.elsevier.com/retrieve/pii/S1053810000904586

Rudny, W. (2013). Organizacja sieciowa jako model biznesu. In W. Rudny & B. Woźniak-Sobczak (Eds.), *Proces kreowania wartości w strukturach sieciowych. Katowice: Wydawnictwo Uniwersytetu Ekonomicznego*.

Rudolf, W. (2010). Koncepcja governance i jej zastosowanie – od instytucji międzynarodowych do niższych szczebli władzy. *Acta Universitatis Lodziensis. Folia Oeconomica*, *245*, 73–82.

Russian Federation Act N° 3085-1 "On Consumer Cooperation in the Russian Federation". (n.d.). Retrieved May, 2016, from http://www.consultant.ru/document/cons_doc_LAW_608/

Russian Federation Federal Act N° 131 "On General Principles of the Organization of Local Self-Government in the Russian Federation". (n.d.). Retrieved November, 2016, from http:// http://www.consultant.ru/document/cons_doc_LAW_44571/

Russian Federation Federal Act N° 14 "On Limited Liability Companies". (n.d.). Retrieved May, 2016, from http://www.consultant.ru/document/cons_doc_LAW_17819/

Russian Federation Federal Act N° 161 "On the State and Municipal Unitary Enterprises". (n.d.). Retrieved May, 2016, from http://www.consultant.ru/document/cons_doc_LAW_39768/

Russian Federation Federal Act N° 208 "On Joint Stock Companies". (n.d.). Retrieved May, 2016, from http://www.consultant.ru/document/cons_doc_LAW_8743/

Russian Federation Federal Act N° 41 "On Production Cooperatives". (n.d.). Retrieved May, 2016, from http://www.consultant.ru/document/cons_doc_LAW_10286/

Rydel, M. & Ronkowski C. (1995). Marketing partnerski. *Marketing i Rynek*, 9.

Rymsza, M. (2003). Aktywna polityka społeczna w teorii i praktyce. In T. Kaźmierczak & M. Rymsza (Eds.), *W stronę aktywnej polityki społecznej* (pp. 19–32). Warszawa: Instytut Spraw Publicznych.

Rymsza, M. (2007). Druga fala ekonomii społecznej w Polsce a koncepcja aktywnej polityki społecznej. In T. Kaźmierczak & M. Rymsza (Eds.), *Kapitał społeczny. Ekonomia społeczna* (pp. 175–192). Warszawa: FISP.

Rymsza, M. (2007). Polityka państwa wobec sektora obywatelskiego w Polsce w latach 1989-2007. In M. Rymsza, G. Markowski, & M. Dudkiewicz (Eds.), *Państwo a trzeci sektor. Prawo i instytucje w działaniu* (pp. 23–35). Warsaw: Instytut Spraw Publicznych.

Rymsza, M. (2013). *Aktywizacja w polityce społecznej W stronę rekonstrukcji europejskich Welfare States?* Warszawa: Wydawnictwo Instytutu Filozofii i Socjologii PAN.

Sabatier, V., Rousselle, T., & Mangematin, V. (2010). From recipe to dinner: Business model portfolio in the European biopharmaceutical industry. *Long Range Planning*, *43*(2/3), 431–447. doi:10.1016/j.lrp.2010.02.001

Saco, R. M., & Goncalves, A. P. (2008). Service Design: An Appraisal. *Design Management Review*, *19*(1), 10–19. doi:10.1111/j.1948-7169.2008.tb00101.x

Sandberg, K. D. (2002). Is It Time to Trade. In *Your Business Model?* Harvard University Update.

Sargeant, A. (1999). *Marketing Management for Non profit Organizations*. Oxford, UK: Oxford University Press.

Saussier, S., & Klien, M. (2014). *Local public enterprises: A taxonomy*. Chaire Economie des Partenariats Public-Privé Institut d'Administration des Entreprises.

Schumpeter, J. A. (1934). The Theory of Economic Development: An Inquiry Into Profits, Capital, Credit, Interest, and the Business Cycle. New Brunswick, NJ: Transaction Publishers.

Scott, J. T. (2015). *The Sustainable Business: A Practitioner's Guide to Achieving Long-Term Profitability and Competitiveness* (2nd ed.). Sheffield, UK: Greenleaf Publishing. doi:10.1007/s13398-014-0173-7.2

Scott, M., Shafera, H., Smitha, J., & Linder, J. C. (2005). The power of business models. *Business Horizons*, *48*(3), 199–207. doi:10.1016/j.bushor.2004.10.014

Šebestová, J., & Palová, Z. (2016). Real Societal Impact of Projects Supported within Human Resources and Employment Operational Programme in Selected Czech Regions. In *Proceedings of the 3rd International Conference on European Integration 2016*. Ostrava: VŠB - Technical University of Ostrava.

Seddon, P. B., Lewis, G. P., & Freeman, P. (2004). The Case for Viewing Business Models as Abstractions of Strategy. Communications of the Association for Information Systems, 13.

Seelos, C., & Mair, J. (2005). Social entrepreneurship: Creating new business models to serve the poor. *Business Horizons*, *48*(3), 241–246. doi:10.1016/j.bushor.2004.11.006

Shane, S. (2004). *Academic Entrepreneurship: University Spinoffs and Wealth Creation*. Cheltenham, UK: Edward Elgar Publishing. doi:10.4337/9781843769828

Shani, D., & Chalasani, S. (2013). Exploiting niches using relationship marketing. *Journal of Services Marketing*.

Sharir, M., & Lerner, M. (2006). Gauging the success of social ventures initiated by individual social entrepreneurs. *Journal of World Business*, *41*(1), 6–20. doi:10.1016/j.jwb.2005.09.004

Sharma, P., & Chrisman, S. J. J. (1999). Toward a reconciliation of the definitional issues in the field of corporate entrepreneurship. *Entrepreneurship Theory and Practice*, *23*(3), 11–27. doi:10.1007/978-3-540-48543-8_4

Shaw, E., & Carter, S. (2007). Social entrepreneurship: Theoretical antecedents and empirical analysis of entrepreneurial processes and outcomes. *Journal of Small Business and Enterprise Development*, *14*(3), 418–434. doi:10.1108/14626000710773529

Sheth, J. N., & Parvatiyar, A. (2000). The domain and conceptual foundations of relationship marketing. Handbook of relationship marketing, 3-38.

Short, C., Moss, T. W., & Lumpkin, G. T. (2009). Research in social entrepreneurship: Past contributions and future opportunities. *Strategic Entrepreneurship Journal*, *3*(2), 161–194. doi:10.1002/sej.69

Sienkiewicz-Małyjurek, K. (2016). Innowacyjność koprodukcji w zarządzaniu publicznym na poziomie lokalnym. *Zeszyty Naukowe Politechniki Śląskiej, Seria: Organizacja i Zarządzanie, 89*.

Siewiorek, A., Gegenfurtner, A., Lainema, T., Saarinen, E., & Lehtinen, E. (2013). The effects of computer-simulation game training on participants opinions on leadership styles. *British Journal of Educational Technology*, *44*(6), 1012–1035. doi:10.1111/bjet.12084

Siewiorek, A., Saarinen, E., Lainema, T., & Lehtinen, E. (2012). Learning leadership skills in a simulated business environment. *Computers & Education*, *58*(1), 121–135. doi:10.1016/j.compedu.2011.08.016

Sinn, H. (1997). Das Selektionsprinzip und der Systemwettbewerb. In A. Oberhauser (Ed.), *Fiskalföderalismus in Europa* (pp. 9–60). Berlin: Duncker & Humblot.

Sirgy, M. J., Efraty, D., Siegel, P., & Dong-Jin, L. (2001). A new measure of quality of work life (QWL) based on need satisfaction and spillover theories. *Social Indicators Research*, *55*(3), 241–302. doi:10.1023/A:1010986923468

Slyvotzky, A.J., Morrison, D.J., & Andelman, B. (2000). *Strefa zysku*. PWE.

Smith, D. (2011). *Functional, Overlapping, Competing Jurisdictions among Fractionalized Agents: Medieval Spain*. Retrieved October, 2015, from http://papers.ssrn.com/sol3/papers.cfm?abstract_id=1937955

Smith, G. (2012). *The Nonprofit Business Model Canvas | Innovative Nonprofit*. Retrieved May 27, 2016, from http://www.innovativenonprofit.com

Smith, G. (2016). *2009), Democratic innovations. Designing Institutions for Citizen Participation* (1st ed.). Cambridge, UK: Cambridge University Press.

Smith, W., Binns, A., & Tushman, M. (2010). Complex Business Models: Managing Strategic Paradoxes Simultaneously. *Long Range Planning*, *43*(2).

Social Innovation Lab. (2013). *Social Business Model Canvas*. Retrieved May 27, 2016, from http://www.socialbusinessmodelcanvas.com

Social Service Europe (SSE). (2012). *Social Innovation: The Role of Social Service Providers*. Retrieved May 20, 2016, from: http://www.socialserviceeurope.eu/#!publications/c1muj

Sørensen, E. (2014). *The metagovernance of public innovation in governance networks*. Paper presented at the University of Bristol Policy & Politics Conference, Bristol, UK.

Sørensen, E., & Torfing, J. (2012). Collaborative Innovation in the Public Sector. *The Innovation Journal, 17*(1), 1–14. Retrieved from https://vpn.utm.my/docview/1362242870?accountid=41678

Sørensen, E., & Torfing, J. (2012). Collaborative Innovation in the Public Sector. *The Innovation Journal*, *17*(1), 1–14.

Spieth, P., Schneckenberg, D., & Matzler, K. (2016). Exploring the linkage between business model (&) innovation and the strategy of the firm. *R&D Management*. http://doi.org/<ALIGNMENT.qj></ALIGNMENT>10.1111/radm.12218

Spieth, P., Schneckenberg, D., & Ricart, J. E. (2014). Business model innovation – state of the art and future challenges for the field. *R & D Management*, *44*(3), 237–247. doi:10.1111/radm.12071

Spindler, J. (1998). *FOCJ – ein Konzept zur Neuordnung der Zusammenarbeit öffentlicher Gebietskörperschaften*. Bern: Haupt.

Spindler, J. (2008a). Die Zweckgemeinde – auch ein waldpolitisches Referenzmodell. *Schweizerische Zeitschrift für das Forstwesen*, *12*(12), 427–434. doi:10.3188/szf.2008.0427

Spindler, J. (2008b). Die Grösse allein ist nicht entscheidend. *Kommunalmagazin*, *3*, 4–6.

Starnawska, M. (2014). Behavior through networking in social entrepreneurship in response to challenging institutional framework – case of 5 social cooperatives. *Problemy Zarzadzania*, *12*(4), 97–116. doi:10.7172/1644-9584.49.6

Starnawska, M. (2014). Zachowanie poprzez sieciowanie w przedsiębiorczości społecznej w odpowiedzi na trudne otoczenie instytucjonalne – przypadek pięciu spółdzielni socjalnych. *Problemy Zarzadzania*, *12*(4), 97–116.

Stubbs, W., & Cocklin, C. (2008). Conceptualizing a „sustainability business model. Organization & Environment, 21(2), 103–127.

Stubbs, W., & Cocklin, C. (2008). Conceptualizing a Sustainability Business Model. *Organization & Environment*, *21*(2), 103–127. doi:10.1177/1086026608318042

Su, K., Li, J., & Fu, H. (2011). Smart city and the applications. In *2011 International Conference on Electronics, Communications and Control, ICECC 2011 - Proceedings* (pp. 1028–1031). http://doi.org/ doi:10.1109/ICECC.2011.6066743

Sudoł, S. (2014). Podstawowe problemy metodologiczne nauk o zarządzaniu. *Organizacja i Kierowanie*, *1*(161), 11–35.

Sułkowski, Ł., & Zawadzki, M. (2016). Corporate University: A Critical Approach. *Entrepreneurship and Management*, *17*(1).

Sundell, A., & Lapuente, V. (2012). Adam Smith or Machiavelli? Political incentives for contracting out local public services. *Public Choice*, *153*(3/4), 469–485. doi:10.1007/s11127-011-9803-1

Svendsen, A. (1998). *The Stakeholder Strategy. Profiting from Collaborative Business Relationships*. San Francisco: Berret-Koehler Publishers, Inc.

Swartout, W., & Van Lent, M. (2003). Making game of system design. *Communications of the ACM*, *46*(7), 32–39. doi:10.1145/792704.792727

Szarfenberg, R. (2007). Rodzaje i formy aktywnej polityki społecznej. In G. Firlit-Fesnak & M. Szylko-Skoczny (Eds.), *Polityka społeczna* (pp. 401–416). Warszawa: PWN.

Szumowski, W. (2014). Zarządzanie publiczne–próba systematyzacji koncepcji. *Nauki o Zarzadzaniu*, *4*(21), 86-98.

Tantalo, C., & Priem, R. L. (2016). Value creation through stakeholder synergy. *Strategic Management Journal*, *37*(2), 314–329. doi:10.1002/smj.2337

Tarnawska, K., & Cwiklicki, M. (2012). Evaluation of The European Social Fund as a Measure To Support Social Innovation in The Public Sector. *Economics and Management*, *17*(1), 237–243. doi:10.5755/j01.em.17.1.2273

Taylor, C. (2003). *Modern Social Imaginaries*. Duke University Press. doi:10.1215/9780822385806

Teach, R. D (1990). Designing Business Simulations. In Guide to Business Gaming and Experiential Learning. Association for Business Simulation and Experiential Learning (ABSEL).

Teece, D. J. (2010). Business Models, Business Strategy and Innovation. *Long Range Planning*, *43*(2-3), 172–194. doi:10.1016/j.lrp.2009.07.003

Teets, J. C. (2012). Reforming Service Delivery in China: The Emergence of a Social Innovation Model. *Journal of Chinese Political Science*, *17*(1), 15–32. doi:10.1007/s11366-011-9176-9

Tellis, W. M. (1997). Introduction to Case Study. *Qualitative Report*, *3*(2), 1–14. PMID:9035020

Ter-Minassian, T. (1997). Intergovernmental Fiscal Relations in a Macroeconomic Perspective: An Overview. In T. Ter-Minassian (Ed.), *Fiscal Federalism in Theory and Practice* (pp. 3–24). Washington, DC: International Monetary Fund.

The evolution of public services from co-production to co-creation and beyond: New Public Management's unfinished trajectory? (2014). *International Journal of Leadership in Public Services*, *10*(2), 91–107.

The Scottish Government. (2011). *A Practical Guide to Forming and Operating Public Social Partnership*. Retrieved June 20, 2016 from: http://www.gov.scot/Resource/Doc/48453/0119024.pdf

Thoelen, A., Cleeren, S., Denis, A., Peters, K., Van Ael, K., & Willems, H. (2015). *Public Service Design A guide for the application of service design in public organisations*. Retrieved from http://www.thespiderproject.eu/wp-content/uploads/2015/09/PSD_manual_UK_LR.pdf

Thompson, J., Alvy, G., & Lees, A. (2000). Social Entrepreneurship – A New Look at The People and The Potential. *Management Decision*, *5*(5), 328–338. doi:10.1108/00251740010340517

Thompson, J., & Doherty, B. (2006). The diverse world of social enterprise: A collection of social enterprise stories. *International Journal of Social Economics*, *33*(5/6), 399–410. doi:10.1108/03068290610660643

Tiebout, Ch. (1956). A Pure Theory of Local Expenditures. *Journal of Political Economy*, *64*(5), 416–424. doi:10.1086/257839

Timmers, P. (1998). *Business models for electronic markets*. European Commission, Directorate-General III.

Timmers, P. (1998). Business Models for Electronic Markets. *Electronic Markets*, *8*(2), 3–8. doi:10.1080/10196789800000016

Timmers, P. (1999). *Electronic Commerce: Strategies and Models from business-to-business trading*. Chichester, UK: Wiley & Sons.

Tolstad, H. K. (2014). Development of rural-tourism experiences through networking: An example from Gudbrandsdalen, Norway. *Norwegian Journal of Geography*, *68*(2), 111–120.

Totawar, A. K., & Nambudiri, R. (2014). Can fairness explain satisfaction? Mediation of quality of work life (QWL) in the influence of organizational justice on job satisfaction. *South Asian Journal of Management*, *21*(2), 101–122.

Tracey, P., Phillips, N., & Haugh, H. (2005). Beyond philanthropy: Community enterprise as a basis for corporate citizenship. *Journal of Business Ethics*, *58*(4), 327–344. doi:10.1007/s10551-004-6944-x

Tracey, P., Phillips, N., & Jarvis, O. (2011). Bridging institutional entrepreneurship and the creation of new organizational forms: A multilevel model. *Organization Science*, *22*(1), 60–80. doi:10.1287/orsc.1090.0522

Truss, A., Marshall, R., & Blair-Stevens, C. (2010). A history of social marketing. In J. French, C. Blair-Stevens, D. McVey, & R. Merritt (Eds.), Social marketing and public health. Theory and practice. Oxford.

Tucker, S. (2014). *Social Innovation for Public Service Excellence*. UNDP Global Centre for Public Service Excellence. Retrieved May 20, 2016, from: http://www.undp.org/content/dam/uspc/docs/GPCSE_Social%20Innovation.pdf

Tung, V. W. S., & Ritchie, B. J. (2011a). Exploring the essence of memorable tourism experience. *Annals of Tourism Research*, *38*(4), 1367–1386. doi:10.1016/j.annals.2011.03.009

Tung, V. W. S., & Ritchie, J. B. (2011b). Investigating the memorable experiences of the senior travel market: An examination of the reminiscence bump. *Journal of Travel & Tourism Marketing*, *28*(3), 331–343. doi:10.1080/10548408.2011.563168

Tvrdoň, M. (2015). Moravskoslezský kraj z pohledu makroekonomických ukazatelů. In *18th International Colloquium on Regional Sciences. Conference Proceedings*. Brno: Masarykova univerzita. doi:10.5817/CZ.MUNI.P210-7861-2015-8

Uljens, M. (1996). On the philosophical foundation of phenomenography. In G. Dall'Alba & B. Hasselgren (Eds.), *Reflections on Phenomenography* (pp. 105–130). Goteborg: Acta Universitatis Gothenburgensis.

Ustawa z dnia 27 kwietnia 2006 roku o spółdzielniach socjalnych (Dziennik Ustaw Nr 94, poz. 651 z późn. Zm).

Valová, G. (2016). *Development of Business Environment and Small Businesses in the Microregion of Valašskomeziříčsko-Kelečsko* (Unpublished master's thesis). Karviná: SU OPF.

Van Berkel, R., & Hornemann Møller, I. (2002). *Active social policies in the EU. Inclusion through participation?* Bristol: Policy Press.

Van de Ven, A. H. (2013). *Engaged Scholarship: A guide for organizational and social research*. Oxford, UK: Oxford University Press.

Van Dooren, W., Bouckaert, G., & Halligan, J. (2015). *Performance Management in the Public Sector* (2nd ed.). London: Routledge.

van Renswoude, K., ten Wolde, A., & Joustra, D. J. (2015). Circular Business Models. Part 1: An introduction to IMSA's circular business model scan. Amsterdam: Academic Press.

Vecchi, V., Brusoni, M., & Borgonovi, E. (2014). Public authorities for entrepreneurship: A management approach to execute competitiveness policies. *Public Management Review*, *16*(2), 256–273. doi:10.1080/14719037.2012.725759

Venkatesh, A., & Vitalari, N. P. (1991). Longitudinal Surveys in Information Systems Research: An Examination of Issues, Methods, and Applications. In *The Information Systems Challenge: Survey Research Methods*. Harvard University Press.

Vidal, I. (2014). The Role of Third Sector in Local Welfare. *International Conference under the Patronage of Municipality of Bologna- Municipality of Ravenna.* Retrieved June 15, 2016, from. https://www.researchgate.net/publication/263442935_The_role_of_Third_Sector_in_Local_Welfare

Vidal, I. (2013). Governance of social enterprises as producers of public services. In P. Valkama, S. J. Bailey, & A.-V. Anttiroiko (Eds.), *Organisational Innovation in Public Services: Forms and Governance*. Basingstoke, UK: Palgrave. doi:10.1057/9781137011848_10

Vigoda-Gadot, E., Shoham, A., Schwabsky, N., & Ruvio, A. (2008). Public sector innovation for Europe: A multinational eight-country exploration of citizens perspectives. *Public Administration*, *86*(2), 307–329. doi:10.1111/j.1467-9299.2008.00731.x

Vis, B., van Kersbergen, K., & Hylands, T. (2011). To What Extent Did the Financial Crisis Intensify the Pressure to Reform the Welfare State? *Social Policy and Administration*, *45*(4), 338–353. doi:10.1111/j.1467-9515.2011.00778.x

Viscio, A. J., & Pasternack, B. A. (1996). Toward a new business model. *Strategic Business*, *2*(1), 125–134.

Viturka, M. (2007). Konkurenceschopnost regionů a možnosti jejího hodnocení. *Politicka Ekonomie*, *55*(5), 637–658. doi:10.18267/j.polek.617

Voelpel, S., Leibold, M., Tekie, E., & von Krogh, G. (2005). Escaping the Red Queen Effect in Competitive Strategy: Sense-testing Business Models. *European Management Journal*, *23*(1), 37–49. doi:10.1016/j.emj.2004.12.008

Vogt, G. (1984). Staatliches Haushaltsrecht. In A. v. Mutius, K. H. Friauf & P. Westermann (Eds.), Handbuch für die öffentliche Verwaltung (pp. 519-584). Neuwied, Darmstadt: Luchterhand.

Wagner, A. (1890). *Finanzwissenschaft*. Leipzig: Winter'sche Verlagshandlung.

Wallerstein, I. (2004). *World-systems Analysis. An Introduction.* Duke University Press.

Walsh, K. (1991). Citizens and consumers: Marketing and public sector management. *Public Money & Management*, *11*(2), 9–16. doi:10.1080/09540969109387649

Walsh, K. (1994). Marketing and public sector management. *European Journal of Marketing*, *28*(3), 63–71. doi:10.1108/03090569410057308

Wang, C.-S., Li, U.-C., & Tzeng, Y.-R. (2015). How to replicate the cognitive process in computer game-based learning units. *Information Technology & People*, *28*(2), 327–343. doi:10.1108/ITP-03-2013-0053

Wang, S.-M. (2014). Public service space remodeling based on service design and behavioral maps. *Journal of Industrial and Production Engineering*, *31*(2), 76–84. doi:10.1080/21681015.2014.887595

Warner, M. E., & Clifton, J. (2014). Marketisation, public services and the city: The potential for Polanyian counter movements. *Cambridge Journal of Regions, Economy and Society*, *7*(1), 45–61. doi:10.1093/cjres/rst028

Warr, P. (2005). Work, well-being, and mental health. In J. Barling, E. K. Kelloway, & M. R. Frone (Eds.), *Handbook of Work Stress* (pp. 547–574). Thousand Oaks, CA: Sage. doi:10.4135/9781412975995.n23

Weber, M. (1947). *The theory of Social and Economic Organization*. New York: Free Press.

Webster, J. G. (2014). *The Marketplace of Attention. Howe Audiences Take Shape in a Digital Age.* Cambridge, MA: The MIT Press.

Whicher, A., Swiatek, P., & Cawood, G. (2013, August). An Overview of Service Design for the Private and Public Sectors. *See Platform.*

Whicher, A., Swiatek, P., & Thurston, P. (2016). Trends in Design and Government in Europe. *Design Management Review*, *27*(1), 44–50. http://doi.org/DOI: 10.1111/drev.10348

Wilhelm, B. (2013). *Inter-Municipal Cooperation (IMC). New Perspectives for IMC in the field of Energy Efficiency (EE).* Retrieved June, 2016, from http://www.sustainable-energybih.org/wp-content/uploads/2013/07/BWilhelm_Presentation-on-IMC.pdf

Williamson, P.J. (2010). Cost Innovation: Preparing for a 'Value-for-Money' Revolution. *Long Range Planning*, *43*(2-3), 343–353. 10.1016/j.lrp.2009.07.008

Wilson, F., & Post, J. E. (2011). Business Models for People, Planet (& Profits): Exploring the Phenomena of Social Business, a Market-based Approach to Social Value Creation. *Small Business Economics*, *40*(3), 715–737. doi:10.1007/s11187-011-9401-0

Wilson, W. (1887). The Study of Administration. *Political Science (Wellington, N.Z.)*, *2*(2), 197–222. doi:10.2307/2139277 PMID:4591257

Windrum, P. (2008). Innovation and entrepreneurship in public services. In P. Windrum & P. Koch (Eds.), *Innovation in Public Sector Services: Entrepreneurship, Creativity and Management* (pp. 3–20). Northampton, MA: Edward Elgar Publishing.

Winkle, C. M., & Lagay, K. (2012). Learning during tourism: The experience of learning from the tourists perspective. *Studies in Continuing Education*, *34*(3), 339–355. doi:10.1080/0158037X.2011.646981

Wirtz, B. W. (2011). Business Model Management: Design – Instruments – Success Factors (1st ed.). Gabler Verlag.

Wokoun, R. (2007). Regionální a strukturální politika (politika soudržnosti) Evropské unie v programovém období 2007-2013. *Urbanismus a územní rozvoj*, (10), 1, 3-7.

Wolińska, I., & Rokicki, B., &Zub, M., &Niedoszewska, A. &Piekutowski, J. (2010). Raport końcowy. *Ocena systemu monitorowania i wartości wskaźnikow PO KL 2007-2013*. Warszawa: PSDB grupa WYG.

World Bank. (2014). *Toolkit Corporate Governance of State-Owned Enterprises*. Retrieved June, 2016, from https://openknowledge.worldbank.org/bitstream/handle/10986/20390/9781464802225.pdf?sequence=1

Wright, T. A., & Bonett, D. G. (2007). Job satisfaction and psychological well-being as non-additive predictors of workplace turnover. *Journal of Management*, *2*(33), 141–160. doi:10.1177/0149206306297582

Wright, T. A., & Cropanzano, R. (2004). The role of psychological well-being in job performance: A fresh look at an age-old quest. *Organizational Dynamics*, *33*(4), 338–351. doi:10.1016/j.orgdyn.2004.09.002

Wronka-Pośpiech, M. (2015). Innowacje społeczne – pojęcie i znaczenie. Studia Ekonomiczne Uniwersytetu Ekonomicznego w Katowicach, 212, 124-136.

Wronka-Pośpiech, M., Frączkiewicz-Wronka, A., & Lara Montero, A. (2015). Innowacje – przesłanki wyłaniania się nowego paradygmatu polityki społecznej. In M. Grewiński & A. Karwacki (Eds.), Innowacyjna polityka społeczna. Warszawa: WSP im. J. Korczaka.

Wronka-Pośpiech, M., & Frączkiewicz-Wronka, A. (2014). The use of ICT for achieving the objectives of the business model – social enterprise perspective. *Polish Journal of Management*, *10*(2), 33–42.

Wronka-Pośpiech, M., & Frączkiewicz-Wronka, A. (2016). Strategic orientation and organisational culture in Polish Public Organisations: Insights from the Miles and Snow typology. *Management*, *20*(1), 126–141. doi:10.1515/manment-2015-0029

Wrzosek, S. (2012). *Administracja publiczna jako przedmiot badań nauki administracji*. Teka Komisji Prawniczej V.

Wygnański, J. (2007). Ekonomizacja organizacji pozarządowych. In J. Hausner (Ed.), *Ekonomia społeczna a rozwój* (pp. 127–156). Kraków: Małopolska Szkoła Administracji Publicznej.

Yeazdanshenas, M. (2014). Designing a Conceptual Framework for Organizational Entrepreneurship in the Public Sector in Iran. *Iranian Journal of Management Studies*, *7*(2), 351–376.

Yin, R. K. (2003). *Case Study Research* (3rd ed.). Thousand Oaks, CA: Sage Publications Inc.

Yin, R. K. (2003). *Case study research: Design and methods*. Thousand Oaks, CA: Sage Publications Inc.

Yip, G. S. (2000). Global Stategy in the Internet Ea. *Business Strategy Review*, *11*(4), 1–14. doi:10.1111/1467-8616.00152

Yoffee, S. F. (2016). Translating Design Policy into a State Design Commission: Maryland. *Design Management Review*, *27*(1), 38–43. http://doi.org/DOI: 10.1111/drev.10347

Yunus, M., Moingeon, B., & Lehman-Ortega, L. (2010). Biulding Social Business Models: Lessons from the Grameen Experience. *Long Range Planning*, *43*(2-3), 308–325. doi:10.1016/j.lrp.2009.12.005

Zainal, Z. (2007). Case study as a research method. *Jurnal Kemanusiaan*, *9*, 1-6.

Zbornik, R. (2012). Tourist destination branding. *Ekonomskog Fakulteta u Istocnom Sarajevu*, *6*, 503–514.

Zerbinati, S., & Souitaris, V. (2005). Entrepreneurship in the public sector: A framework of analysis in European local governments. *Entrepreneurship & Regional Development*, *17*(1), 43–64. doi:10.1080/0898562042000310723

Zhiyong, L., & Rosenbloom, D. H. (1992).. . *Public Administration Review*, *52*(6), 535–537. doi:10.2307/977163

Zimmermann, H. (1999). *Kommunalfinanzen*. Baden-Baden: Nomos.

Zolberg, V. (1980). Displayed Art and Performed Music: Selective Innovation and the Structure of Artistic Media. *The Sociological Quarterly*, *21*(2), 219–231. doi:10.1111/j.1533-8525.1980.tb00606.x

Zott, C., Amit, R., & Massa, L. (2010). *The Business Model: Theoretical Roots, Recent Developments, and Future Research*. IESE. Retrieved from http://www.iese.edu/research/pdfs/DI-0862-E.pdf

Zott, Ch., Amit, R., Massa L. (2011). The Business Model: Recent Developments and Future Research. *Journal of Management*. doi:10.1177/0149206311406265

Zott, C., & Amit, R. (2007). Business model design and the performance of entrepreneurial firms. *Organization Science*, *18*(2), 181–199. doi:10.1287/orsc.1060.0232

Zott, C., & Amit, R. (2010). Business model design: An activity system perspective. *Long Range Planning*, *43*(2–3).

Zott, Ch., & Amit, R. (2008). The fit between product market strategy and business model: Implications for firm performance. *Strategic Management Journal*, *29*(1), 1–26. doi:10.1002/smj.642

БутоваТ. В.Пухова,М. М.Щукин,И. А. (2013). Проблемы и перспективы становления института межмуниципального сотрудничества в России. *Управленческие науки*, *3*(8), 4-15.

Винницкий, А. В. (2010). Участие публичных образований в имущественных отношениях: проблемы сбалансированности административно-правового и гражданско-правового регулирования. *Административное и муниципальное право*, *11*, 82-91.

ВинницкийА. В. (2013). *Публичная собственность*. Москва: Статут.

Гельман, В. (2006). Возвращение Левиафана? Политика рецентрализации в современной России. *Политические исследования*, *2*, 90-109.

Ирискина, О. В. (2010). Современные формы управления межмуниципальным сотрудничеством в Российской Федерации. *Известия Саратовского университета. Новая серия. Серия Социология. Политология*, *2*(10), 60–65.

Истомина, Н. А. (2015). Среднесрочное бюджетное планирование в регионе: этапы эволюции и перспективы развития. *Вестник ВГУ. Серия: экономика и управление*, *2*, 77-82.

ЛадыгинВ. В. (2011). Возможности создания муниципального бизнес-инкубатора с использованием механизмов межмуниципальной кооперации. *Вопросы государственного и муниципального управления, 2*, 25-37.

Ляпин, И. Ф. (2015). Конституционные особенности децентрализации государственной власти в российской федерации. *Вестник Нижегородского университета им. Н.И. Лобачевского*, *1*, 152–154.

Петров, Н. (2000). Федерализм по-российски. *Pro et Contra*, *5*(1), 7–32.

Проект Федерального закона N° 252441-6 «О публично-правовых компаниях в Российской Федерации и о внесении изменений в отдельные законодательные акты Российской Федерации». (n.d.). Retrieved June, 2016, from http://www.consultant.ru/law/hotdocs/46731.html

About the Contributors

Saioa Arando is Research Coordinator at Mondragon Innovation & Knowledge (MIK), the research centre at Mondragon University's Faculty of Business, and simuntaneously the Coordinator of " People in cooperation: leadership and ownership" research area. She completed here doctorate at the University of Deusto in Economics. In 2006, she was awarded the Vanek-Hovat Prize for best paper by a young economist at the IAFEP International Conference. She is actively collaborating in several research projects with national and international researches and her work has received funding from different public institutions. Her papers have been published at journals such as "ILR Review ", "JEBO" or "Revesco".

Mariia Chebotareva received her MA in Economics from Voronezh State University, Russia in 2014. In 2014 she won admission to Voronezh State University as a doctorial student and got her Aurora Grant to study abroad. Mariia Chebotareva is currently a second year PhD student, she writes on local public goods provision and municipal cooperation at both Voronezh State University and University of Tartu.

Aldona Frączkiewicz-Wronka, professor, Head of the Department of Public Management and Social Sciences at the University of Economics in Katowice. Her research interests focus on issues related to the management of public sector organizations, head of research team in many research projects funded by the Ministry of Science and Higher Education as well as the European Union, author of numerous publications on the functioning of the public sector in Poland, has experience with a proven track record for successfully training managers in the health care sector and public administration, expert for the National Foresight Programme "Poland 2000", Technology Foresight for development of public services in the GOP, has designed and facilitated courses for broad range of clients i.e. public entities, such as the Ministry of Science and Higher Education, Ministry of Regional Development, Ministry of Health, the Marshal's Office in Katowice and others.

Peter Friedrich studied Economics at University of Münster. He received his master diploma in 1964, 1964-1974 he was an assistant at chair of public finance (prof. Timm) and lecturer at University of Münster. In 1969 Prof. P. Friedrich received Doktor (PhD) at University of Münster (Investment Theory for Municipal Enterprices), 1971-1973 he received research grant of German Research Association (DFG). In 1973 he was habilitated. In 1974 he was a Dozent at University of Münster, 1974-1978 Chair (full professor) of Public Finance at University of Siegen, 1978-1988 Chair of Public Finance at University of Bamberg,1988-2006 Chair of Public Finance at University of Federal Armed Forces at München. Prof. P. Friedrich retired in 2006, from 2006 to 2009 he was an extraordinary professor of public finance at University of Tartu, 2006-2014 he was a professor at High Banking School Torun/Poland, since 2009 he is a senior researcher at the University of Tartu. Since 1977 to 2007 member of the editorial board of the "Journal of Public and Nonprofit Services. From 1973 to 2006 he was Co-editor of Schriften zur öffentlichen Verwaltung und öffentlichen Wirtschaft - Series on Public Administration and Public Economics- Nomos-Verlagsgesellschaft, Baden-Baden and Berliner Wissenschaftsverlag, Berlin.

Adam Jabłoński is an associate professor in The Department of Management at the University of Dąbrowa Górnicza (Wyższa Szkoła Biznesu w Dąbrowie Górniczej). He is also Vice-President of the Board of a reputable management consulting company "OTTIMA plus" Ltd. of Katowice, and President of the "Southern Railway Cluster" Association of Katowice, which supports development in railway transport and the transfer of innovation, as well as cooperation with European railway clusters (as a member of the European Railway Clusters Initiative). He holds a postdoctoral degree in Economic Sciences, specializing in Management Science. Having worked as a management consultant since 1997, he has broadened his experience and expertise through co-operation with a number of leading companies in Poland and abroad. He is the author of a variety of studies and business analyses on business models, value management, risk management, the balanced scorecard and corporate social responsibility. He has also written and co-written several monographs and over 100 scientific articles in the field of management, published both in Poland and abroad. Adam's academic interests focus on the issues of modern and efficient business model design, including Sustainable Business Models and the principles of company value building strategy that includes the rules of Corporate Social Responsibility.

Marko Kesti works as an adjunct professor of Administrative Science in University of Lapland, Finland. Kesti completed his Ph.D. in 2012 and is now leading the research program dealing with human capital productivity. He has written several books at his research subject and is active in presenting research findings at conferences and seminars. Kesti is utilizing his research in helping organizations to analyze and improve their human capital performance. His analyzing tools and development methods are widely in use at Finnish business and municipal organizations.

Terhi Kesti (M.Soc.Sc. (Man.), eMBA, BIT, BBA, B.Com.) has 20 years' experience in productive leadership and HR-management in IT-business. She knows well challenges of supervisor's tasks. During 2014-2016 she updated her knowledge of leadership in master's program in Lapland's University in Finland. Her pro gradu thesis concerned experiences of playing and learning by Productive Leadership simulation game (published in Finnish). Her special interests are situation sensible leadership, productivity, and gamification. She has been a presenter at leadership and gamification seminars and workshop facilitator in game playing sessions. Her presentations are geared to helping supervisors to become more credible by recognizing the efforts of their leadership actions.

Barbara Kożuch - Professor of Economic Sciences at Jagiellonian University in Kraków and specialist in Organization Theory and Public Management. Principal fields of interest and specialization: foundations of organizational sciences, systems and methods of management, project management and human resource management, multicultural management, public management, including conditions of efficient functioning of the public sector, particularly in regards to change management in health care, education, culture, and public safety. The Editor-in-Chief of the "International Journal of Contemporary Management" and a member of Scientific Board in Journals, i.e. "Journal of Intercultural Management". A member of the Committee of Organization and Management of the Polish Academy of Sciences (2007-2015). She is well-renowned in Poland and abroad for her expertise in public management issues. She has been the keynote speaker at various international conferences.

Jaana Leinonen is a senior lecturer of administrative science at the University of Lapland. Her research interests cover a wide range of topics including public management, local government, political leadership, social media, grounded theory, and information management.

Slawomir J. Magala, until Dec. 2015 professor of cross-cultural management, department of organizational sciences and human resource management at the Rotterdam School of Management, Erasmus University. Involved in global consulting and training experiences in post-communist and pre-capitalist countries including China, India, Estonia, Kazakhstan and Namibia, he wrote on "Cross Cultural Competence" (Routledge,2005), the meaning of meaning in organizations and in academic communities (The Management of Meaning in Organizations, Palgrave, 2009) and on „Ethical Control and Cultural Change. In Cultural Dreams Begin Organizational Responsibilities" (Journal of Public Affairs, 10/3, 2010). Interested in visual sociology and multimedia aesthetics, he had translated Susan Sontag into Polish and wrote „Perplexing Images. Relational Identities In Cultural Tempospaces" (In; Lowe, Sid, ed., Managing In Changing Times. A Guide for the Perplexed Manager, Sage, 2010). In 2011 the Polish translation of his 1981 book published under the name Stanislaw Starski "Class Struggle in Classless Poland" (South End Press, Boston) had been re-published in Gdansk by the European Solidarity Center.

Zuzana Palová is a student of doctoral study program Business Economics and Management at Silesian University in Opava, Faculty of Business Administration in Karvina. She works in non-profit organization Unika Centrum, o.p.s. as financial and project manager. She is participating at the Faculty of Business Administration in researches especially in the field of regional disparities and social innovations.

Jarmila Šebestová, Assistant Professor, is an experienced researcher dealing with Small Businesses. She has participated in a number of international projects, including IPREG—Innovative Policy Research for Economic Growth and the E-WORLD project (International Entrepreneurs Network). She is Vice President of the European Council of Small Businesses for the Czech Republic. Research interests: small business performance, small business dynamics, social entrepreneurship.

Robert Seliga is a doctor of economic sciences. His research interests include organization and management, and in particular: public marketing, relationship marketing and social marketing. Currently an an assistant professor of the University of Social Sciences, Vice-Dean of the Department at the University of Social Sciences in London and Director of Science at the University of Social Sciences in Łodz. He serves as the associate editor of the Journal of Intercultural Management, editor of the Entrepreneurship and Management Journal.

Beata Maria Staszewska is a Ph.D. Candidate at Kozminski University - Warsaw. She holds Masters degree in Psychology from the Maria-Curie Skłodowska University. She also completed Postgraduate Management Studies from the Lublin University of Technology and the University of Illinois at Urbana-Champaign. She was a business trainer, strategy consultant and an author of methodologies, which created possibilities for social research using ICT systems (including Equal Programme). Currently, works as CEO of Recreation Park Zoom Nature. She specializes in management strategy, human resource management and industrial psychology, and studies leadership, dynamic capabilities and knowledge management. Her current areas of strategic management interest are a business model and business model development. She is particularly interested in the application of the management strategy theory into business practice.

Monika Tkacz, research assistant in MIK (Mondragon Innovation & Knowledge), Spain. Her research is focused on entrepreneurship, new business models and innovation in social and public sector. From the beginning of her career she has been actively contributing to numerous social projects aimed at seeking effective solutions for socioeconomic issues, participant of start-up pitches for emerging social enterprises (Startery Central and Easter Europe in Warsaw), co-author of publications on cooperative business model and innovation in social economy area, team member of women empowerment movement Aldazank in Basque Country in Spain.

Martyna Wronka-Pośpiech, PhD, Assistant Professor in Management at University of Economics in Katowice, Department of Entrepreneurship and Management Innovation. Her research is focused on entrepreneurship, innovation and strategic management in social and public sector. From the beginning of her university career she is additionally engaged in social projects aiming at stimulating and activating people into action. Therefore, she has over 7 years of experience with a proven track record for successfully empowering trainees working in nonprofit sector through counselling, consulting, coaching and developing strategy.

Index

I

K

L

M

N

O

P

Q

R

S

T

V

www.ingramcontent.com/pod-product-compliance
Lightning Source LLC
LaVergne TN
LVHW081252011026
849162LV00006BA/1507